OUT OF THE WOODS

OUT OF THE WOODS

ESSAYS IN ENVIRONMENTAL HISTORY

Edited by Char Miller and Hal Rothman

UNIVERSITY OF PITTSBURGH PRESS

Published by the University of Pittsburgh Press, Pittsburgh, Pa. 15261
Copyright © 1997, University of Pittsburgh Press
Manufactured in the United States of America
Printed on acid-free paper
10 9 8 7 6 5 4 3 2 1

LIBRARY OF CONGRESS CATALOGING-IN-PUBLICATION DATA
Out of the woods : essays in environmental history / edited by Char Miller
    and Hal Rothman.
            p.   cm.
        A compilation of essays from the first 19 volumes of Environmental history review and
    its predecessor Environmental review.
        Includes index.
        ISBN 0-8229-3982-7 (alk. paper). — ISBN 0-8229-5631-4 (pbk. : alk.
    paper)
        1. Environmental sciences—History.   I. Miller, Char, 1951– .
    II. Rothman, Hal, 1958– .
    GE50.O98   1997
    363.7—dc21   97-4747

A CIP catalog record for this book is available from the British Library.

William Cronon's "The Trouble with Wilderness: Or, Getting Back to the Wrong Nature" is
from *Uncommon Ground: Toward Reinventing Nature* by William Cronon. Copyright © 1995 by
William Cronon. Reprinted by permission of W. W. Norton Company, Inc.

for Frank Lubbock Miller III (1917–1995)
and Lauralee, Talia, and Brent Rothman

Contents

Acknowledgments

No project of this kind is ever without its problems and dilemmas. We are indebted to so many for minimizing the bumps along the way, but particularly to Evelyn Luce, secretary of the History Department at Trinity University, who diligently typed the many essays for which electronic versions were not available. In many other organizational ways she helped keep this manuscript on track, as did Eunice Herrington, senior secretary of the Trinity History Department, and Gary Kates, the chair. We also could not have pursued this book without the cooperation of its many contributors, of course, all of whom were happy their work would find a new audience! To the anonymous readers for the University of Pittsburgh Press, whose insightful criticism helped make this a better volume, and to the marvelous staff at the press who have done so much to further its publication, we are indeed thankful.

We owe a great deal to our respective families, naturally. Char Miller's graciously affected not to notice the vast stacks of manuscripts that seemed to fill up the house; he learned this storage technique from his late father, from whom he also learned to love and value the study of the past. Hal Rothman wishes to thank his wife Lauralee, daughter Talia, and son Brent, whose birth coincided with the idea for this book, for having given freely of the time he should have shared with them so that he could complete this and other projects; his love and gratitude for them knows no bounds.

Introduction

While a graduate student at Yale University in the late 1960s, Donald Worster was hazed by classmates puzzled by his fascination with the hitherto unknown field of environmental history; his peers, Worster would recall nearly thirty years later, regarded him and his scholarly interest in the relationship between humanity and the natural world as "something of a joke." Did his historical approach mean, one fellow student asked while sharing a table in the Yale Commons, that Worster intended to look at history from the point of view of bears?

It was hard to grasp what Worster and the handful of pioneers who launched the new field of environmental history in the early 1970s sought to accomplish, if only because they seemed to propose that scholars needed to step outside the usual frame of the study of human experience to work and write within this new arena of scholarly endeavor. They appeared to promote a methodological perspective that sought, consciously or not, to fulfill Aldo Leopold's dictum that we will never fully grasp the essential importance of wildness to civilization until we begin to "think like a mountain." Adopting Leopold's vision, however, was (and remains) as complicated as it would be to adapt to a bear's perspective. Realized or not, the formulation remains a part of the complex challenge facing those committed to studying a realm beyond the norm of human sensibilities.[1]

Out of the Woods chronicles the evolution of this and other intellectual commitments of environmental history by gathering together essays from the first nineteen volumes of *Environmental History Review* and its predecessor, *Environmental Review*. It brings together articles that reflect the changing status of the field over two decades, its shifting emphases, issues, and perspectives. At the same time, we have included works that have asked compelling questions and thereby led historians to follow up on these important insights. Because not all the questions once raised continue to frame environmental scholarship, *Out of the Woods* serves as a kind of extended historiographical essay, its coverage ranging from the field's inception and awkward beginnings in the late 1960s to its current status as a more sophisticated and interdisciplinary academic enterprise.

The roots of this historiographical field stem from several distinct sources. One is the influential Annales school of French scholarship. Annalistes such as Marc Bloch, Lucian Febvre, and, later, Fernand Braudel pioneered aspects of what would become environmental history into their work; Bloch's studies of French peasants, Febrve's social geographies, and Braudel's inspired explication of the Mediterranean world emphasized the importance nature played in shaping human behavior; Braudel in particular explored nature's influence on local cultures, and the influence these in turn had on their national societies. The physical world humans occupy helps construct their profound sense of place, and historians of the environment are fundamentally historians of that sense and of those places.

This perception was reinforced through the scholarly insights of Walter Prescott Webb, who, by framing the critical significance of aridity to the American west added an environmental determinism to the historiography. To him, frontiers were not only zones of human contact between what he called savagery and civilization, but sites of powerful collisions between ways of life from divergent climates, a grand sweep of an argument that his positivist, triumphalist baggage constrained. James Malin later extended this argument; like Webb, he was obsessed with aridity, and pressed Webb's paradigmatic thinking about frontiers into the realm of ecological theory. By modern standards, Malin was quirky. An unreconstructed social Darwinist, he saw in nature the justification of the free market principles of Adam Smith that had been buried beneath the Industrial Revolution. If at their core these two influential thinkers reflected the importance of the environment in human history, and insisted that it was a legitimate force in causative explanations for human behavior or action, their insistence was so firm, their faith so absolute as to make their argument inflexible, even dogmatic. It would be left to Donald Worster and a later generation of environmental historians to transform this determinism into a fluid, more malleable form.

The "new" social history that emerged in the 1960s and 1970s was another antecedent of modern environmental history and has come to dominate American historical scholarship. Its emphasis on the lives of ordinary people—workers, husbands and wives, the poor, and others—and its "bottoms-up" focus meshed with and reinforced the tendency among the then-nascent environmental histo-rians to utilize a grass roots approach to their work; for them, local history was the bedrock of the new discipline. Rooted in the idea of place, they built from the ground up. It seemed almost natural, too, to examine the relationship between these newly enfranchised people, the institutions in which they labored, and the landscapes through which they moved; this early insight into the interconnect-edness of humanity and place now serves as an antecedent for bioregionalism.

These writers and their ideas, then, constituted a basis from which the mod-ern field of environmental history emerged in the late 1960s and early 1970s. However, two scholars' different perspectives laid down the field's original con-tours. One of these was Samuel P. Hays, whose influential *Conservation and the Gospel of Efficiency* (1959) established the political tradition in environmental his-tory. In the nearly thirty years between its publication and that of his no less influential *Beauty, Health, and Permanence* (1987), which chronicled the historic transformation from conservation to environmentalism, he probed deeply into the political processes, federal legislation, and policy that made up the environ-mental revolution—work that has spawned a generation of successors. At the same time, Roderick Nash added new energy to the study of the intellectual his-tory of the environment. His *Wilderness and the American Mind* (1967), which built on the tradition of such scholarship as Han Huth's *Nature and the American* (1957) and Arthur A. Ekirch Jr.'s *Man and Nature in America* (1963), resonated more fully with a broad-based readership in part because of its insights, in part because of its timing: Nash published his book at precisely the moment when the environment as an idea—which was after all the subject of *Wilderness and the American Mind*—became a driving force in American culture.

This was the time, then, when the development of an environmental histori-ography took off. In addition to the influences of Hays and Nash, other seminal works appeared, including J. Donald Hughes's *In the House of Stone and Light: A Human History of the Grand Canyon* (1967) and Donald Worster's *American Envi-ronmentalism* (1973). Amid this was a resurgence of interest in the nature writings of Henry David Thoreau and Ralph Waldo Emerson and the republication of Aldo Leopold's *Sand County Almanac* (1968) and Charles A. Reich's stunningly popular *The Greening of America* (1970), to cite but a few of the major works that found their way onto best-seller lists and into student backpacks. A natural con-sequence of this expanded reading list was the appearance of college courses in the study of the environment. Among those introducing the fledgling curriculum were Nash at the University of California, Santa Barbara, and Worster at Yale,

Brandeis, and Hawaii, Manoa. These academic pioneers had begun to stake their claim amid course catalogs and curricula.[2]

This new discipline's parameters soon became clear, in the classic pattern of academic organizational strategy, when in 1973 initial conversations occurred about the need for a journal and a professional society devoted to environmental history. These discussions were the brainchild of the convivial John Opie, now at the New Jersey Institute of Technology, but then affiliated with Duquesne University. At the 1973 American Studies Association meeting in San Antonio, Texas, Opie took Worster for a stroll along the city's famed Riverwalk, and when they sat down on a park bench, they began to hatch ideas for how to organize historians interested in environmental issues. Into their discussions they drew fellow conventioneers Susan Flader of the University of Missouri and Stephen J. Pyne, then a graduate student at the University of Texas; together, this small cadre combined with other scholars and met at the end of the conference to sketch out a formal structure of what would become known as the American Society for Environmental History. There was, Don Worster would remember, a "universal feeling of enthusiasm."[3]

This momentum carried forward to the Organization of American Historians meeting in Denver in April 1974, where a session on the teaching of environmental history was held. John Opie recalled the pleasure of discovering at that meeting "that each of us was not the only one teaching the subject," a sense of collaboration that confirmed the felt need for a new society, with its own journal, the *Environmental Review*, and an internal, and more informal, form of communication, the "Environmental History Newsletter." The subdiscipline now had a structure, a future.[4]

One who helped to chart the society's and the discipline's future course was John Opie, who served as the *Environmental Review*'s first editor, and thereby left an indelible imprint on the field. His energy carried the journal through its earliest and hardest moments, and in 1983 he was succeeded by Don Hughes, whom Opie described as a "friend and co-conspirator." Hughes, on the faculty at the University of Denver, brought a flair and graciousness to the journal, inaugurating, for instance, its first special number, dedicated to the issue of Native Americans and the environment and edited by Richard White, then a promising young scholar at the University of Utah. By the time Hughes chose to step down at the end of a three-year term in 1985, *Environmental Review* had become a quarterly journal and had made its mark. To say that for the first generation of environmental historians there was no more important publishing outlet than *Environ-*

mental Review is to grossly understate its centrality. Without the journal, the opportunities to contribute to this form of historiographical research simply would not have existed.

The journal's central place in the field has only been enhanced during its second decade of existence. Through William Robbins, who succeeded Hughes with the first issue of 1986—a term that lasted into 1988—the journal began to accommodate new voices and views, a process that John Opie, who returned for a second term as editor in the middle of 1988, embraced, and that current editor Hal Rothman has continued to promote.[5]

This inclusivity and expansiveness has been one of the hallmarks of environmental history, as the chapters in *Out of the Woods* suggest. In the journal's first decade, for example, environmental scholarship tended to focus on the political history of the conservation movement, or on the concept of wilderness, foci that led historians to assess the damage that human society had wrought upon the landscape, a view animated by the political energies of the day. Often, these analyses were marked by an urgent sense of advocacy, as their authors believed they had an obligation to alert the public to the character and consequences of ecological devastation, historical and contemporary. The framework by which environmental historians have interpreted their field has continued to be shaped by questions concerning humanity's impact on the earth. While this assessment came to include what Alfred Crosby called humanity's "portmanteau biota"—the collection of animals, microbes, and plants that accompany us as we move across the globe—even these accoutrements tended to confine scholarship to but one realm, that which Donald Worster has defined as the agro-ecological dimension.

That emphasis has begun to change over the past fifteen years or so, as environmental historians have increasingly widened their frames of reference, absorbing the perceptions of urban history and pollution studies, as well as those of environmental justice and its corollary, eco-racism (two perspectives that reveal the continuing presence of political advocacy within environmental scholarship). Historians have also applied more theoretical constructs to the field, ranging from the adoption of chaos theory and Gaian perspectives to the methodological insights of race, class, and gender. Many of these have begun to transform how we conceive of the environment, a self-critical reflectiveness that has become typical of academic disciplines in the waning years of the twentieth century. *Out of the Woods* reflects these influences, too: its physical layout and organizational structure are emblematic of the shifts in the kind and focus of the scholarly agenda.

As these and other intellectual innovations alter the manner in which environmental historians conceive of and argue about the world, they will have ramifications for public discourse: because of its interdisciplinary character, the field has always had the ability to challenge, if not change, the way other historians, policy makers, political activists, and the judiciary think, and then act, on these thoughts. Those who practice environmental history will know that they have been successful if these actions incorporate, even if only infrequently, a bear's point of view.

IDEAS MATTER

Environmental history did not burst onto the intellectual scene in the early 1970s complete with theoretical models or precise methodological foci: it was, and is, a field of scholarship the very nature of which seems to defy theoretical impress. Take as an example one of its many charges, the attempt to reconstruct and then interpret humanity's shifting relationship to the physical world in which it lives. To pursue this kind of analysis, scholars must necessarily be flexible toward and responsive to an ever-changing set of historical variables and cultural perspectives. They, like the environment they study, have not been, nor can they remain, static.

This is as true for the questions environmental historians have asked as it is for the methodologies they have employed, and the ideas they have developed: each has varied greatly from one scholar to the next, from one time to another. Perhaps this fluidity represents, as Carolyn Merchant implies in "The Theoretical Structure of Ecological Revolutions," an early stage in the field's intellectual development. Explicit analyses of the theories underpinning scholarly perspectives, then, may be a matter of maturation, one confirmation of which is that the three chapters in this section were originally published during the journal's second decade. As it turns out, each is concerned with the evolution of nature as an idea, with the shifts in the human representation of the nonhuman world. Merchant probes the major ecological transformations in

New England land and life between the seventeenth and nineteenth centuries, a probe that examines the interplay between
ecological alterations, modes of production and reproduction,
and the impact these have had on human consciousness. A
self-consciousness is central to Donald Worster's "The Ecology of Order and Chaos," a close reading of the changes in
meaning imputed to the word *ecology* and the intellectual paradigms it has generated: from the orderly vision of Frederick
Clement, for whom ecological change was rational, predictable, and thus knowable, to the emerging nonlinear conceptions of a chaotic universe that is perhaps beyond our comprehension. In such a conception of the world, Worster
wonders, what "is there to love or preserve"?

An answer to this query unfolds in William Cronon's "The
Trouble with Wilderness," a controversial challenge to the
prevailing and romantic distinction Americans have long held
between wild and domestic landscapes. The trouble, he argues, lies in our insistence that wilderness is "the best antidote
to our human selves," a refuge from ourselves. But this conception of wilderness is a "cultural invention" of the very civilization we apparently hope to escape: "We mistake ourselves
when we suppose that wilderness can be the solution to our
culture's problematic relationships with the nonhuman world,
for wilderness is no small part of the problem." Only when we
learn to perceive wildness as "humane as it is natural" will we
be able to "get on with the unending struggle to live rightly in
the world."

The Ecology of Order and Chaos

DONALD WORSTER

The science of ecology has had a popular impact unlike that of any other academic field of research. Consider the extraordinary ubiquity of the word itself: it has appeared in the most everyday places and the most astonishing, on Day-Glo T-shirts, in corporate advertising, and on bridge abutments. It has changed the language of politics and philosophy—springing up in a number of countries are political groups that are self-identified as "Ecology Parties." Yet who ever proposed forming a political party named after comparative linguistics or advanced paleontology? On several continents we have a philosophical movement termed "Deep Ecology," but nowhere has anyone announced a movement for "Deep Entomology" or "Deep Polish Literature." Why has this funny little word, ecology, coined by an obscure nineteenth-century German scientist, acquired so powerful a cultural resonance, so widespread a following?

Behind the persistent enthusiasm for ecology, I believe, lies the hope that this science can offer a great deal more than a pile of data. It is supposed to offer a pathway to a kind of moral enlightenment that we can call, for the purposes of simplicity, "conservation." The expectation did not originate with the public but first appeared among eminent scientists within the field. For instance, in his 1935 book *Deserts on the March*, the noted University of Oklahoma, and later Yale, botanist Paul Sears urged Americans to take ecology seriously, promoting it in their universities and making it part of their governing process. "In Great Britain," he pointed out, the ecologists are being consulted at every step in planning the proper utilization of those parts of the Empire not yet settled, thus ... ending the era of haphazard exploitation. There are hopeful, but all too few signs that our own national government realizes the part which ecology must play in a permanent program.[1]

Sears recommended that the United States hire a few thousand ecologists at the county level to advise citizens on questions of land use and thereby bring an end to environmental degradation; such a brigade, he thought, would put the whole nation on a biologically and economically sustainable basis.

In a 1947 addendum to his text, Sears added that ecologists, acting in the public interest, would instill in the American mind that "body of knowledge," that "point of view, which peculiarly implies all that is meant by conservation."[2] In other words, by the time of the 1930s and 1940s, ecology was being hailed as a much needed guide to a future motivated by an ethic of conservation. And conservation for Sears meant restoring the biological order, maintaining the health of the land and thereby the well-being of the nation, pursuing by both moral and technical means a lasting equilibrium with nature.

While we have not taken to heart all of Sears's suggestions—have not yet put any ecologists on county payrolls, with an office next door to the tax collector and sheriff—we have taken a surprisingly long step in his direction. Every day in some part of the nation, an ecologist is at work writing an environmental impact report or monitoring a human disturbance of the landscape or testifying at a hearing.

Twelve years ago I published a history, going back to the eighteenth century, of this scientific discipline and its ideas about nature.[3] The conclusions in that book still strike me as being, on the whole, sensible and valid: that this science has come to be a major influence on our perception of nature in modern times; that its ideas, on the other hand, have been reflections of ourselves as much as objective apprehensions of nature; that scientific analysis cannot take the place of moral reasoning; that science, including the science of ecology, promotes, at least in some of its manifestations, a few of our darker ambitions toward nature and therefore itself needs to be morally examined and critiqued from time to time. Ecology, I argued, should never be taken as an all-wise, always trustworthy guide. We must be willing to challenge this authority, and indeed challenge the authority of science in general; not be quick to scorn or vilify or behead, but simply, now and then, to question. During the period since my book was published, there has accumulated a considerable body of new thinking and new research in ecology. In this essay I mean to survey some of that recent thinking, contrasting it with its predecessors, and to raise a few of the same questions I did before. Part of my ar-

DONALD WORSTER

gument will be that Paul Sears would be astonished, and perhaps dismayed, to hear the kind of advice that ecological experts have to give these days. Less and less do they offer, or even promise to offer, what he would consider to be a program of moral enlightenment of conservation in the sense of a restored equilibrium between humans and nature.

There is a clear reason for that outcome, I will argue, and it has to do with drastic changes in the ideas that ecologists hold about the structure and function of the natural world. In Sears's day ecology was basically a study of equilibrium, harmony, and order; it had been so from its beginnings. Today, however, in many circles of scientific research, it has become a study of disturbance, disharmony, and chaos, and coincidentally or not, conservation is often not even a remote concern.

At the time *Deserts on the March* appeared in print, and through the time of its second and even third edition, the dominant name in the field of American ecology was that of Frederic L. Clements, who more than any other individual introduced scientific ecology into our national academic life. He called his approach "dynamic ecology," meaning it was concerned with change and evolution in the landscape. At its heart Clements's ecology dealt with the process of vegetational succession—the sequence of plant communities that appear on a piece of soil, newly made or disturbed, beginning with the first pioneer communities that invade and get a foothold.[4] Here is how I have defined the essence of the Clementsian paradigm:

> Change upon change became the inescapable principle of Clements's science. Yet he also insisted stubbornly and vigorously on the notion that the natural landscape must eventually reach a vaguely final climax stage. Nature's course, he contended, is not an aimless wandering to and fro but a steady flow toward stability that can be exactly plotted by the scientist.[5]

Most interestingly, Clements referred to that final climax stage as a "superorganism," implying that the assemblage of plants had achieved the close integration of parts, the self-organizing capability, of a single animal or plant. In some unique sense, it had become a live, coherent thing, not a mere collection of atomistic individuals, and exercised some control over the nonliving world around it, as organisms do.

Until well after World War II Clements's climax theory dominated ecological thought in this country.[6] Pick up almost any textbook in the field written forty, or even thirty, years ago, and you will likely find mention of the climax. It was this theory that Paul Sears had studied and took to be the core lesson of ecology that his county ecologists should teach their fellow citizens: that nature tends toward a climax state and that, as far as practicable, they should learn to respect and preserve it. Sears wrote that the chief work of the scientist ought to be to show "the unbalance which man has produced on this continent" and to lead people back to some approximation of nature's original health and stability.[7]

But then, beginning in the 1940s, while Clements and his ideas were still in the ascendent, a few scientists began trying to speak a new vocabulary. Words like "energy flow," "trophic levels," and "ecosystem" appeared in the leading journals, and they indicated a view of nature shaped more by physics than botany. Within another decade or two nature came to be widely seen as a flow of energy and nutrients through a physical or thermodynamic system. The early figures prominent in shaping this new view included C. Juday, Raymond Lindeman, and G. Evelyn Hutchinson. But perhaps its most influential exponent was Eugene P. Odum, hailing from North Carolina and Georgia, discovering in his southern saltwater marshes, tidal estuaries, and abandoned cotton fields the animating, pulsating force of the sun, the global flux of energy. In 1953 Odum published the first edition of his famous textbook, *The Fundamentals of Ecology*.[8] In 1966 he became president of the Ecological Society of America.

By now anyone in the United States who regularly reads a newspaper or magazine has come to know at least a few of Odum's ideas, for they furnish the main themes in our popular understanding of ecology, beginning with the sovereign idea of the ecosystem. Odum defined the ecosystem as "any unit that includes all of the organisms (i.e., the "community") in a given area interacting with the physical environment so that a flow of energy leads to clearly defined trophic structure, biotic diversity, and material cycles (i.e., exchange of materials between living and nonliving parts) within the system."[9] The whole earth, he argued, is organized into an interlocking series of such "ecosystems," ranging in size from a small pond to so vast an expanse as the Brazilian rainforest.

What all those ecosystems have in common is a "strategy of development," a kind of game plan that gives nature an overall direction. That strategy is, in Odum's words, "directed toward achieving as large and diverse an organic structure as is possible within the limits set by the available energy input and the prevailing physical conditions of existence."[10] Every single ecosystem, he believed, is either moving toward or has already achieved that goal. It is a clear, coherent, and easily observable strategy; and it ends in the happy state of order.

Nature's strategy, Odum added, leads finally to a world of mutualism and cooperation among the organisms inhabiting an area. From an early stage of competing against one another, they evolve toward a more symbiotic relationship.

They learn, as it were, to work together to control their surrounding environment, making it more and more suitable as a habitat, until at last they have the power to protect themselves from its stressful cycles of drought and flood, winter and summer, cold and heat. Odum called that point "homeostasis." To achieve it, the living components of an ecosystem must evolve a structure of interrelatedness and cooperation that can, to some extent, manage the physical world—manage it for maximum efficiency and mutual benefit.

I have described this set of ideas as a break from the past, but that is misleading. Odum may have used different terms than Clements, may even have had a radically different vision of nature at times; but he did not repudiate Clements's notion that nature moves toward order and harmony. In the place of the theory of the "climax" stage he put the theory of the "mature ecosystem." His nature may have appeared more as an automated factory than as a Clementsian super-organism, but like its predecessor it tends toward order.

The theory of the ecosystem presented a very clear set of standards as to what constituted order and disorder, which Odum set forth in the form of a "tabular model of ecological succession." When the ecosystem reaches its end point of homeostasis, his table shows, it expends less energy on increasing production and more on furnishing protection from external vicissitudes: that is, the biomass in an area reaches a steady level, neither increasing nor decreasing, and the emphasis in the system is on keeping it that way—on maintaining a kind of no-growth economy. Then the little, aggressive, weedy organisms common at an early stage in development (the r-selected species) give way to larger, steadier creatures (K-selected species), who may have less potential for fast growth and explosive reproduction but also better talents at surviving in dense settlements and keeping the place on an even keel.[11] At that point there is supposed to be more diversity in the community—i.e., a greater array of species. And there is less loss of nutrients to the outside; nitrogen, phosphorous, and calcium all stay in circulation within the ecosystem rather than leaking out. Those are some of the key indicators of ecological order, all of them susceptible to precise measurement. The suggestion was implicit but clear that if one interfered too much with nature's strategy of development, the effects might be costly: a serious loss of nutrients, a decline in species diversity, an end to biomass stability. In short, the ecosystem would be damaged.

The most likely source of that damage was no mystery to Odum: it was human beings trying to force up the production of useful commodities and stupidly risking the destruction of their life support system.

Man has generally been preoccupied with obtaining as much "production" from the landscape as possible, by developing and maintaining early succession-al types of ecosystems, usually monocultures. But, of course, man does not live by food and fiber alone; he also needs a balanced CO_2-O_2 atmosphere, the climatic buffer provided by oceans and masses of vegetation, and clean (that is, unproductive) water for cultural and industrial uses. Many essential life-cycle resources, not to mention recreational and esthetic needs, are best provided man by the less "productive" landscapes. In other words, the landscape is not just a supply depot but is also the *oikos*—the home—in which we must live.[12]

Odum's view of nature as a series of balanced ecosystems, achieved or in the making, led him to take a strong stand in favor of preserving the landscape in as nearly natural a condition as possible. He suggested the need for substantial restraint on human activity—for environmental planning "on a rational and scientific basis." For him as for Paul Sears, ecology must be taught to the public and made the foundation of education, economics, and politics; America and other countries must be "ecologized."

Of course not everyone who adopted the ecosystem approach to ecology ended up where Odum did. Quite the contrary, many found the ecosystem idea a wonderful instrument for promoting global technocracy. Experts familiar with the ecosystem and skilled in its manipulation, it was hoped in some quarters, could manage the entire planet for improved efficiency. "Governing" all of nature with the aid of rational science was the dream of these ecosystem technocrats.[13] But technocratic management was not the chief lesson, I believe, the public learned in Professor Odum's classroom; most came away devoted, as he was, to preserving large parts of nature in an unmanaged state and sure that they had been given a strong scientific rationale, as well as knowledge base, to do it. We must defend the world's endangered ecosystems, they insisted. We must safeguard the integrity of the Greater Yellowstone ecosystem, the Chesapeake Bay ecosystem, the Serengeti ecosystem. We must protect species diversity, biomass stability, and calcium recycling. We must make the world safe for K-species.[14]

That was the rallying cry of environmentalists and ecologists alike in the 1960s and early 1970s, when it seemed that the great coming struggle would be between what was left of pristine nature, delicately balanced in Odum's beautifully rational ecosystems, and a human race bent on mindless, greedy destruction. A decade or two later the situation has changed considerably. There are still environmental threats around, to be sure, and they are more dangerous than ever. The newspapers inform us of continuing disasters like the massive 1989 oil spill in Alaska's Prince William Sound, and reporters persist in using words like "ecosystem" and "balance" and "fragility" to describe such disasters. So do many scientists, who continue to acknowledge their theoretical indebtedness to Odum. For instance, in a recent British poll, 447 ecologists out of 645 questioned ranked the "ecosystem" as one of the most important concepts their discipline has contributed to our understanding of the natural world; indeed, "ecosystem" ranked first on their list, drawing more votes than nineteen other leading concepts.[15] But all the same, and despite the persistence of environmental problems, Odum's ecosystem is no longer the main theme in research or teaching in the science. A survey of recent ecology textbooks shows that the concept is not even mentioned in one leading work and has a much diminished place in the others.[16]

Ecology is not the same as it was. A rather drastic change has been going on in this science of late—a radical shifting away from the thinking of Eugene Odum's generation, away from its assumptions of order and predictability, a shifting toward what we might call a new ecology of chaos.

In July 1973, the *Journal of the Arnold Arboretum* published an article by two scientists associated with the Massachusetts Audubon Society, William Drury and Ian Nisbet, and it challenged Odum's ecology fundamentally. The title of the article was simply "Succession," indicating that old subject of observed sequences in plant and animal associations. With both Frederic Clements and Eugene Odum, succession had been taken to be the straight and narrow road to equilibrium. Drury and Nisbet disagreed completely with that assumption. Their observations, drawn particularly from northeastern temperate forests, strongly suggested that the process of ecological succession does not lead anywhere. Change is without any determinable direction and goes on forever, never reaching a point of stability. They found no evidence of any progressive development in nature: no progressive increase over time in biomass stabilization, no progressive diversification of species, no progressive movement toward a greater cohesiveness in plant and animal communities, nor toward a greater success in regulating the environment. Indeed, they found none of the criteria Odum had posited for mature ecosystems. The forest, they insisted, no matter what its age, is nothing but an erratic, shifting mosaic of trees and other plants. In their words, "most of the phenomena of succession should be understood as resulting from the differential growth, differential survival, and perhaps differential dispersal of species adapted to grow at different points on stress gradients."[17] In other words, they could see lots of individual species, each doing its thing, but they could locate no emergent collectivity, nor any strategy to achieve one.

Prominent among their authorities supporting this view was the nearly forgotten name of Henry A. Gleason, a taxonomist who, in 1926, had challenged Frederic Clements and his organismic theory of the climax in an article entitled, "The Individualistic Concept of the Plant Association." Gleason had argued that we live in a world of constant flux and impermanence, not one tending toward Clements's climaxes. There is no such thing, he argued, as balance or equilibrium or steady-state. Each and every plant association is nothing but a temporary gathering of strangers, a clustering of species unrelated to one another, here for a brief while today, on their way somewhere else tomorrow. "Each . . . species of plant is a law unto itself," he wrote.[18] We look for cooperation in nature and we find only competition. We look for organized wholes, and we can discover only loose atoms and fragments. We hope for order and discern only a mishmash of conjoining species, all seeking their own advantage in utter disregard of others.

Thanks in part to Drury and Nisbet, this "individualistic" view was reborn in the mid-1970s and, during the past decade, it became the core idea of what some scientists hailed as a new, revolutionary paradigm in ecology. To promote it, they attacked the traditional notion of succession; for to reject that notion was to reject the larger idea that organic nature tends toward order. In 1977 two more biologists, Joseph Connell and Ralph Slatyer, continued the attack, denying the old claim that an invading community of pioneering species, the first stage in Clements's sequence, works to prepare the ground for its successors, like a group of Daniel Boones blazing the trail for civilization. The first comers, Connell and Slatyer maintained, manage in most cases to stake out their claims and successfully defend them; they do not give way to a later, superior group of colonists. Only when the pioneers die or are damaged by natural disturbances, thus releasing the resources they have monopolized, can latecomers find a foothold and get established.[19]

As this assault on the old thinking gathered momentum, the word "disturbance" began to appear more frequently in the scientific literature and be taken far more seriously. Disturbance was not a common subject in Odum's heyday, and it almost never appeared in combination with the adjective "natural." Now, however, it was as though scientists were out looking strenuously for signs of disturbance in nature—especially signs of disturbance that were not caused by humans—and they were finding it everywhere. During the past decade those new ecologists succeeded in leaving little tranquility in primitive nature. Fire is one of the most common disturbances they noted. So is wind, especially in the form of violent hurricanes and tornadoes. So are invading populations of microorganisms and pests and predators. And volcanic eruptions. And invading ice sheets of the Quaternary Period. And devastating droughts like that of the 1930s in the American West. Above all, it is these last sorts of disturbances, caused by the restlessness of climate, that the new generation of ecologists has emphasized. As one of the most influential of them, Professor Margaret Davis of the University of Minnesota, has written: "For the last 50 years or 500 or 1,000—as long as anyone would claim for ecological time—there has never been an interval when temperature was in a steady state with symmetrical fluctuations about a mean. . . . Only on the longest time scale, 100,000 years, is there a tendency toward cyclical variation, and the cycles are asymmetrical, with a mean much different from today."[20]

One of the most provocative and impressive expressions of the new post-Odum ecology is a book of essays edited by S. T. A. Pickett and P. S. White, *The Ecology of Natural Disturbance and Patch Dynamics* (1985). I submit it as symptomatic of much of the thinking going on today in the field. Though the final section of the book does deal with ecosystems, the word has lost much of its former meaning and implications. Two of the authors in fact open their contribution with a complaint that many scientists assume that "homogeneous ecosystems are a reality," when in truth "virtually all naturally occurring and man-disturbed ecosystems are mosaics of environmental conditions." "Historically," they write, "ecologists have been slow to recognize the importance of disturbances and the heterogeneity they generate." The reason for this slowness? "The majority of both theoretical and empirical work has been dominated by an equilibrium perspective."[21] Repudiating that perspective, these authors take us to the tropical forests of South and Central America and to the Everglades of Florida, showing us instability on every hand: a wet, green world of continual disturbance—or as they prefer to say, "of perturbations." Even the grasslands of North America, which inspired Frederic Clements's theory of the climax, appear in this collection as regularly disturbed environments. One paper describes them as a "dynamic, fine-textured mosaic" that is constantly kept in upheaval by the workings of badgers, pocket gophers, and mound-building ants, along with fire, drought, and eroding wind and water.[22] The message in all these papers is consistent: The climax notion is dead, the ecosystem has receded in usefulness, and in their place we have the idea of the lowly patch. Nature should be regarded as a landscape of patches, big and little, patches of all textures and colors, a patchwork quilt of living things, changing continually through time and space, responding to an unceasing barrage of perturbations. The stitches in that quilt never hold for long.

Now, of course, scientists have known about gophers and winds, the Ice Age and droughts for a considerable time. Yet heretofore they have not let those disruptions spoil their theories of balanced plant and animal associations, and we must ask why that was so. Why did Clements and Odum tend to dismiss such forces as climatic change, at least of the less catastrophic sort, as threats to the order of nature? Why have their successors, on the other hand, tended to put so much emphasis on those same changes, to the point that they often see nothing but instability in the landscape?

One clue comes from the fact that many of these disturbance boosters are not and have never been ecosystem scientists; they received their training in the subfield of population biology and reflect the growing confidence, methodological maturity, and influence of that subfield.[23] When they look at a forest, the population ecologists see only the trees.

See them and count them—so many white pines, so many hemlocks, so many maples and birches. They insist that if we know all there is to know about the individual species that constitute a forest, and can measure their lives in precise, quantitative terms, we will know all there is to know about that forest. It has no "emergent" or organismic properties. It is not some whole greater than the sum of its parts, requiring "holistic" understanding. Outfitted with computers that can track the life histories of individual species, and chart the rise and fall of populations, they have brought a degree of mathematical precision to ecology that is awesome to contemplate. And what they see when they look at population histories for any patch of land is wildly swinging oscillations. Populations rise and populations fall, like stock market prices, auto sales, and hemlines. We live, they insist, in a nonequilibrium world.[24]

There is another reason for the paradigmatic shift I have been describing, though I suggest it quite tentatively and can offer only sketchy evidence for it. For some scientists, a nature characterized by highly individualistic associations, constant disturbance, and incessant change may be more ideologically satisfying than Odum's ecosystem, with its stress on cooperation, social organization, and environmentalism. A case in point is the very successful popularizer of contemporary ecology, Paul Colinvaux, author of *Why Big Fierce Animals Are Rare* (1978). His chapter on succession begins with these lines: "If the planners really get hold of us so that they can stamp out all individual liberty and do what they like with our land, they might decide that whole counties full of inferior farms should be put back into forest." Clearly, he is not enthusiastic about land-use planning or forest restoration. And he ends that same chapter with these remarkably revealing and self-assured words:

We can now . . . explain all the intriguing, predictable events of plant successions in simple, matter of fact, Darwinian ways. Everything that happens in successions comes about because all the different species go about earning their livings as best they may, each in its own individual manner. What look like community properties are in fact the summed results of all these bits of private enterprise.[25]

Apparently, if this example is any indication, the social Darwinists are back on the scene, and at least some of them are ecologists, and at least some of their opposition to Odum's science may have to do with a revulsion toward its political implications, including its attractiveness for environmentalists. Colinvaux is very clear about the need to get some distance between himself and groups like the Sierra Club.

I am not alone in wondering whether there might be a deeper, half-articulated ideological motive generating the new direction in ecology. The Swedish historian of science, Thomas Söderqvist, in his recent study of ecology's development in his country, concludes that the present generation of evolutionary ecologists seem to do ecology for fun only, indifferent to practical problems, including the salvation of the nation. They are mathematically and theoretically sophisticated, sitting indoors calculating on computers, rather than traveling out in the wilds. They are individualists, abhorring the idea of large-scale ecosystem projects. Indeed, the transition from ecosystem ecology to evolutionary ecology seems to reflect the generational transition from the politically conscious generation of the 1960s to the 'yuppie' generation of the 1980s.[26]

That may be an exaggerated characterization, and I would not want to apply it to every scientist who has published on patch dynamics or disturbance regimes. But it does draw our attention to an unmistakable attempt by many ecologists to disassociate themselves from reform environmentalism and its criticisms of human impact on nature.

I wish, however, that the emergence of the new post-Odum ecology could be explained so simply in those two ways: as a triumph of reductive population dynamics over holistic consciousness, or as a triumph of social Darwinist or entrepreneurial ideology over a commitment to environmental preservation. There is, it seems, more going on than that, and it is going on all through the natural sciences—biology, astronomy, physics—perhaps going on through all modern technological societies. It is nothing less than the discovery of chaos. Nature, many have begun to believe, is *fundamentally* erratic, discontinuous, and unpredictable. It is full of seemingly random events that elude our models of how things are supposed to work. As a result, the unexpected keeps hitting us in the face.

Clouds collect and disperse, rain falls or doesn't fall, disregarding our careful weather predictions, and we cannot explain why. Cars suddenly bunch up on the freeway, and the traffic controllers fly into a frenzy. A man's heart beats regularly year after year, then abruptly begins to skip a beat now and then. A ping pong ball bounces off the table in an unexpected direction. Each little snowflake falling out of the sky turns out to be completely unlike any other. Those are ways in which nature seems, in contrast to all our previous theories and methods, to be chaotic. If the ultimate test of any body of scientific knowledge is its ability to predict events, then all the sciences and pseudo-sciences—physics, chemistry, climatology, economics, ecology—fail the test regularly. They all have been announcing laws, designing models, predicting what an individual atom or person is supposed to do; and now, increasingly, they are beginning to confess that the world never quite behaves the way it is supposed to do.

Making sense of this situation is the task of an altogether new kind of inquiry calling itself the science of chaos. Some say it portends a revolution in thinking equivalent to quantum mechanics or relativity. Like those other twentieth-century revolutions, the science of chaos rejects tenets going back as far as the days of Sir Isaac Newton. In fact, what is occurring may be not two or three separate revolutions but a single revolution against all the principles, laws, models, and applications of classical science, the science ushered in by the great Scientific Revolution of the seventeenth century.[27] For centuries we have assumed that nature, despite a few appearances to the contrary, is a perfectly predictable system of linear, rational order. Give us an adequate number of facts, scientists have said, and we can describe that order in complete detail—can plot the lines along which everything moves and the speed of that movement and the collisions that will occur. Even Darwin's theory of evolution, which in the last century challenged much of the Newtonian worldview, left intact many people's confidence that order would prevail at last in the evolution of life; that out of the tangled history of competitive struggle would come progress, harmony, and stability. Now that traditional assumption may have broken down irretrievably. For whatever reason, whether because empirical data suggests it or because extrascientific cultural trends do—the experience of so much rapid social change in our daily lives—scientists are beginning to focus on what they had long managed to avoid seeing. The world is more complex than we ever imagined, they say, and indeed, some would add, ever can imagine.[28]

Despite the obvious complexity of their subject matter, ecologists have been among the slowest to join the cross-disciplinary science of chaos. I suspect that the influence of Clements and Odum, lingering well into the 1970s, worked against the new perspective, encouraging faith in linear regularities and equilibrium in the interaction of species. Nonetheless, eventually there arrived a day of conversion. In 1974 the Princeton mathematical ecologist Robert May published a paper with the title, "Biological Populations with Nonoverlapping Generations: Stable Points, Stable Cycles, and Chaos."[29] In it he admitted that the mathematical models he and others had constructed were inadequate approximations of the ragged life histories of organisms. They did not fully explain, for example, the aperiodic outbreaks of gypsy moths in eastern hardwood forests or the Canadian lynx cycles in the subarctic. Wildlife populations do not follow some simple Malthusian pattern of increase, saturation, and crash.

More and more ecologists have followed May and begun to try to bring their subject into line with chaotic theory. William Schaefer is one of them; though a student of Robert MacArthur, a leader of the old equilibrium school, he has been lately struck by the same anomaly of unpredictable fluctuations in populations as May and others. Though taught to believe in the "so-called 'Balance of Nature,'" he writes, ". . . the idea that populations are at or close to equilibrium," things now are beginning to look very different.[30] He describes himself has having to reach far across the disciplines, to make connections with concepts of chaos in the other natural sciences, in order to free himself from his field's restrictive past.

The entire study of chaos began in 1961, with efforts to simulate weather and climate patterns on a computer at MIT. There, meteorologist Edward Lorenz came up with his now famous "Butterfly Effect," the notion that a butterfly stirring the air today in a Beijing park can transform storm systems next month in New York City. Scientists call this phenomenon "sensitive dependence on initial conditions." What it means is that tiny differences in input can quickly become substantial differences in output. A corollary is that we cannot know, even with all our artificial intelligence apparatus, every one of the tiny differences that have occurred or are occurring at any place or point in time; nor can we know which tiny differences will produce which substantial differences in output. Beyond a short range, say, of two or three days from now, our predictions are not worth the paper they are written on.

The implications of this "Butterfly Effect" for ecology are profound. If a single flap of an insect's wings in China can lead to a torrential downpour in New York, then what might it do to the Greater Yellowstone Ecosystem? What can ecologists possibly know about all the forces impinging on, or about to impinge on, any piece of land? What can they safely ignore and what must they pay attention to?

Ecology of Order and Chaos   15

What distant, invisible, minuscule events may even now be happening that will change the organization of plant and animal life in our back yards? This is the predicament, and the challenge, presented by the science of chaos, and it is altering the imagination of ecologists dramatically.

John Muir once declared, "when we try to pick out anything by itself, we find it hitched to everything else in the universe."[31] For him, that was a manifestation of an infinitely wise plan in which everything functioned with perfect harmony. The new ecology of chaos, though impressed like Muir with interdependency, does not share his view of "an infinitely wise plan" that controls and shapes everything into order. There is no plan, today's scientists say, no harmony apparent in the events of nature. If there is order in the universe—and there will no longer be any science if all faith in order vanishes—it is going to be much more difficult to locate and describe than we thought.

For Muir, the clear lesson of cosmic complexity was that humans ought to love and preserve nature just as it is. The lessons of the new ecology, in contrast, are not at all clear. Does it promote, in Ilya Prigogine and Isabelle Stenger's words, "a renewal of nature," a less hierarchical view of life, and a set of "new relations between man and nature and between man and man?"[32] Or does it increase our alienation from the world, our withdrawal into post-modernist doubt and self-consciousness? What is there to love or preserve in a universe of chaos? How are people supposed to behave in such a universe? If such is the kind of place we inhabit, why not go ahead with all our private ambitions, free of any fear that we may be doing special damage? What, after all, does the phrase "environmental damage" mean in a world of so much natural chaos? Does the tradition of environmentalism to which Muir belonged, along with so many other nature writers and ecologists of the past like Paul Sears, Eugene Odum, Aldo Leopold, and Rachel Carson make sense any longer? I have no space here to attempt to answer those questions or to make predictions but only issue a warning that they are too important to be left for scientists alone to answer. Ecology today, no more than in the past, can not be assumed to be all-knowing or all-wise or eternally true.

Whether they are true or false, permanent or passingly fashionable, it does seem entirely possible that these changes in scientific thinking toward an emphasis on chaos will not produce any easing of the environmentalist's concern. Though words like ecosystem or climax may fade away and some new vocabulary take their place, the fear of risk and danger will likely become greater than ever. Most of us are intuitively aware, whether we can put our fears into mathematical formulae or not, that the technological power we have accumulated is *destructive-*

ly chaotic; not irrationally, we fear it and fear what it can to do us as well as the rest of nature.[33] It may be that we moderns, after absorbing the lessons of today's science, find we cannot love nature quite so easily as Muir did; but it may also be that we have discovered more reason than ever to respect it—to respect its baffling complexity, its inherent unpredictability, its daily turbulence. And to flap our own wings in it a little more gently.

The Theoretical Structure of
Ecological Revolutions

CAROLYN MERCHANT

Environmental history has reached a point in its evolution in which explicit attention to the theories that underlie its various interpretations is called for. Theories about the social construction of science and nature that have emerged over the past decade in the wake of Thomas Kuhn's *Structure of Scientific Revolutions* is one such approach. It accepts the relativist stance toward science set forth in the first edition of his book. (Kuhn backed away from that position toward a view of the progress of knowledge in a second edition.) Marxist theories that attempt to understand history as constructions of the material-social world existing in particular times and places provide a second influence. The theory of ecological revolutions that follows draws on social construction approaches and uses New England as a case study.[1]

Two major transformations in New England land and life took place between 1600 and 1860. The first, a colonial ecological revolution, occurred during the seventeenth century and was externally generated. It resulted in the collapse of indigenous Indian ecologies and the incorporation of a European ecological complex of animals, plants, pathogens, and people. It was legitimated by a set of symbols that placed cultured Europeans above wild nature, other animals, and "beastlike savages." It substituted a visual for an oral consciousness and an image of nature as female and subservient to a transcendent male God for an animistic fabric of symbolic exchanges between people and nature.

The second transformation, a capitalist ecological revolution, took place roughly between the American Revolution and about 1860. That second revolution was internally generated and resulted in the reintroduction of soil nutrients and native species. It demanded an economy of increased human labor, land management, and a legitimating mechanistic science. It split human consciousness into a disembodied analytic mind and a romantic emotional sensibility.

My thesis is that ecological revolutions are major transformations in human relations with nonhuman nature. They arise from changes, tensions, and contradictions that develop between a society's mode of production and its ecology, and between its modes of production and reproduction. Those dynamics in turn support the acceptance of new forms of consciousness, ideas, images, and world views. The course of the colonial and capitalist ecological revolutions in New England may be understood through a description of each society's production, reproduction, and forms of consciousness, the processes by which they broke down, and an analysis of the new relations between the emergent colonial or capitalist society and nonhuman nature.

Two frameworks of analysis offer springboards for discussing the structure of such ecological revolutions. In *The Structure of Scientific Revolutions* (first edition), Thomas Kuhn approached major transformations in scientific consciousness from a perspective internal to the workings of science and the community of scientists. One of the strengths of Kuhn's provocative account is its recognition of stable world views in science that exist for relatively long periods but are rapidly transformed during times of crisis and stress. One of its limitations is its failure to incorporate an interpretation of social forces external to the daily activities of science practitioners in their laboratories and field stations. Social and economic circumstances affect internal developments in scientific theories, at least indirectly. A viewpoint that incorporates social, economic, and ecological changes is required for a more complete understanding of scientific change.

A second approach to revolutionary transformations is that of Karl Marx and Friedrich Engels. According to their base/superstructure theory of history, social revolutions begin in the economic base of a particular social formation and result in a fairly rapid transformation of the legal, political, and ideological superstructure. In the most succinct statement of his theory of history, Marx wrote:

At a certain stage of their development, the material productive forces of society come in conflict with the existing relations of production. . . . Then begins an epoch of social revolution. With the change of the economic foundation the entire immense super-structure is more or less rapidly transformed.[2]

One weakness of that approach is the determinism Marx assigns to the economic base and the sharp demarcation between base and superstructure. But its strength lies in its view of society and change. If a society at a given time can be understood as a mutually supportive structure of dynamically interacting parts, then the process of its breakdown and transformation to a new whole can be described. Both Kuhn's theory of scientific revolution and Marx's theory of social revolution are starting points for a theory of ecology and history.

Science and history are both social constructions. Science is an ongoing negotiation with nonhuman nature for what counts as reality. Scientists socially construct nature, representing it differently in different historical epochs. Those social constructions change during scientific evolutions. Historians also socially construct the past in accordance with concepts relevant to the historian's present. History is thus a continuing negotiation between the historian and historical sources. Ecology is a particular twentieth-century construction of nature relevant to the concerns of environmental historians.

A scientific world view answers three key questions:

(1) What is the world made of? (the ontological question)
(2) How does change occur? (the historical question)
(3) How do we know? (the epistemological question)

World views such as animism, Aristotelianism, mechanism, and quantum field theory construct answers to these fundamental questions differently.

Environmental history poses similar questions:

(1) What concepts describe the world?
(2) What is the process by which change occurs?
(3) How does a society know the natural world?

The concepts most useful for this approach to environmental history are ecology, production, reproduction, and consciousness. Because of the differences in the immediacy of impact of production, reproduction, and consciousness on nonhuman nature, a structured, leveled framework of analysis is needed. This framework provides the basis for an understanding of stability as well as evolutionary change and transformation. Although change may occur at any level, ecological revolutions are characterized by major alterations at all three levels.

Widening tensions between the requirements of ecology and production in a given habitat and between production and reproduction initiate those changes. Those dynamics in turn lead to transformations in consciousness and legitimating world views.

Since the Scientific Revolution of the seventeenth century, the West has seen nature primarily through the spectacles of mechanistic science. Matter is dead and inert, remaining at rest or moving with uniform velocity in a straight line unless acted on by external forces. Change comes from outside as in the operation of a machine. The world is a clock, adjustable by human clock makers; nature is passive and manipulable.

An ecological approach to history asserts the idea of nature as a historical actor. It challenges the mechanistic tradition by focusing on the interchange of energy, materials, and information among living and nonliving beings in the natural environment. Nonhuman nature is not passive, but an active complex that participates in change over time and responds to human-induced change. Nature is a whole of which humans are only one part. We interact with plants, animals, and soils in ways that sustain or deplete local habitats, but through science and technology, we have greater power to alter the whole in a short period of time.

But like the mechanistic paradigm, the ecological paradigm is a socially constructed theory. Although it differs from mechanism by taking relations, context, and networks into consideration, it has no greater or lesser claim to ultimate truth than do earlier paradigms. Both mechanism and ecology construct their theories through a socially sanctioned process of problem identification, selection and deselection of particular "facts," inscription of the selected facts into texts, and the acceptance of a constructed order of nature by the scientific community. But laboratory and field ecology merge through the replication of laboratory conditions in the field. Farm, field, and forest are viewed as an ecological whole that includes both nonhuman nature and the human designer. The ecological approach of the twentieth century, like the earlier mechanistic one, has resulted from a socially constructed set of experiences sanctioned by scientific authority and a set of social practices and policies.[3]

Production is the human counterpart of "nature's" activity. The need to produce subsistence to reproduce human energy on a daily basis connects human communities with their local environments. Production for subsistence (or use) from the elements (or resources) of nature and the production of surpluses for market exchange are the primary ways in which humans interact directly with the local habitat. An ecological perspective unites the laws of nature with the processes of production through exchanges of energy. All animals, plants, and minerals are energy niches involved in the actual exchange of energy, materials, and information. The relation between human beings and the nonhuman world is reciprocal; when humans alter their surroundings, "nature" responds to those changes through ecological laws.

Production is the extraction, processing, and exchange of nature's parts as resources. In traditional cultures exchanges are often gifts or symbolic alliances while in market societies they are exchanged as commodities. For much of Western history, humans have produced and bartered food, clothing, and shelter primarily within the local community to reproduce daily life. But when commodities are marketed for profit, as in capitalist societies, they are often removed from the local habitat to distant places and exchanged for money. Marx and Engels distinguished between use-value production, or production for subsistence, and production for profit. When people "exploit" nonhuman nature, they do so in one of two ways: they either make immediate or personal use of it for subsistence, or they exchange its products as commodities for personal profit or gain.

New England is a significant historical example because several types of production evolved within the bounds of its present geographical area. Native Americans engaged primarily in gathering and hunting in the north and in horticulture in the south. Colonial Americans combined mercantile trade in natural resources with subsistence-oriented agriculture. The market and transportation revolutions of the nineteenth century initiated the transition to capitalist production. Historical bifurcation points within the evolutionary process can be identified roughly between 1600 and 1675 (the colonial ecological revolution) and between 1775 and 1860 (the capitalist ecological revolution).

To continue over time, life must be reproduced from generation to generation. The habitat is populated and repopulated with living organisms of all kinds. Biologically, all species must reproduce themselves inter-generationally. For humans, reproduction is both biological and social. Each adult generation must maintain itself, its parents, and its offspring so that human life may continue. And each individual must reproduce its own energy and that of its offspring (intra-generationally) on a daily basis through gathering, growing, or preparing food. Socially, humans must reproduce future laborers by passing on family and community norms. And they must reproduce and maintain the larger social order through the structures of governance and laws (such as property inheritance) and the ethical codes that reinforce behavior. Thus, although production is twofold—oriented toward subsistence use or market exchange—reproduction is fourfold, having both biological and social articulations.

Reproduction is the biological and social process through which humans are born, nurtured, socialized, and governed. Through reproduction sexual relations are legitimated, population sizes and family relationships are maintained, and property and inheritance practices are reinforced. In subsistence-oriented economies, production and reproduction are united in the maintenance of the local community. Under capitalism production and reproduction separate into two different spheres.

Claude Meillassoux's *Maidens, Meal, and Money* (1981) best explains the necessary connections between biological and social reproduction in subsistence economies. Production, he argues, exists for the sake of reproduction; the production and exchange of human energy are the keys to the reproduction of human life. Food must be extracted or produced to maintain the daily energy of producing adults, to maintain the energy of the children who will be the future producers, and to maintain that of the elders, the past producers. In this way reproducing life on a daily (intragenerational) basis through energy is linked directly to the intergenerational reproduction of the human species.[4]

Although the biological reproduction of life is possible only through the necessary connections between inter- and intragenerational reproduction, the community as a self-perpetuating unit is maintained by social reproduction. In addition, the political, legal, or governmental structures that maintain the mode of production will play the role of reproducing the social whole.[5]

Whereas Meillassoux was interested primarily in the concept of reproduction in subsistence societies, sociologist Abby Peterson examined the gender-sex dimension in politics to formulate an analysis of reproduction in capitalist societies. Under capitalism, the division of labor between the sexes has meant that men bear the responsibility for and dominate the production of exchange commodities, while women bear responsibility for reproducing the workforce and social relations. Peterson argues:

> Women's responsibility for reproduction includes both the biological reproduction of the species (intergenerational reproduction) and the intragenerational reproduction of the work force through unpaid labor in the home. Here too is included the reproduction of social relations—socialization.[6]

Under capitalist patriarchy, reproduction is subordinate to production.

Meillassoux's and Peterson's work offers an approach by which the analysis of reproduction can be advanced beyond demography to include daily life and the community itself. The sphere of reproduction is fourfold, having two biological and two social manifestations: (1) the intergenerational reproduction of the species (both human and nonhuman), (2) the intragenerational reproduction of daily life, (3) the reproduction of social norms within the family and community, and (4) the reproduction of the legal-political structures that maintain social order within the community and the state. The fourfold sphere of reproduction exists in a dynamic relationship with the twofold (subsistence or market-oriented) sphere of production.

Production and reproduction are in dynamic tension. When reproductive patterns are altered, as in population growth or changes in property inheritance, production is affected. Conversely, when production changes, as in the addition or depletion of resources or in technological innovation, reproductive structures are altered. A dramatic change at the level of either reproduction or production can alter the dynamic between them, resulting in a major transformation of the social whole.

Socialist-feminists have further elaborated the interaction between production and reproduction. In a 1976 article, "The Dialectics of Production and Reproduction in History," Renaté Bridenthal argues that changes in production give rise to changes in reproduction, creating tensions between them. For example, the change from an agrarian to an industrial capitalist economy—one that characterized the capitalist ecological revolution—can be described in terms of tensions, contradictions, and synthesis within the gender roles associated with production and reproduction. In the agrarian economy of colonial America, production and reproduction were symbiotic. Women participated in both spheres because the production and reproduction of daily life were centered in the household and domestic communities. Likewise, men working in barns and fields and women working in farmyards and farmhouses socialized children into production. But with industrialization, the production of items such as textiles and shoes moved out of the home into the factory, while farms became specialized and mechanized. Production became more public, reproduction more private, leading to their social and structural separation. For working-class women, the split between production and reproduction imposed a double burden of wage labor and housework; for middle-class women, it led to enforced idleness as "ladies of leisure."[7]

In New England the additional tensions between the requirements of intergenerational reproduction and those of subsistence production in rural areas also stimulated the capitalist ecological revolution. A partible system of patriarchal inheritance meant that farm sizes decreased after three or four generations to the point that not all sons inherited enough land to reproduce the subsistence system. The tensions between the requirements of subsistence-oriented production (a large family labor force) and social reproduction through partible inheritance (all sons must inherit farms) helped create a supply of landless sons, wage laborers for the transition to capitalist agriculture. The requirements of reproduction in its fourfold sense, therefore, came into conflict with the requirements of subsistence-oriented (use-value) production, stimulating a movement toward capital-intensive market production.

Consciousness is the totality of one's thoughts, feelings, and impressions, the awareness of one's acts and volitions. Group consciousness is a collective awareness by an aggregate of individuals. Both environments and culture shape individual and group consciousness. In different historical epochs, particular characteristics dominate a society's consciousness. Those forms of consciousness, through which the world is perceived, understood, and interpreted, are socially constructed and subject to change.

A society's symbols and images of nature express its collective consciousness. They appear in mythology, cosmology, science, religion, philosophy, language, and art. Scientific, philosophical, and literary texts are sources of the ideas and images used by controlling elites whereas rituals, festivals, songs, and myths provide clues to the consciousness of ordinary people. Ideas, images, and metaphors legitimate human behavior toward nature and are translated into action through ethics, morals, and taboos. According to Charles Taylor, particular intellectual frameworks give rise to a certain range of normative variations and not others, because their related values are not accidental. When sufficiently powerful, world views and their associated values can override social changes. But if they are weak, they can be undermined. A tribe of New England Indians or a community of colonial Americans may have a religious world view that holds it together for many decades while its economy is gradually changing. But eventually with the acceleration of commercial change, ideas that had formerly existed on the periphery, or among selected elites, may become dominant if they support and legitimate the new economic directions.[8]

For Native American cultures, consciousness was an integration of all the bodily senses in sustaining life. In that mimetic consciousness culture was transmitted intergenerationally through imitation in song myth, dance, sport, gathering, hunting, and planting. Aural/oral transmission of tribal knowledge through myth and transactions between animals, Indians, and neighboring tribes produced sustainable relations between the human and the nonhuman worlds. The primal gaze of locking eyes between hunter and hunted initiated the moment of ordained killing when the animal gave itself up so that the Indian could survive. (The very meaning of the gaze stems from the intent look of expectancy when a deer first sees a fire, becomes aware of a scent, or looks into the eyes of a pursuing hunter.) For Indians engaged in an intimate survival relationship with nature, sight, smell, sound, taste, and touch were all of equal importance, integrated in a total participatory consciousness.[9]

When Europeans took over Native American habitats during the colonial ecological revolution, vision became dominant within the mimetic fabric. Although imitative, oral, face-to-face transactions still guided daily life for most colonial settlers and Indians, Puritan eyes turned upward toward a transcendent God who sent down his word in written form in the Bible. Individual Protestants learned to read so that they could interpret God's word for themselves. The biblical word in turn legitimated the imposition of agriculture and artifact in the new land. The objectifying scrutiny of fur trader, lumber merchant, and banker who viewed nature as resource and commodity submerged the primal gaze of the Indians. Treaties and property relations that extracted land from the Indians were codified in writing. Alphanumeric literacy became central to religious expression, social survival, and upward mobility.[10]

The Puritan imposition of a visually oriented consciousness was shattering to the continuation of Indian animism and ways of life. With the commercializing of the fur trade and the missionary efforts of Jesuits and Puritans, a society in which humans, animals, plants, and rocks were equal subjects was changed to one dominated by transcendent vision in which human subjects were separate from resource objects. That change in consciousness characterized the colonial ecological revolution.

The rise of an analytical, quantitative consciousness was a feature of the capitalist ecological revolution. Capitalist ecological relations emphasized efficient management and control of nature. With the development of mechanistic science and its use of perspective diagrams, visualization was integrated with numbering. The superposition of scientific, quantitative approaches to nature and its resources characterized the capitalist ecological revolution. Through education, analytic consciousness expanded beyond that of dominant elites to include most ordinary New Englanders.

Viewed as a social construction, "nature" (as it was conceptualized in each social epoch—Indian, colonial, and capitalist) is not some ultimate truth that was gradually discovered through the scientific processes of observation, experimentation, and mathematics. Rather, it was a relative, changing structure of human representations of "reality." Ecological revolutions are processes through which different societies change their relationship to nature. They arise from tensions between production and ecology, and between production and reproduction. The results are new constructions of nature, both materially and in human consciousness.

The Trouble with Wilderness: Or, Getting Back to the Wrong Nature

WILLIAM CRONON

The time has come to rethink wilderness.

This will seem a heretical claim to many environmentalists, since the idea of wilderness has for decades been a fundamental tenet—indeed, a passion—of the environmental movement, especially in the United States. For many Americans, wilderness stands as the last remaining place where civilization, that all too human disease, has not fully infected the earth. It is an island in the polluted sea of urban-industrial modernity, the one place we can turn for escape from our own too-muchness. Seen in this way, wilderness presents itself as the best antidote to our human selves, a refuge we must somehow recover if we hope to save the planet. As Henry David Thoreau once famously declared, "In Wildness is the preservation of the World."[1]

But is it? The more one knows of its peculiar history, the more one realizes that wilderness is not quite what it seems. Far from being the one place on earth that stands apart from humanity, it is quite profoundly a human creation—indeed, the creation of very particular human cultures at very particular moments in human history. It is not a pristine sanctuary where the last remnant of an untouched, endangered, but still transcendent nature can for at least a little while longer be encountered without the contaminating taint of civilization. Instead, it is a product of that civilization, and could hardly be contaminated by the very

28

stuff of which it is made. Wilderness hides its unnaturalness behind a mask that is all the more beguiling because it seems so natural. As we gaze into the mirror it holds up for us, we too easily imagine that what we behold is Nature when in fact we see the reflection of our own unexamined longings and desires. For this reason, we mistake ourselves when we suppose that wilderness can be the solution to our culture's problematic relationships with the nonhuman world, for wilderness is itself no small part of the problem.

To assert the unnaturalness of so natural a place will no doubt seem absurd or even perverse to many readers, so let me hasten to add that the nonhuman world we encounter in wilderness is far from being merely our own invention. I celebrate with others who love wilderness the beauty and power of the things it contains. Each of us who has spent time there can conjure images and sensations that seem all the more hauntingly real for having engraved themselves so indelibly on our memories. Such memories may be uniquely our own, but they are also familiar enough to be instantly recognizable to others. Remember this? The torrents of mist shoot out from the base of a great waterfall in the depths of a Sierra canyon, the tiny droplets cooling your face as you listen to the roar of the water and gaze up toward the sky through a rainbow that hovers just out of reach. Remember this too: looking out across a desert canyon in the evening air, the only sound a lone raven calling in the distance, the rock walls dropping away into a chasm so deep that its bottom all but vanishes as you squint into the amber light of the setting sun. And this: the moment beside the trail as you sit on a sandstone ledge, your boots damp with the morning dew while you take in the rich smell of the pines, and the small red fox—or maybe for you it was a raccoon or a coyote or a deer—that suddenly ambles across your path, stopping for a long moment to gaze in your direction with cautious indifference before continuing on its way. Remember the feelings of such moments, and you will know as well as I do that you were in the presence of something irreducibly nonhuman, something profoundly Other than yourself. Wilderness is made of that too.

And yet: what brought each of us to the places where such memories became possible is entirely a cultural invention. Go back 250 years in American and European history, and you do not find nearly so many people wandering around remote corners of the planet looking for what today we would call "the wilderness experience." As late as the eighteenth century, the most common usage of the word "wilderness" in the English language referred to landscapes that generally carried adjectives far different from the ones they attract today. To be a wilderness then was to be "deserted," "savage," "desolate," "barren"—in short, a "waste," the word's nearest synonym. Its connotations were anything but positive, and the emotion one was most likely to feel in its presence was "bewilderment"—or terror.[2]

Many of the word's strongest associations then were biblical, for it is used over and over again in the King James Version to refer to places on the margins of civilization where it is all too easy to lose oneself in moral confusion and despair. The wilderness was where Moses had wandered with his people for forty years, and where they had nearly abandoned their God to worship a golden idol.[3] "For Pharaoh will say of the Children of Israel," we read in Exodus, "They are entangled in the land, the wilderness hath shut them in."[4] The wilderness was where Christ had struggled with the devil and endured his temptations: "And immediately the Spirit driveth him into the wilderness. And he was there in the wilderness for forty days tempted of Satan; and was with the wild beasts; and the angels ministered unto him."[5] The "delicious Paradise" of John Milton's Eden was surrounded by a "steep wilderness, whose hairy sides / Access denied" to all who sought entry.[6] When Adam and Eve were driven from that garden, the world they entered was a wilderness that only their labor and pain could redeem. Wilderness, in short, was a place to which one came only against one's will, and always in fear and trembling. Whatever value it might have arose solely from the possibility that it might be "reclaimed" and turned toward human ends—planted as a garden, say, or a city upon a hill.[7] In its raw state, it had little or nothing to offer civilized men and women.

But by the end of the nineteenth century, all this had changed. The wastelands that had once seemed worthless had for some people come to seem almost beyond price. That Thoreau in 1862 could declare wildness to be the preservation of the world suggests the sea change that was going on. Wilderness had once been the antithesis of all that was orderly and good—it had been the darkness, one might say, on the far side of the garden wall—and yet now it was frequently likened to Eden itself. When John Muir arrived in the Sierra Nevada in 1869, he would declare, "No description of Heaven that I have ever heard or read of seems half so fine."[8] He was hardly alone in expressing such emotions. One by one, various corners of the American map came to be designated as sites whose wild beauty was so spectacular that a growing number of citizens had to visit and see them for themselves. Niagara Falls was the first to undergo this transformation, but it was soon followed by the Catskills, the Adirondacks, Yosemite, Yellowstone, and others. Yosemite was deeded by the United States government to the

State of California in 1864 as the nation's first wildland park, and Yellowstone became the first true national park in 1872.[9]

By the first decade of the twentieth century, in the single most famous episode in American conservation history, a national debate had exploded over whether the city of San Francisco should be permitted to augment its water supply by damming the Tuolumne River in Hetch Hetchy Valley, well within the boundaries of Yosemite National Park. The dam was eventually built, but what today seems no less significant is that so many people fought to prevent emerging movement to preserve wilderness. Fifty years earlier, such opposition would have been unthinkable. Few would have questioned the merits of "reclaiming" a wasteland like this in order to put it to human use. Now the defenders of Hetch Hetchy attracted widespread national attention by portraying such an act not as improvement or progress but as desecration and vandalism. Lest one doubt that the old biblical metaphors had been turned completely on their heads, listen to John Muir attack the dam's defenders. "Their arguments," he wrote, "are curiously like those of the devil, devised for the destruction of the first garden—so much of the very best Eden fruit going to waste; so much of the best Tuolumne water and Tuolumne scenery going to waste."[10] For Muir and the growing number of Americans who shared his views, Satan's home had become God's own temple.

The sources of this rather astonishing transformation were many, but for the purposes of this essay they can be gathered under two broad headings: the sublime and the frontier. Of the two, the sublime is the older and more pervasive cultural construct, being one of the most important expressions of that broad transatlantic movement we today label as romanticism; the frontier is more peculiarly American, though it too had its European antecedents and parallels. The two converged to remake wilderness in their own image, freighting it with moral values and cultural symbols that it carries to this day. Indeed, it is not too much to say that the modern environmental movement is itself a grandchild of romanticism and post-frontier ideology, which is why it is no accident that so much environmentalist discourse takes its bearings from the wilderness these intellectual movements helped create. Although wilderness may today seem to be just one environmental concern among many, it in fact serves as the foundation for a long list of other such concerns that on their face seem quite remote from it. That is why its influence is so pervasive and, potentially, so insidious.

To gain such remarkable influence, the concept of wilderness had to become loaded with some of the deepest core values of the culture that created and idealized it: it had to become sacred. This possibility had been present in wilderness

even in the days when it had been a place of spiritual danger and moral temptation. If Satan was there, then so was Christ, who had found angels as well as wild
beasts during His sojourn in the desert. In the wilderness the boundaries between
human and nonhuman, between natural and supernatural, had always seemed
less certain than elsewhere. This was why the early Christian saints and mystics
had often emulated Christ's desert retreat as they sought to experience for themselves the visions and spiritual testing He had endured. One might meet devils
and run the risk of losing one's soul in such a place, but one might also meet God.
For some that possibility was worth almost any price.

By the eighteenth century this sense of the wilderness as a landscape where the
supernatural lay just beneath the surface was expressed in the doctrine of the *sub-*
lime, a word whose modern usage has been so watered down by commercial hype
and tourist advertising that it retains only a dim echo of its former power.[11] In the
theories of Edmund Burke, Immanuel Kant, William Gilpin, and others, sublime
landscapes were those rare places on earth where one had more chance than elsewhere to glimpse the face of God.[12] Romantics had a clear notion of where one
could be most sure of having this experience. Although God might, of course,
choose to show Himself anywhere, He would most often be found in those vast,
powerful landscapes where one could not help feeling insignificant and being reminded of one's own mortality. Where were these sublime places? The eighteenth-century catalog of their locations feels very familiar, for we still see and value landscapes as it taught us to do. God was on the mountain top, in the chasm,
in the waterfall, in the thundercloud, in the rainbow, in the sunset. One has only
to think of the sites that Americans chose for their first national parks—Yellowstone, Yosemite, Grand Canyon, Rainier, Zion—to realize that virtually all of
them fit one or more of these categories. Less sublime landscapes simply did not
appear worthy of such protection; not until the 1940s, for instance, would the
first swamp be honored, in Everglades National Park, and to this day there is no
national park in the grasslands.[13]

Among the best proofs that one had entered a sublime landscape was the
emotion it evoked. For the early romantic writers and artists who first began to
celebrate it, the sublime was far from being a pleasurable experience. The classic
description is that of William Wordsworth as he recounted climbing the Alps
and crossing the Simplon Pass in his autobiographical poem *The Prelude.* There,
surrounded by crags and waterfalls, the poet felt himself literally to be in
the presence of the divine—and experienced an emotion remarkably close to
terror:

The immeasurable height
Of woods decaying, never to be decayed,
The stationary blasts of waterfalls,
And in the narrow rent at every turn
Winds thwarting winds, bewildered and forlorn,
The torrents shooting from the clear blue sky,
The rocks that muttered close upon our ears,
Black drizzling crags that spake by the way-side
As if a voice were in them, the sick sight
And giddy prospect of the raving stream,
The unfettered clouds and region of the Heavens,
Tumult and peace, the darkness and the light—
Were all like workings of one mind, the features
Of the same face, blossoms upon one tree;
Characters of the great Apocalypse,
The types and symbols of Eternity,
Of first, and last, and midst, and without end.[14]

This was no casual stroll in the mountains, no simple sojourn in the gentle lap of nonhuman nature. What Wordsworth described was nothing less than a religious experience, akin to that of the Old Testament prophets as they conversed with their wrathful God. The symbols he detected in this wilderness landscape were more supernatural than natural, and they inspired more awe and dismay than joy or pleasure. No mere mortal was meant to linger long in such a place, so it was with considerable relief that Wordsworth and his companion made their way back down from the peaks to the sheltering valleys.

Lest you suspect that this view of the sublime was limited to timid Europeans who lacked the American know-how for feeling at home in the wilderness, remember Henry David Thoreau's 1846 climb of Mount Katahdin in Maine. Although Thoreau is regarded by many today as one of the great American celebrators of wilderness, his emotions about Katahdin were no less ambivalent than Wordsworth's about the Alps.

It was vast, Titanic, and such as man never inhabits. Some part of the beholder, even some vital part, seems to escape through the loose grating of his ribs as he ascends. He is more lone than you can imagine. . . . Vast, Titanic, inhuman Nature has got him at disadvantage, caught him alone, and pilfers him of some of his divine faculty. She does not smile on him as in the plains. She seems to say sternly, why came ye here before your time? This ground is not prepared for you. Is it not enough that I smile in the valleys? I have never made this soil for thy feet, this air for thy breathing, these rocks for thy neighbors. I cannot pity nor fondle thee here, but forever relentlessly drive thee hence to where I *am* kind. Why seek me where I have not called thee, and then complain because you find me but a stepmother?[15]

This is surely not the way a modern backpacker or nature lover would describe Maine's most famous mountain, but that is because Thoreau's description owes as much to Wordsworth and other romantic contemporaries as to the rocks and clouds of Katahdin itself. His words took the physical mountain on which he stood and transmuted it into an icon of the sublime: a symbol of God's presence on earth. The power and the glory of that icon were such that only a prophet might gaze on it for long. In effect, romantics like Thoreau joined Moses and the children of Israel in Exodus when "they looked toward the wilderness, and behold, the glory of the Lord appeared in the cloud."[16]

But even as it came to embody the awesome power of the sublime, wilderness was also being tamed—not just by those who were building settlements in its midst but also by those who most celebrated its inhuman beauty. By the second half of the nineteenth century, the terrible awe Wordsworth and Thoreau regarded as the appropriately pious stance to adopt in the presence of their mountain top God was giving way to a much more comfortable, almost sentimental demeanor. As more and more tourists sought out the wilderness as a spectacle to be looked at and enjoyed for its great beauty, the sublime in effect became domesticated. The wilderness was still sacred, but the religious sentiments it evoked were more those of a pleasant parish church than those of a grand cathedral or a harsh desert retreat. The writer who best captures this late romantic sense of a domesticated sublime is undoubtedly John Muir, whose descriptions of Yosemite and the Sierra Nevada reflect none of the anxiety or terror one finds in earlier writers. Here he is, for instance, sketching on North Dome in Yosemite Valley:

> No pain here, no dull empty hours, no fear of the past, no fear of the future. These blessed mountains are so compactly filled with God's beauty, no petty personal hope or experience has room to be. Drinking this champagne water is pure pleasure, so is breathing the living air, and every movement of limbs is pleasure, while the body seems to feel beauty when exposed to it as it feels the campfire or sunshine, entering not by the eyes alone, but equally through all one's flesh like radiant heat, making a passionate ecstatic pleasure glow not explainable.

The emotions Muir describes in Yosemite could hardly be more different from Thoreau's on Katahdin or Wordsworth's on the Simplon Pass. Yet all three men are participating in the same cultural tradition and contributing to the same myth: the mountain as cathedral. The three may differ about the way they choose to express their piety—Wordsworth favoring an awe-filled bewilderment, Thoreau a stern loneliness, Muir a welcome ecstasy—but they agree completely about the church in which they prefer to worship. Muir's closing words on North Dome diverge from his older contemporaries only in mood, not in their ultimate content:

> Perched like a fly on this Yosemite dome, I gaze and sketch and bask, oftentimes settling down into dumb admiration without definite hope of ever learning much, yet with the longing, unresting effort that lies at the door of hope, humbly prostrate before the vast display of God's power, and eager to offer self-denial and renunciation with eternal toil to learn any lesson in the divine manuscript.[17]

Muir's "divine manuscript" and Wordsworth's "Characters of the great Apocalypse" were in fact pages from the same holy book. The sublime wilderness had ceased to be a place of satanic temptation and become instead a sacred temple, much as it continues to be for those who love it today.

But the romantic sublime was not the only cultural movement that helped transform wilderness into a sacred American icon during the nineteenth century. No less important was the powerful romantic attraction of primitivism, dating back at least to Rousseau—the belief that the best antidote to the ills of an overly refined and civilized modern world was a return to simpler, more primitive living. In the United States, this was embodied most strikingly in the national myth of the frontier. The historian Frederick Jackson Turner wrote in 1893 the classic academic statement of this myth, but it had been part of American cultural traditions for well over a century. As Turner described the process, easterners and European immigrants, in moving to the wild unsettled lands of the frontier, shed the trappings of civilization, rediscovered their primitive racial energies, reinvented direct democratic institutions, and thereby reinfused themselves with an energy, an independence, and a creativity that were the source of American democracy and national character. Seen in this way, wild country became a place not just of religious redemption but of national renewal, the quintessential location for experiencing what it meant to be an American.

One of Turner's most provocative claims was that by the 1890s the frontier was passing away. Never again would "such gifts of free land offer themselves" to the American people. "The frontier has gone," he declared, "and with its going has closed the first period of American history."[18] Built into the frontier myth from its very beginning was the notion that this crucible of American identity was temporary and would pass away. Those who have celebrated the frontier have almost always looked backward as they did so, mourning an older, simpler, truer world that is about to disappear forever. That world and all of its attractions, Turner said, depended on free land—on wilderness. Thus, in the myth of the vanishing frontier lay the seeds of wilderness preservation in the United States, for if wild land had been so crucial in the making of the nation, then surely one must save its last remnants as monuments to the American past—and as an insurance policy to protect its future. It is no accident that the movement to set aside national parks and wilderness areas began to gain real momentum at precisely the time that laments about the passing frontier reached their peak. To protect wilderness was in a very real sense to protect the nation's most sacred myth of origin.

Among the core elements of the frontier myth was the powerful sense among certain groups of Americans that wilderness was the last bastion of rugged individualism. Turner tended to stress communitarian themes when writing frontier history, asserting that Americans in primitive conditions had been forced to band together with their neighbors to form communities and democratic institutions. For other writers, however, frontier democracy for communities was less compelling than frontier freedom for individuals.[19] By fleeing to the outer margins of settled land and society—so the story ran—an individual could escape the confining strictures of civilized life. The mood among writers who celebrated frontier individualism was almost always nostalgic; they lamented not just a lost way of life but the passing of the heroic men who had embodied that life. Thus Owen Wister in the introduction to his classic 1902 novel *The Virginian* could write of "a vanished world" in which "the horseman, the cow-puncher, the last romantic figure upon our soil" rode only "in his historic yesterday" and would "never come again." For Wister, the cowboy was a man who gave his word and kept it ("Wall Street would have found him behind the times"), who did not talk lewdly to women ("Newport would have thought him old-fashioned"), who worked and played hard, and whose "ungoverned hours did not unman him."[20] Theodore Roosevelt wrote with much the same nostalgic fervor about the "fine, manly qualities" of the "wild rough-rider of the plains." No one could be more heroically masculine, thought Roosevelt, or more at home in the western wilderness:

> There he passes his days, there he does his life-work, there, when he meets death, he faces it as he has faced many other evils, with quiet, uncomplaining fortitude. Brave, hospitable, hardy, and adventurous, he is the grim pioneer of our race; he prepares the way for the civilization from before whose face he must himself disappear. Hard and dangerous though his existence is, it has yet a wild attraction that strongly draws to it his bold, free spirit.[21]

This nostalgia for a passing frontier way of life inevitably implied ambivalence, if not downright hostility, toward modernity and all that it represented. If one saw the wild lands of the frontier as freer, truer, and more natural than other, more modern places, then one was also inclined to see the cities and factories of urban-industrial civilization as confining, false, and artificial. Owen Wister looked at the post-frontier "transition" that had followed "the horseman of the plains," and did not like what he saw: "a shapeless state, a condition of men and manners as unlovely as is that moment in the year when winter is gone and spring not come, and the face of Nature is ugly."[22] In the eyes of writers who shared Wister's distaste for modernity, civilization contaminated its inhabitants and absorbed them into the faceless, collective, contemptible life of the crowd. For all of its troubles and dangers, and despite the fact that it must pass away, the frontier had been a better place. If civilization was to be redeemed, it would be by men like the Virginian who could retain their frontier virtues even as they made the transition to post-frontier life.

The mythic frontier individualist was almost always masculine in gender: here, in the wilderness, a man could be a real man, the rugged individual he was meant to be before civilization sapped his energy and threatened his masculinity. Wister's contemptuous remarks about Wall Street and Newport suggest what he and many others of his generation believed—that the comforts and seductions of civilized life were especially insidious for men, who all too easily became emasculated by the femininizing tendencies of civilization. More often than not, men who felt this way came, like Wister and Roosevelt, from elite class backgrounds. The curious result was that frontier nostalgia became an important vehicle for expressing a peculiarly bourgeois form of antimodernism. The very men who most benefited from urban-industrial capitalism were among those who believed they must escape its debilitating effects. If the frontier was passing, then men who had the means to do so should preserve for themselves some remnant of its wild landscape so that they might enjoy the regeneration and renewal that came from sleeping under the stars, participating in blood sports, and living off the land. The frontier might be gone, but the frontier experience could still be had if only wilderness were preserved.

Thus the decades following the Civil War saw more and more of the nation's wealthiest citizens seeking out wilderness for themselves. The elite passion for wild land took many forms: enormous estates in the Adirondacks and elsewhere (disingenuously called "camps" despite their many servants and amenities), cattle ranches for would-be rough riders on the Great Plains, guided big game hunting trips in the Rockies, and luxurious resort hotels wherever railroads pushed their way into sublime landscapes. Wilderness suddenly emerged as the landscape of choice for elite tourists, who brought with them strikingly urban ideas of the countryside through which they traveled. For them, wild land was not a site for productive labor and not a permanent home; rather, it was a place of recreation. One went to the wilderness not as a producer but as a consumer, hiring guides and other backcountry residents who could serve as romantic surrogates for the rough riders and hunters of the frontier if one was willing to overlook their new status as employees and servants of the rich.

In just this way, wilderness came to embody the national frontier myth, standing for the wild freedom of America's past and seeming to represent a highly attractive natural alternative to the ugly artificiality of modern civilization. The irony, of course, was that in the process wilderness came to reflect the very civilization its devotees sought to escape. Ever since the nineteenth century, celebrating wilderness has been an activity mainly for well-to-do city folks. Country people generally know far too much about working the land to regard unworked land as their ideal. In contrast, elite urban tourists and wealthy sportsmen projected their leisure-time frontier fantasies onto the American landscape and so created wilderness in their own image.

There were other ironies as well. The movement to set aside national parks and wilderness areas followed hard on the heels of the final Indian wars, in which the prior human inhabitants of these areas were rounded up and moved onto reservations. The myth of the wilderness as "virgin," uninhabited land had always been especially cruel when seen from the perspective of the Indians who had once called that land home. Now they were forced to move elsewhere, with the result that tourists could safely enjoy the illusion that they were seeing their nation in its pristine, original state, in the new morning of God's own creation.[23] Among the things that most marked the new national parks as reflecting a post-frontier consciousness was the relative absence of human violence within their boundaries. The actual frontier had often been a place of conflict, in which invaders and invaded fought for control of land and resources. Once set aside within the fixed and carefully policed boundaries of the modern bureaucratic state, the wilderness lost its savage image and became safe: a place more of reverie than of revulsion or fear. Meanwhile, its original inhabitants were kept out by dint of force, their earlier uses of the land redefined as inappropriate or even illegal. To this day, for instance, the Blackfeet continue to be accused of "poaching" on the lands of Glacier National Park that originally belonged to them and that were never ceded by treaty.

The removal of Indians to create an "uninhabited wilderness"—uninhabited as never before in the human history of the place—reminds us just how invented, just how constructed, the American wilderness really is. To return to my opening argument: there is nothing natural about the concept of wilderness. It is entirely a creation of the culture that holds it dear, a product of the very history it seeks to deny. Indeed, one of the most striking proofs of the cultural invention of wilderness is its thoroughgoing erasure of the history from which it sprang. In virtually all of its manifestations, wilderness represents a flight from history. Seen as the original garden, it is a place outside of time, from which human beings had to be ejected before the fallen world of history could properly begin. Seen as the frontier, it is a savage world at the dawn of civilization, whose transformation represents the very beginning of the national historical epic. Seen as the bold landscape of frontier heroism, it is the place of youth and childhood, into which men escape by abandoning their pasts and entering a world of freedom where the constraints of civilization fade into memory. Seen as the sacred sublime, it is the home of a God who transcends history by standing as the One who remains untouched and unchanged by time's arrow. No matter what the angle from which we regard it, wilderness offers us the illusion that we can escape the cares and troubles of the world in which our past has ensnared us.[24]

This escape from history is one reason why the language we use to talk about wilderness is often permeated with spiritual and religious values that reflect human ideals far more than the material world of physical nature. Wilderness fulfills the old romantic project of secularizing Judeo-Christian values so as to make a new cathedral not in some petty human building but in God's own creation, Nature itself. Many environmentalists who reject traditional notions of the Godhead and who regard themselves as agnostics or even atheists nonetheless express feelings tantamount to religious awe when in the presence of wilderness— a fact that testifies to the success of the romantic project. Those who have no difficulty seeing God as the expression of our human dreams and desires nonetheless have trouble recognizing that in a secular age Nature can offer precisely the same sort of mirror.

Thus it is that wilderness serves as the unexamined foundation on which so many of the quasi-religious values of modern environmentalism rest. The critique of modernity that is one of environmentalism's most important contributions to the moral and political discourse of our time more often than not appeals, explicitly or implicitly, to wilderness as the standard against which to measure the failings of our human world. Wilderness is the natural, unfallen antithesis of an unnatural civilization that has lost its soul. It is a place of freedom in which we can recover the true selves we have lost to the corrupting influences of our artificial lives. Most of all, it is the ultimate landscape of authenticity. Combining the sacred grandeur of the sublime with the primitive simplicity of the frontier, it is the place where we can see the world as it really is, and so know ourselves as we really are—or ought to be.

But the trouble with wilderness is that it quietly expresses and reproduces the very values its devotees seek to reject. The flight from history that is very nearly the core of wilderness represents the false hope of an escape from responsibility, the illusion that we can somehow wipe clean the slate of our past and return to the tabula rasa that supposedly existed before we began to leave our marks on the world. The dream of an unworked natural landscape is very much the fantasy of people who have never themselves had to work the land to make a living— urban folk for whom food comes from a supermarket or a restaurant instead of a field, and for whom the wooden houses in which they live and work apparently have no meaningful connection to the forests in which trees grow and die. Only people whose relation to the land was already alienated could hold up wilderness as a model for human life in nature, for the romantic ideology of wilderness leaves precisely nowhere for human beings actually to make their living from the land.

This, then, is the central paradox: wilderness embodies a dualistic vision in which the human is entirely outside the natural. If we allow ourselves to believe that nature, to be true, must also be wild, then our very presence in nature represents its fall. The place where we are is the place where nature is not. If this is so— if by definition wilderness leaves no place for human beings, save perhaps as contemplative sojourners enjoying their leisurely reverie in God's natural cathedral—then also by definition it can offer no solution to the environmental and other problems that confront us. To the extent that we celebrate wilderness as the measure with which we judge civilization, we reproduce the dualism that sets humanity and nature at opposite poles. We thereby leave ourselves little hope of discovering what an ethical, sustainable, *honorable* human place in nature might actually look like.

Worse: to the extent that we live in an urban-industrial civilization but at the same time pretend to ourselves that our *real* home is in the wilderness, to just that extent we give ourselves permission to evade responsibility for the lives we actually lead. We inhabit civilization while holding some part of ourselves—what we imagine to be the most precious part—aloof from its entanglements. We work our nine-to-five jobs in its institutions, we eat its food, we drive its cars (not least to reach the wilderness), we benefit from the intricate and all too invisible networks with which it shelters us, all the while pretending that these things are not an essential part of who we are. By imagining that our true home is in the wilderness, we forgive ourselves the homes we actually inhabit. In its flight from history, in its siren song of escape, in its reproduction of the dangerous dualism that sets human beings outside of nature—in all of these ways, wilderness poses a serious threat to responsible environmentalism at the end of the twentieth century.

By now I hope it is clear that my criticism in this essay is not directed at wild nature per se, or even at efforts to set aside large tracts of wild land, but rather at the specific habits of thinking that flow from this complex cultural construction called "wilderness." It is not the things we label as wilderness that are the problem—for nature and large tracts of the natural world *do* deserve protection—but rather what we ourselves mean when we use that label. Lest one doubt how pervasive these habits of thought actually are in contemporary environmentalism, let me list some of the places where wilderness serves as the ideological underpinning for environmental concerns that might otherwise seem quite remote from it. Defenders of biological diversity, for instance, although sometimes appealing to more utilitarian concerns, often point to "untouched" ecosystems as the best and richest repositories of the undiscovered species we must certainly try to protect. Although at first blush an apparently more "scientific" concept than wilderness, biological diversity in fact invokes many of the same sacred values, which is why organizations like the Nature Conservancy have been so quick to employ it as an alternative to the seemingly fuzzier and more problematic concept of wilderness. There is a paradox here, of course. To the extent that biological diversity (indeed, even wilderness itself) is likely to survive in the future only by the most vigilant and self-conscious management of the ecosystems that sustain it, the ideology of wilderness is potentially in direct conflict with the very thing it encourages us to protect.[25]

The most striking instances of this have revolved around "endangered species," which serve as vulnerable symbols of biological diversity while at the same time standing as surrogates for wilderness itself. The terms of the Endangered Species Act in the United States have often meant that those hoping to defend pristine wilderness have had to rely on a single endangered species like the spotted owl to gain legal standing for their case—thereby making the full power of sacred land inhere in a single numinous organism whose habitat then becomes the object of intense debate about appropriate management and use. The ease with which anti-environmental forces like the Wise Use Movement have attacked single-species preservation efforts suggests the vulnerability of strategies like these.

Perhaps partly because our own conflicts over such places and organisms have become so messy, the convergence of wilderness values with concerns about biological diversity and endangered species has helped produce a deep fascination for remote ecosystems, where it is easier to imagine that nature might somehow be "left alone" to flourish by its own pristine devices. The classic example is the tropical rain forest, which since the 1970s has become the most powerful modern icon of unfallen, sacred land—a veritable Garden of Eden—for many Americans and Europeans. And yet protecting the rain forest—as a growing number of environmentalists now realize—all too often means protecting it from the people who live there. Those who seek to preserve such "wilderness" from the activities of native peoples run the risk of reproducing the same tragedy—being forcibly removed from an ancient home—that befell American Indians. Third World countries face massive environmental problems and deep social conflicts, but these are not likely to be solved by a cultural myth that encourages us to "preserve" peopleless landscapes that have not existed in such places for millennia. At its worst, as environmentalists are beginning to realize, exporting American notions of wilderness in this way can become an unthinking and self-defeating form of cultural imperialism.[26]

Perhaps the most suggestive example of the way that wilderness thinking can underpin other environmental concerns has emerged in the recent debate about "global change." In 1989 the journalist Bill McKibben published a book entitled *The End of Nature,* in which he argued that the prospect of global climate change as a result of unintentional human manipulation of the atmosphere means that nature as we once knew it no longer exists.[27] Whereas earlier generations inhabited a natural world that remained more or less unaffected by their actions, our own generation is uniquely different. We and our children will henceforth live in a biosphere completely altered by our own activity, a planet in which the human and the natural can no longer be distinguished because the one has overwhelmed the other. In McKibben's view, nature has died, and we are responsible for killing it. "The planet," he declares, "is utterly different now."[28]

But such a perspective is possible only if we accept the wilderness premise that nature, to be natural, must also be pristine—remote from humanity and untouched by our common past. In fact, everything we know about environmental history suggests that people have been manipulating the natural world on various scales for as long as we have a record of their passing. Moreover, we have unassailable evidence that many of the environmental changes we now face also occurred quite apart from human intervention at one time or another in the earth's past.[29] The point is not that our current problems are trivial, or that our devastating effects on the earth's ecosystems should be accepted as inevitable or "natural." It is rather that we seem unlikely to make much progress in solving these problems if we hold up to ourselves as the mirror of nature a wilderness we ourselves cannot inhabit.

To do so is merely to take to a logical extreme the paradox that was built into wilderness from the beginning: if nature dies because we enter it, then the only way to save nature is to kill ourselves. The absurdity of this proposition flows from the underlying dualism it expresses. Not only does it ascribe greater power to humanity than we in fact possess—physical and biological nature will surely survive in some form or another long after we ourselves have gone the way of all flesh—but in the end it offers us little more than a self-defeating counsel of despair. The tautology gives us no way out: if wild nature is the only thing worth saving, and if our mere presence destroys it, then the sole solution to our own unnaturalness, the only way to protect sacred wilderness from profane humanity, would seem to be suicide. It is not a proposition that seems likely to produce very positive or practical results.

And yet radical environmentalists and deep ecologists all too frequently come close to accepting this premise as a first principle. When they express, for instance, the popular notion that our environmental problems began with the invention of agriculture, they push the human fall from natural grace so far back into the past that all of civilized history becomes a tale of ecological declension. Earth First! founder Dave Foreman captures the familiar parable succinctly when he writes,

> Before agriculture was midwifed in the Middle East, humans were in the wilderness. We had no concept of "wilderness" because everything was wilderness and *we were a part of it*. But with irrigation ditches, crop surpluses, and permanent villages, we became *apart from* the natural world. . . . Between the wilderness that created us and the civilization created by us grew an ever-widening rift.[30]

In this view the farm becomes the first and most important battlefield in the long war against wild nature, and all else follows in its wake. From such a starting place, it is hard not to reach the conclusion that the only way human beings can hope to live naturally on earth is to follow the hunter-gatherers back into a wilderness Eden and abandon virtually everything that civilization has given us. It may indeed turn out that civilization will end in ecological collapse or nuclear disaster, whereupon one might expect to find any human survivors returning to a way of life closer to that celebrated by Foreman and his followers. For most of us, though, such a debacle would be cause for regret, a sign that humanity had failed to fulfill its own promise and failed to honor its own highest values—including those of the deep ecologists.

In offering wilderness as the ultimate hunter-gatherer alternative to civilization, Foreman reproduces an extreme but still easily recognizable version of the myth of frontier primitivism. When he writes of his fellow Earth Firsters that "we believe we must return to being animal, to glorying in our sweat, hormones, tears, and blood" and that "we struggle against the modern compulsion to become dull, passionless androids," he is following in the footsteps of Owen Wister.[31] Although his arguments give primacy to defending biodiversity and the autonomy of wild nature, his prose becomes most passionate when he speaks of preserving "the wilderness experience." His own ideal "Big Outside" bears an uncanny resemblance to that of the frontier myth: wide open spaces and virgin land with no trails, no signs, no facilities, no maps, no guides, no rescues, no modern equipment. Tellingly, it is a land where hardy travelers can support themselves by hunting with "primitive weapons (bow and arrow, atlatl, knife, sharp rock)."[32] Foreman claims that "the primary value of wilderness is not as a proving ground for young Huck Finns and Annie Oakleys," but his heart is with Huck and Annie all the same. He admits that "preserving a quality wilderness experience for the human visitor, letting her or him flex Paleolithic muscles or seek visions, remains a tremendously important secondary purpose."[33] Just so does Teddy Roosevelt's rough rider live on in the greener garb of a new age.

However much one may be attracted to such a vision, it entails problematic consequences. For one, it makes wilderness the locus for an epic struggle between malign civilization and benign nature, compared with which all other social, political, and moral concerns seem trivial. Foreman writes, "The preservation of wildness and native diversity is *the* most important issue. Issues directly affecting only humans pale in comparison."[34] Presumably so do any environmental problems whose victims are mainly people, for such problems usually surface in landscapes that have already "fallen" and are no longer wild. This would seem to exclude from the radical environmentalist agenda problems of occupational health and safety in industrial settings, problems of toxic waste exposure on "unnatural" urban and agricultural sites, problems of poor children poisoned by lead exposure in the inner city, problems of famine and poverty and human suffering in the "overpopulated" places of the earth—problems, in short, of environmental justice. If we set too high a stock on wilderness, too many other corners of the earth become less than natural and too many other people become less than human, thereby giving us permission not to care much about their suffering or their fate.

It is no accident that these supposedly inconsequential environmental problems mainly affect poor people, for the long affiliation between wilderness and wealth means that the only poor people who count when wilderness is *the* issue are hunter-gatherers, who presumably do not consider themselves to be poor in the first place. The dualism at the heart of wilderness encourages its advocates to conceive of its protection as a crude conflict between the "human" and the "nonhuman"—or, more often, between those who value the nonhuman and those who do not. This in turn tempts one to ignore crucial differences *among* humans and the complex cultural and historical reasons why different peoples may feel very differently about the meaning of wilderness.

Why, for instance, is the "wilderness experience" so often conceived as a form of recreation best enjoyed by those whose class privileges give them the time and resources to leave their jobs behind and "get away from it all"? Why does the protection of wilderness so often seem to pit urban recreationists against rural people who actually earn their living from the land (excepting those who sell goods and services to the tourists themselves)? Why in the debates about pristine natural areas are "primitive" peoples idealized, even sentimentalized, until the moment they do something unprimitive, modern, and unnatural, and thereby fall from environmental grace? What are the consequences of a wilderness ideology that devalues productive labor and the very concrete knowledge that comes from working the land with one's own hands?[35] All of these questions imply conflicts among different groups of people, conflicts that are obscured behind the deceptive clarity of "human" vs. "nonhuman." If in answering these knotty questions we resort to so simplistic an opposition, we are almost certain to ignore the very subtleties and complexities we need to understand.

But the most troubling cultural baggage that accompanies the celebration of wilderness has less to do with remote rain forests and peoples than with the ways we think about ourselves—we American environmentalists who quite rightly worry about the future of the earth and the threats we pose to the natural world. Idealizing a distant wilderness too often means not idealizing the environment in which we actually live, the landscape that for better or worse we call home. Most of our most serious environmental problems start right here, at home, and if we are to solve those problems, we need an environmental ethic that will tell us as much about *using* nature as about *not* using it. The wilderness dualism tends to cast any use as *ab*-use, and thereby denies us a middle ground in which responsible use and nonuse might attain some kind of balanced, sustainable relationship. My own belief is that only by exploring this middle ground will we learn ways of imagining a better world for all of us: humans and nonhumans, rich people and poor, women and men, First Worlders and Third Worlders, white folks and people of color, consumers and producers—a world better for humanity in all of its diversity and for all the rest of nature too. The middle ground is where we actually live. It is where we—all of us, in our different places and ways—make our homes.

That is why when I think of the times I myself have come closest to experiencing what I might call the sacred in nature, I find myself remembering not some remote wilderness but places much closer to home. I think, for instance, of a small pond near my house where water bubbles up from limestone springs to feed a series of pools that rarely freeze in winter and so play home to waterfowl that stay here for the protective warmth even on the coldest of winter days, gliding silently through steaming mists as the snow falls from gray February skies. I think of a November evening long ago when I found myself on a Wisconsin hilltop in rain and dense fog, only to have the setting sun break through the clouds to cast an otherworldly golden light on the misty farms and woodlands below, a scene so unexpected and joyous that I lingered past dusk so as not to miss any part of the gift that had come my way. And I think perhaps most especially of the blown-out, bankrupt farm in the sand country of central Wisconsin where Aldo Leopold and his family tried one of the first American experiments in ecological restoration, turning ravaged and infertile soil into carefully tended ground where the human and the nonhuman could exist side by side in relative harmony. What I celebrate about such places is not *just* their wildness, though that certainly is among their most important qualities; what I celebrate even more is that they remind us of the wildness in our own backyards, of the nature that is all around us if only we have eyes to see it.

Indeed, my principal objection to wilderness is that it may teach us to be dismissive or even contemptuous of such humble places and experiences. Without our quite realizing it, wilderness tends to privilege some parts of nature at the expense of others. Most of us, I suspect, still follow the conventions of the romantic sublime in finding the mountain top more glorious than the plains, the ancient forest nobler than the grasslands, the mighty canyon more inspiring than the humble marsh. Even John Muir, in arguing against those who sought to dam his beloved Hetch Hetchy valley in the Sierra Nevada, argued for alternative dam sites in the gentler valleys of the foothills—a preference that had nothing to do with nature and everything with the cultural traditions of the sublime.[36] Just as problematically, our frontier traditions have encouraged Americans to define "true" wilderness as requiring very large tracts of roadless land what Dave Foreman calls "The Big Outside." Leaving aside the legitimate empirical question in conservation biology of how large a tract of land must be before a given species can reproduce on it, the emphasis on big wilderness reflects a romantic frontier belief that one hasn't really gotten away from civilization unless one can go for days at a time without encountering another human being. By teaching us to fetishize sublime places and wide-open country, these peculiarly American ways of thinking about wilderness encourage us to adopt too high a standard for what counts as "natural." If it isn't hundreds of square miles big, if it doesn't give us God's-eye views or grand vistas, if it doesn't permit us the illusion that we are alone on the planet, then it really isn't natural. It's too small, too plain, or too crowded to be *authentically* wild.

In critiquing wilderness as I have done in this essay, I'm forced to confront my own deep ambivalence about its meaning for modern environmentalism. On the one hand, one of my own most important environmental ethics is that people should always be conscious that they are part of the natural world, inextricably tied to the ecological systems that sustain their lives. Any way of looking at nature that encourages us to believe we are separate from nature—as wilderness tends to do—is likely to reinforce environmentally irresponsible behavior. On the other hand, I also think it no less crucial for us to recognize and honor nonhuman nature as a world we did not create, a world with its own independent, nonhuman reasons for being as it is. The autonomy of nonhuman nature seems to me an indispensable corrective to human arrogance. Any way of looking at nature that helps us remember—as wilderness also tends to do—that the interests of people are not necessarily identical to those of every other creature or of the earth itself is likely to foster *responsible* behavior. To the extent that wilderness has served as an important vehicle for articulating deep moral values regarding our obligations and responsibilities to the nonhuman world, I would not want to jettison the contributions it has made to our culture's ways of thinking about nature.

If the core problem of wilderness is that it distances us too much from the very things it teaches us to value, then the question we must ask is what it can tell us about *home*, the place where we actually live. How can we take the positive values we associate with wilderness and bring them closer to home? I think the answer to this question will come by broadening our sense of the otherness that wilderness seeks to define and protect. In reminding us of the world we did not make, wilderness can teach profound feelings of humility and respect as we confront our fellow beings and the earth itself. Feelings like these argue for the importance of self-awareness and self-criticism as we exercise our own ability to transform the world around us, helping us set responsible limits to human mastery—which without such limits too easily become human hubris. Wilderness is the place where, symbolically at least, we try to withhold our power to dominate. Wallace Stegner once wrote of

> the special human mark, the special record of human passage, that distinguishes man from all other species. It is rare enough among men, impossible to any other form of life. *It is simply the deliberate and chosen refusal to make any marks at all*. . . . We are the most dangerous species of life on the planet, and every other species, even the earth itself, has cause to fear our power to exterminate. But we are also the only species which, when it chooses to do so, will go to great effort to save what it might destroy.[37]

The myth of wilderness, which Stegner knowingly reproduces in these remarks, is that we can somehow leave nature untouched by our passage. By now it should be clear that this for the most part is an illusion. But Stegner's deeper message then becomes all the more compelling. If living in history means that we cannot help leaving marks on a fallen world, then the dilemma we face is to decide what kinds of marks we wish to leave. It is just here that our cultural traditions of wilderness remain so important. In the broadest sense, wilderness teaches us to ask whether the Other must always bend to our will, and, if not, under what circumstances it should be allowed to flourish without our intervention. This is surely a question worth asking about everything we do, and not just about the natural world.

When we visit a wilderness area, we find ourselves surrounded by plants and animals and physical landscapes whose otherness compels our attention. In forcing us to acknowledge that they are not of our making, that they have little or no need of our continued existence, they recall for us a creation far greater than our own. In the wilderness, we need no reminder that a tree has its own reasons for being, quite apart from us. The same is less true in the gardens we plant and tend ourselves: there it is far easier to forget the otherness of the tree.[38] Indeed, one could almost measure wilderness by the extent to which our recognition of its otherness requires a conscious, willed act on our part. The romantic legacy means that wilderness is more a state of mind than a fact of nature, and the state of mind that today most defines wilderness is *wonder*. The striking power of the wild is that wonder in the face of it requires no act of will, but forces itself upon us—as an expression of the nonhuman world experienced through the lens of our cultural history—as proof that ours is not the only presence in the universe.

Wilderness gets us into trouble only if we imagine that this experience of wonder and otherness is limited to the remote corners of the planet, or that it somehow depends on pristine landscapes we ourselves do not inhabit. Nothing could be more misleading. The tree in the garden is in reality no less other, no less worthy of our wonder and respect, than the tree in an ancient forest that has never known an ax or a saw—even though the tree in the forest reflects a more intricate web of ecological relationships. The tree in the garden could easily have sprung from the same seed as the tree in the forest, and we can claim only its location and perhaps its form as our own. Both trees stand apart from us; both share our common world. The special power of the tree in the wilderness is to remind us of this fact. It can teach us to recognize the wildness we did not see in the tree we planted in our own backyard. By seeing the otherness in that which is most unfamiliar, we can learn to see it too in that which at first seemed merely ordinary. If wilderness can do this—if it can help us perceive and respect a nature we had forgotten to recognize as natural—then it will become part of the solution to our environmental dilemmas rather than part of the problem.

This will only happen, however, if we abandon the dualism that sees the tree in the garden as artificial—completely fallen and unnatural—and the tree in the wilderness as natural—completely pristine and wild. Both trees in some ultimate sense are wild; both in a practical sense now depend on our management and care. We are responsible for both, even though we can claim credit for neither. Our challenge is to stop thinking of such things according to a set of bipolar moral scales in which the human and the nonhuman, the unnatural and the natural, the fallen and the unfallen, serve as our conceptual map for understanding and valuing the world. Instead, we need to embrace the full continuum of a natural landscape that is also cultural, in which the city, the suburb, the pastoral, and the wild each has its proper place, which we permit ourselves to celebrate without needlessly denigrating the others. We need to honor the Other within and the Other next door as much as we do the exotic Other that lives far away—a lesson that applies as much to people as it does to (other) natural things. In particular, we need to discover a common middle ground in which all of these things, from the city to the wilderness, can somehow be encompassed in the word "home." Home, after all, is the place where finally we make our living. It is the place for which we take responsibility, the place we try to sustain so we can pass on what is best in it (and in ourselves) to our children.[39]

The task of making a home in nature is what Wendell Berry has called "the forever unfinished lifework of our species." "The only thing we have to preserve nature with," he writes, "is culture; the only thing we have to preserve wildness with is domesticity."[40] Calling a place home inevitably means that we will *use* the nature we find in it, for there can be no escape from manipulating and working and even killing some parts of nature to make our home. But if we acknowledge the autonomy and otherness of the things and creatures around us—an autonomy our culture has taught us to label with the word "wild"—then we will at least think carefully about the uses to which we put them, and even ask if we should use them at all. Just so can we still join Thoreau in declaring that "in Wildness is the preservation of the World," for wildness (as opposed to wilderness) can be found anywhere: in the seemingly tame fields and woodlots of Massachusetts, in the cracks of a Manhattan sidewalk, even in the cells of our own bodies. As Gary Snyder has wisely said, "A person with a clear heart and open mind can experience the wilderness anywhere on earth. It is a quality of one's own consciousness. The planet is a wild place and always will be."[41] To think ourselves capable of causing "the end of nature" is an act of great hubris, for it means forgetting the wildness that dwells everywhere within and around us.

Learning to honor the wild—learning to remember and acknowledge the autonomy of the other—means striving for critical self-consciousness in all of our actions. It means that deep reflection and respect must accompany each act of use, and means too that we must always consider the possibility of nonuse. It means looking at the part of nature we intend to turn toward our own ends and asking whether we can use it again and again and again—sustainably—without its being diminished in the process. It means never imagining that we can flee into a mythical wilderness to escape history and the obligation to take responsibility for our own actions that history inescapably entails. Most of all, it means practicing remembrance and gratitude, for thanksgiving is the simplest and most basic of ways for us to recollect the nature, the culture, and the history that have come together to make the world as we know it. If wildness can stop being (just) out there and start being (also) in here, if it can start being as humane as it is natural, then perhaps we can get on with the unending task of struggling to live rightly in the world—not just in the garden, not just in the wilderness, but in the home that encompasses them both.

PLACE SETTINGS

We know ourselves in part by the land on which we live. Its shapes and contours mold our existence, our sense of place. But place is also what we make of it: we help structure specific forms of landscape—whether rural or suburban or urban, agrarian or industrial—and then impute meaning to them, helping us to understand why we live as we do. Our relationship to these terrains is like that of an ocean wave, whose watery form is partly determined by the very landmass its grinding power will transform.

The importance of this reciprocity emerges in the following chapters which address the nature of landscapes. Each appropriately asserts the preeminence of place, each digs through what geographer I. G. Simmons calls its "historical depth" to uncover the multiple layers of interaction between humanity and its particular bioregion. Simmons opens "The Earliest English Cultural Landscapes" with a beguiling observation that from his study window he can see "a patch of land . . . that has been in the last 10,000 years a tundra slope, a dense deciduous forest, agricultural land, a coal-mine yard, a clay pit, and grazing land," an opening that he then uses to pull us back across the millennia so as to examine what these evolving landscapes can tell us about how its ancient peoples' lives and labors were etched into the terrain.

Less old, but no less revealing, is Robert MacCameron's careful recovery of the significance of changes in colonial New

Mexico's rugged landscape. Among other things, it reminds us of the limitations of historians' assumptions about the relative impact that different economic systems can have on the environment. Well before the introduction of capitalism, both Spanish mercantilism and indigenous subsistence economies had decided impacts on the region and human behavior. The need to learn to read such landmarks, whether subtle or otherwise, is something emphasized in John Stilgoe's early contribution to *Environmental Review,* *"Landschaft* and Linearity: Two Archetypes of Landscape." He captures the clash between these two competing perspectives when he observes expatriate Henry James's bewildered reactions to the explosive changes wrought by the Industrial Revolution in the United States. Stilgoe suggests why it is that we, like James, can become strangers in a once-known land: our "visions of what the built is" molds our "dreams of what it ought to be," a response that can prevent "men from realizing—and from loving—the space they inhabit."

The Earliest Cultural Landscapes of England

I. G. SIMMONS

Even to the casual observer, the English landscape has historical depth. Every piece of terrain speaks of a history in which environmental processes, whether natural or human-induced, have changed the face of the land more than once. Sometimes remnants of an earlier state show through, like a palimpsest; in other places the latest environment seems all-pervasive and appears to blot out all that went before it. But that is rarely the case, because somewhere in a sediment or an archive or a memory will be interpretable knowledge of the history of that piece of terrain.[1]

From my study window, for instance, I can see a patch of land about two hectares in size that has been in the last 10,000 years a tundra slope, a dense deciduous forest, agricultural land, a coal-mine yard, a clay pit, and grazing land. The area now is well on its way to becoming deciduous woodland once again, because it has been designated as an area of landscape and recreational value; this symbolizes the major phases of the evolution of the English landscape, which are well known to the educated visitor and native alike. Since 1939 the major characteristic of public policy has been planning: local and national authorities have had powers to control most forms of public development in both town and country, with some long-standing exceptions like farming and very recent changes in inner-city areas of near-dereliction. This recent phase was in some ways a reaction to the changes initiated in the nineteenth century when industrialization and urban growth took place in an environment of laissez-faire economics, a process that produced a lot of wealth and a great number of unwanted environmental changes.

Before industrialization, a change of immense significance had been enclosure, when the open fields inherited from medieval agriculture were parceled out to individual landowners. Those changes produced some of the most characteristic features of the English rural scene: the relatively small, rectangular fields with hedgerow boundaries spiked with oaks and ashes at irregular intervals and interspersed with patches of coppice woodland. Many of the farmhouses in the fields (and now offering such good bargain bed-and-breakfast accommodations for the traveler) were moved there from the villages as part of the enclosure movement.

The agglomeration of holdings for the very rich meant the chance to acquire large acreages upon which to build stately homes and to set them in a landscape park, employing the likes of Humphrey Repton or "Capability" Brown to lay out a landscape built upon an earlier foundation of agricultural colonization of dense oak and lime woodland. Domesday Book (1086) shows a mosaic of farmland in common ownership, large tracts of forest, villages and hamlets, and of course towns, some of which had castles and cathedrals. The latter were often symbols of Norman domination because the Conquest of 1066 had brought a nationwide bureaucracy unrivaled even by the earlier Romans whose impact upon the native landscape was probably more limited than their ability to make roads and take baths.[2]

What happened before then? We see in our present landscapes the great Iron Age hillforts like Maiden Castle in Dorset and even earlier monuments such as Stonehenge and Avebury (both in Wiltshire and both dating in large measure from the late Neolithic and early Bronze Age). But of what environmental events and processes are these stones mementoes? Beyond question it is the coming of agriculture (circa 3500 B.C.E.) and its firm establishment on the Atlantic fringe of Eurasia, a great distance and a long time after the domestication of barley and wheat in Southwest Asia (circa 7000 B.C.E.).

But we know that humans lived in England long before then; hence, can an environmental story be told for an even earlier, pre-agricultural period? Surprisingly, the paleoecological evidence for conditions in the Pleistocene and early Holocene areas occupied by human societies is reasonably good, as is the archaeological evidence for Paleolithic people (to 10,000 B.C.E.) and Mesolithic groups (10,000 to 3500 B.C.E.). There is a large corpus of evidence from archaeozoology and from paleoecology that relates both to phyto- and zoogeography and to the conditions and changes of human existence at that time. Thus it is possible to discuss, albeit not always conclusively, whether the hunter-gatherer occupation of England in the intervals of glaciation and in the immediate postglacial period laid a foundation for the cultural landscape of the country.[3]

There is a great amount of evidence for geological, climatic, vegetational, and faunal (including human) conditions in both the glacial and interglacial periods of the Pleistocene. But the number of sites where there is sufficient evidence to indicate whether human groups exerted any lasting influence upon the ecosystem are very few. Most of the localities investigated might display human remains, or mammal bones, or implements, or muds with pollen and spores, or gravel river terraces, but not more than one or two of these. Some sites in East Anglia, therefore, that include long records of floristic change during an interglacial period are of great interest. That is especially so when human tools are present as well.

The sites discussed in this essay are from the second of the major interglacial periods of the British Pleistocene, referred to as the Hoxnian after its type-site at Hoxne in Suffolk, and correlated with the Holstein interglacial period of continental Europe. They date from the Middle Pleistocene, circa 330,000 years ago. The type-site consists of a sequence of lake sediments in a hollow within a till sheet created by the preceding episode of glaciation. Clays and muds predominate and they preserve pollen, seeds, fruits, and other organic debris as well as inorganic materials. At Mark's Tey in Essex, the sequence is even more evident, and there are also several other sites in England where parts of the interglacial succession are found. In its full form, a progression can be traced from ice through tundra to coniferous forest and then to temperate deciduous forest at the warmest part of the interval, and then a recession through the phases to tundra again.

At Hoxne and Mark's Tey, however, calculations from deposition rates and other internal factors show that the zone containing evidence for temperate forest (a stable set of ecosystems lasting at least 2,700 years) has an interruption of approximately 350 years in the sequence. The pollen of the trees least tolerant of cooler conditions diminished rapidly (especially that of hazel, *Corylus avellana)* and were replaced by those of grasses and other herbaceous plants. The tree alder *(Alnus glutinosa),* tolerant of waterlogged conditions, suffered less, therefore it is assumed that the forest away from the lakeside suffered a rapid and sustained recession. Because the same phase turns up in two places sixty kilometers apart, the calamity must have been widespread.

But it is interesting to note that at both Mark's Tey and Hoxne charcoal is found in the lake deposits at the appropriate horizons; moreover at Hoxne (though nowhere else), flint implements of the Acheulian type associated with *Homo erectus* are also found in the deposits.[4] This is the *prima facie* evidence of early environmental impact: remains of human cultures, traces of fire, diminution of trees and their replacement by grasses and herbs.

The interpretations of this phase, however, are not uniform. The botanists who carried out the original work are inclined to look for an explanation in terms of natural phenomena: Charles Turner suggests that there must have been a major forest fire at Mark's Tey after which it took 350 years for the forest to recover. Climatic change is generally ruled out because the vegetation is not the type that occurs with cooling. On the other hand, Turner believes that large herbivorous mammals may have played a significant part in maintaining the open ground once it was created. He points to the example of how hippos grazing on grassland by the side of a lake would transfer large quantities of graminaceous pollen to the water via their semi-liquid excreta. Turner, therefore, accepts the idea of an initial forest fire due to natural causes followed by herbivore pressure.[5] Another site in Norfolk shows the recession of forest but here spruce maintains its frequencies, suggesting that fire was not an important factor (there was no charcoal at this site, either). Hence for botanists, a relationship between Paleolithic hominids and environmental change is only a tenuous possibility.

By contrast, archaeologists are much more willing to see the hand of humans in the environmental changes recorded in the deposits. They acknowledge the lack of consensus: Alex Morrison argues that "whether or not" the human element was present "cannot be demonstrated with any certainty." But Derek A. Roe admits that such a fire may have had purely natural causes but believes that it "is open to us to reflect on the habitual carelessness of *Homo sapiens* with regard to fire and to view his predecessors with dark suspicion." Finally, J. J. Wymer points to the coincidence at Hoxne of the level at which hunters began to occupy the lakeside and the charcoal found in the lake muds. He concludes that the "earliest example of man's impact on the environment may be contained . . . [in] the lake beds at Hoxne, Suffolk."[6]

Until more detailed work on the deposits provides better data about the exact course of the events during the Hoxnian interglacial period, we can form only an interim judgment. Nevertheless, a marker has been put down for the possibility of an English cultural landscape 330,000 years ago. If it existed, it was totally obliterated by the succeeding ice sheets and left no remnant in today's visible landscape.

Between 10,000 and 8000 B.C.E. the ice withdrew, though not in a single, smooth recession. Upper Paleolithic cultures (of *Homo sapiens*) occupied the warmer places in the south, and although some charcoals have been found in caves, no extrapolations about the environment can be made from those. The only suggestion of an ecological impact is in the apparent selective dependence upon the wild horse and the reindeer for food and raw materials.[7] But in the absence of good data on the populations of both humans and wild animals, it seems unlikely that landscapes of this Late Devensian period had anything but the most transient of humanized elements.

The climate ameliorated, however, between 8000 and 5500 B.C.E., in the period of time known as the Holocene. Until circa 5500 England was physically joined to the continent, so the trees that had "over-wintered" in southern Europe and the Caucasus gradually returned until the climax vegetation of a mixed deciduous forest was established by about 6000 B.C.E. Hunter-gatherers still inhabited the country, and they remained until the coming of agriculture (circa 3500 B.C.E. and after). Those Mesolithic cultures and their lithic remains are found in most parts of England, both upland and lowland.

Apart from the stone tools, other remains are scarce except where an accident of preservation in waterlogged conditions has left other material, such as at Star Carr in Yorkshire. We have to interpret the environmental impact of those folk from the processes of change in vegetation revealed by the techniques of Quaternary ecology. Thus, the Mesolithic occupants of England in the earlier part of that period would have lived in a park-tundra environment that was rapidly becoming pine, birch, and hazel woodland. Mixed deciduous woodland dominated by oak and lime, but with elm and alder as important constituents, succeeded that landscape. Even in the Atlantic climate, the forest colonized the uplands; as a consequence the tree line was well above the 900-foot level of today (although exotic conifers grow much higher on the slopes), reaching above 3,000 feet in some areas. As the tundra fauna of wild horses and reindeer diminished, forest animals like the red and roe deer and wild pigs replaced them; a few wild oxen also remained.[8]

The impact of humans on the environmental systems of the assembly phase of the deciduous forest is not easily detectable. At Star Carr, the birch woodland surrounding the habitation (circa 7600 B.C.E.) seems to have suffered some clearance, being replaced by hazel and open ground with ruderal plants like the nettle and species of the goosefoot family. Not far away, at Kildale on the North Yorkshire Moors, the skeleton of a wild ox was encased in deposits of the eighth millenium B.C.E.; the bone level also included charcoal and pollens of disturbed ground, as well as the birch heath that was the "natural" vegetation of the time. If fire was used to run animals to their death, it is likely that vegetation successions were deflected, some perhaps permanently. Charcoal, pollens of open ground and broken soils, and even silt in-wash stripes in valley deposits are evidence that sporadic interruptions of the forest succession took place throughout that ecological period on the North Yorkshire Moors. There is nothing to suggest the continuous roar of fire, but as it were, the occasional whiff of smoke hinting at a battle for survival being fought some distance away.[9]

The botanist Alan G. Smith has looked at the forest-assembly phase in some detail, using evidence from Ireland and the rest of the British Isles. He notes that hazel becomes rapidly abundant in the postglacial period, but after the high forest of oak and lime is in place, hazel is common only after the opening of the forest and after fire, because it readily stump-sprouts after burning. Smith suggests that its high profile during the forest-assembly phase in part could be due to the increased incidence of fire that resulted from human presence. He extends the argument to alder as well, showing that in areas of Mesolithic occupation there were often disturbances leading to gaps in the forest cover; the pollen of open-ground plants like plantain, the spores of bracken fern, and the pollen of the ash *(Fraxinus)*, which is intolerant of shade, testify to that possibility. But alder pollen increases in frequency after such discontinuities; therefore, its establishment and expansion may be a result of environmental change brought about by human groups. The evidence shows that the forest cover of the mid-Holocene period may not be as pristine as it first appears from the paleoecological record.[10]

For the period of the established high forest and its relationship with the later mesolithic people, the evidence is much greater in volume and more substantive. In part that is a climatic accident, because the insulation of Britain from the continent (circa 5000 B.C.E.) led to increased wetness and to more peat growth. Those deposits in turn have preserved the evidence for human/environment relations in the period 5500 to 3500 B.C.

Analysis of the subfossil pollen and other plant remains (and occasionally animal bones) from peats, lake muds, and other organic deposits for this period reveals a particular pattern. There is a high frequency of charcoal in the deposits which comes in two forms; either or both may be present at any one site. One type consists of large chunks (one to five centimeters) that are angular in shape and are usually identifiable to species, often a tree such as birch or alder. The other type is microscopic (less than one millimeter) and rounded in shape; it is often called "microcharcoal" in the literature describing its occurrence. The incidence of microcharcoal in those profiles is usually in the form of a constant "background rainout" through the period, with peaks at intervals when the quantity shoots up markedly. The larger pieces of charcoal are found less frequently, but when they are detected, they form continuous spreads of a thick layer of material (up to five centimeters), sometimes over an area as large as 400-by-60 meters. The former type generally is interpreted as evidence of burning somewhere within the catchment zone of the sampled profile but not adjacent to it. But the continuity of the deposited flecks infers that the burning was continuous throughout the period with a radius of perhaps five kilometers from the investigated site. The more chunky material cannot be interpreted as being other than direct evidence of a burn near at hand because large and angular pieces of charcoal could not have traveled very far. In some peat deposits where perhaps two meters of deposit refer to the mid-Holocene, there can be two or three such layers of charcoal-rich peat.

If we look at the pollen and spore content of those layers and the intervening peats that lack the large charcoal fragments (but not the soot), a clear pattern emerges. At the level of charcoal pieces, the vegetation is clearly being changed, because the tree pollen content is markedly less. Oak, alder, and lime seem to be the main losers, as if they had quite suddenly ceased to contribute to the local pollen rain. In their place is a variety of plants. The first group are shrubby species that are known to be intolerant of dense shade but that would flourish if the forest canopy were opened up. Preeminent among these is the hazel. This plant is known to flower prolifically if given more light and it will sprout vigorously from a burned stump; hence it could easily have benefited from the presence of fewer trees. The pollen of other shrubs like the rowan and the bramble and of small trees like birch and ash also appear at those horizons. A second group is those plants of the forest floor where the canopy is incomplete and the soils have a tendency toward acidity: the bracken fern and heather. Once bracken is established it tends to be a permanent feature of the ecology unless it is shaded out by overgrowing scrub or forest. The fern maintains its position by exuding chemicals toxic to other plants, by smothering them with the density of its canopy of fronds, and, when it dies, by the matted blanket of slowly rotting leafy material. Both bracken and heather can survive fire by sprouting from undamaged underground parts.

Another category of plants belongs to open grassy swards, containing the pollen of grasses and of grassland herbs, including plantain and some species, such as cow-wheat *(Melampyrum)*, that are characteristic of forest floors not long after the canopy has been broken. Some herbs that tend to follow fire, like the small sorrel *Rumex acetosella*, are also found, along with the wormwood, *Artemisia*. Occasionally there is evidence of bare and broken soils: silt is found in the peat profile, and the pollen of plants such as nettles *(Urtica)* testifies to the breaking of the soil surface.

Each of those phases is a few centimeters thick in the actual deposits. But what happened between those ecological time spans? Something of a reversal of the process, it appears. The trees gradually come back and the open-ground plants yield to them: the forest has regenerated, although not quite into a virgin condition. Some bracken usually persists and the variety of trees suggests a more open canopy than before. Alder may be replaced with willow in places where the soil seems to have turned more acid as a result of the forest recession.

In the absence of fire, therefore, the forest returns in a slightly different shape, but not everywhere. On the uplands above 1,100 feet where there is a small basin or a gentle slope, the forest may not return at all. In its place there is peat of wet conditions and of very acid surroundings. The cotton-sedge *Eriophorum* and the bog-moss *Sphagnum* are examples. Those plants can receive adequate nutrition from the minerals present in rainwater (of which at these altitudes there is at least forty-five inches per year) and can grow quickly on the corpses of their immediate ancestors. The peat spreads from its initial foci over all the terrain that does not have slopes beyond two or three degrees and blankets out the detail of the original topography. For that reason, it is known to ecologists as "blanket peat." In that scenario, trees were basically living water pumps; once removed, their action in evaporating water from their surfaces and transpiring it from their leaves ceased and the soils became waterlogged. In those conditions the only plants able to survive were those able to live under nutrient-poor conditions; thus the whole soil-plant-water system shifted toward a state of greater acidity.[11]

The environment of the Mesolithic hunter-gatherers was one in which the closed canopy deciduous high forest was subject to openings caused by fire, after which some regeneration took place. The replacement forest was different in species composition and form, and on the uplands peat replaced it in some topographic situations. But were hunters or natural causes responsible for the fires? It is apparent that oak forest (especially in high rainfall areas) is about as likely to burst into flames after a lightning strike as a sackful of wet socks—and even if lightning does strike more than once in the same place (as Damon Runyon might have put it), three times is beyond coincidence.[12] Add to that paleoenvironmental evidence the use of fire by near-recent hunter-gatherers in many parts of the world, and there is a good circumstantial case for the deliberate and controlled use of fire by Mesolithic folk as a tool of environmental management.

The purposes of such management can be relatively easily explained (recall the assembly of dense deciduous forest after a period of open park-tundra and pine-hazel woodland and the changing fauna that accompanied it). The greater the high forest, the less habitat space for two important food resources. The first of these are food plants, in particular the hazel. The nuts of this shrub turn up in almost every archaeological excavation of the period and were clearly a valuable food item: they store well and are high in fats and minerals. A woodland with lots of "edge" where hazel could flower abundantly and produce heavy crops would be valuable real estate. But no hunter is by choice a vegetarian all the time and the same argument, *mutatis mutandis,* applies to deer.

Two relevant species are the roe deer *(Capreolus capreolus)* and the red deer *(Cervus elephas);* both would find more food in a mosaic environment of forest diversified by dense shrubs and adjacent grassland. Open areas would also attract wild pigs rooting for soil invertebrates.[13] Scholars in this field are not sure how those clearings originated: Mesolithic people may have seized upon existing glades such as occur when a tree dies or is windthrown; and in the uplands the location of some of the burned areas is adjacent to springs. An opening with plenty of browse for deer but near water and with shrubs to afford concealment for the hunters sounds like a recipe for ensuring that venison will be on the menu. It so happens that in such areas the springs are close to the boundary between the dense oak and lime forest of lower altitudes and the more open oak, birch, and pine woodland higher up. There too, circumstances would make the preservation of more open areas feasible.

Two additional pieces of paleoenvironmental evidence support this model of managing the forest edge for subsistence. On the southwestern upland of Dartmoor, the first fires appear in the record just as the forest is beginning to cover the highest parts of the upland terrain, and Canute-like, it appears as if there was an attempt to halt and turn back the tide of trees. In several other places change coincides with the decline of elm pollen, which in western Europe marks the establishment of agriculture (and a "Neolithic" culture). In most deposits, the regional microcharcoal influx diminishes or even stops, although there are forest clearances giving local charcoal deposits. By this time, however, they have cereal pollen in them. Together with that phenomenon, a few sites indicate the reestablishment of trees, sometimes on top of a meter or two of peat. The evidence implies that with the coming of agriculture as a major means of subsistence there was no longer a need to apply fire to the landscape so vigorously; hence some trees were able to recolonize formerly open areas.[14]

By comparison with the earlier Mesolithic and Paleolithic, there is considerable evidence to show that the earliest cultural landscapes were formed in the period 5500–3500 B.C.E. That scholars will produce better data for earlier times is very likely, therefore these judgments are provisional. But for the moment, the first well-documented English cultural landscape is represented by the following: the conversion of mixed deciduous forest into a mosaic of high forest, open-canopy woodland; and grassy clearings with fringes of scrub and bracken fern and patches of wet sedge and peat bog. Among those, bands of seasonal food collectors moved with no knowledge of crop-based agriculture.

The coming of agriculture was one of the great turning points in western Europe. For those people there was no gradual emergence of specialized collection, *in situ* husbandry, incipient domestication, and then full domestication with its accompanying land-use system. As agriculture spread westward, the full suite of phenomena presumably came as a package, even if accessory hunting persisted. We do not know precisely why agriculture was adopted (nor is it relevant to this essay); but we can say that although it marked a major phase shift in a cultural and economic sense, it need not have been so ecologically. The mosaic of small clearings that were abandoned and reverted to trees is the model for the earliest agriculture in western Europe; and it was a movable set of occasions, with clearings abandoned as soil fertility fell or weeds became too bothersome.[15] Because of a lack of evidence, we do not know whether every part of England was subject to those processes. But it is certainly true of the uplands and of the light soil areas of the lowlands (with the creation of heaths and the blowing of degraded soils). That leaves only the clay lowlands for another generation of paleoenvironmentalists to investigate. The story for Wales is much the same; that for Scotland beyond the Highland boundary is considerably different, with southern Scotland being more like northern England in its environmental history.

It is interesting to note that in the uplands the spreads of blanket peat that resulted from human intervention in the ecosystems of the Mesolithic are still there. They have been added to laterally because the process of tree removal (leading to paludification) continued into later prehistory and even historic times. The spreads of peat also gained vertical increments, because until the beginning of acid rain in the nineteenth century, there was nothing much to stop them growing. But the core area of blanket peat growing around and uphill from the springheads can still be seen. It represents both the first cultural landscape and in large measure it is a part of today's landscape as well. The depth of its visual history is even greater than most observers suppose, and its appreciation is open to anybody with the right knowledge, a 1:50,000 map and, although less easily obtained, a clear day.[16]

Landschaft and Linearity
Two Archetypes of Landscape

JOHN R. STILGOE

Older Americans dislike the man-made environment. They call it monotonous, homogenized, commercial, or chaotic. Their adjectives presume a standard of judgment with which contemporary "built space" is compared and found vexingly inferior. The standard is a remembered spatial order so complete and so perfectly reflective of the good life that it survives as an unqualified archetype in the national memory.[1] It is the primary essence of landscape.

Landscape is a slippery word. In the sixteenth century *landschaft* defined a compact territory extensively modified by permanent inhabitants. A *landschaft* was not a town exactly, nor a manor or a village, but a self-sufficient, fully realized construct of fields, paths, and clustered structures encircled by unimproved forest or marsh.[2] The modern word is transformed, abused in phrases like "the landscape of injustice," which deny its ancient relations to land and to shape.[3] Vestiges of the archetypal *landschaft* endured well into the nineteenth century across Europe and the United States, and while the form no longer orders space or language, its memory controls men's imaginations.

Landschaft (or *landskap* or *landschap*) made perfect sense to Germans, Danes, and Dutchmen accustomed to compact, discrete space, but seventeenth-century English merchants and sea captains smitten with Dutch scenery painting took home only its sound. Thus, *landschap* entered the English language as *landskip*, and referred at first only to the pictures imported from Holland. Very quickly, though, it was used in new ways. Soon it defined any natural or rural view that approximated those painted by the Dutch, but by mid-eighteenth century signified the ornamental gardens of great country estates. Gardeners, reshaping fields and woods according to picturesque standards, made it a verb, and artists made it a synonym for *depict*.[4] Implicit in every definition is the old-country awareness of knowable space, however, and the contemporary critic reviewing "the moral and intellectual landscape" vaguely apprehends it.[5]

Landscape endures and thrives because the English language is particularly unsuited to topographical discussion. In the early years of the seventeenth century, the English countryside changed so dramatically that men abandoned their traditional spatial vocabulary. *Vill*, for example, once defined something like a *landschaft*, a collection of dwellings and other structures crowded together within a circle of pastures, meadows, planting fields, and woodlots. Like the Anglo-Saxon *tithing*, and like *landschaft*, it connoted the inhabitants of the place and their obligations to one another and to their land. By 1650, however, *village* had supplanted *vill*; the encircling ring of improved land and its involved maintenance were far less important to men moving freely from one cluster of dwellings to the next. Like *hamlet* and *town*, *village* soon defined only the built-up nucleus, not the surrounding fields or the intricate web of interpersonal association implicit in earlier arrangements of space.[6] Traders and other travelers thought in spaces larger than vills, and seized on *landscape* to define their vague perceptions of places now dependent on roads and long-distance commerce. In their eyes, a landscape was an extensive, cultivated expanse dotted with villages, towns, and cities; it was best seen from a mountain top, and best depicted in a painting or on a map, not in prose.

Travelers found it difficult to describe landscapes because their view was either too broad or too narrow. From a distance many structures and land forms seemed insignificant; up close they were extremely complex. Eighteenth- and nineteenth-century spatial description falls into two categories: the sweeping catalogs of large regions written by observers confined to well-traveled roads, and the intimate depictions of compact places penned by residents or long-time visitors.[7] Until late in the nineteenth century, the built environment easily accommodated the two schools of observers and descriptions multiplied. But suddenly, at the close of the century in the United States and a few years later in Britain, the rate of spatial change accelerated and observers learned that old perceptions no longer applied to space transformed beyond comprehension.

Henry James made the discovery for himself, when he returned to the United States in 1904, after an absence of twenty-four years. Twelve months later he left for England, defeated among other things by space he no longer recognized. Despite his misgivings, he published *The American Scene* in 1906.[8] It is a cryptic travelogue, filled with misadventure, disappointment, and disgust. Unwittingly, he had arrived in the United States near the end of a forty-year transformation. The nation was changing from a pedestrian to a vehicular orientation.

The American Scene is crucial because it identifies the transformation. James did not understand the changes he described, but his book enumerates the most significant of the time, and orders them against a backdrop of traditional space. James oscillated between backward areas and the most up-to-date cities, between Berkshire farmhouses and Manhattan hotels. He landed in New York, was astonished by its skyline, and fled after several days to the quiet of New Hampshire, Cape Cod, and Cambridge. From there, his wits collected, he returned to New York and plunged into the Bowery, Lower East Side, and Manhattan, striving to experience a city wholly reshaped. After several exhausting weeks he retreated to the unchanged stillness of Newport, and after a second rest set out for Boston. His last exploratory thrust into the booming resort areas of Florida forced him home to England for a final recuperation. American space overtaxed James's powers of observation and ability to control composition. In the end, description and analysis proved impossible.

James was secure in compact places—the abandoned farms of New Hampshire or the fishing villages of Cape Cod or the small towns of the Berkshires, each with its elm-lined street, white-painted houses, and quiet. In such places, he remarked, "the scene is everywhere the same; whereby tribute is always ready and easy, and you are spared all shocks of surprise and saved any extravagance of discrimination." Harvard College drowsing in the early September sunlight, Newport deserted in off-season, George Washington's monumental Mount Vernon, and "the old Spanish Fort, the empty, sunny, grassy shell by the low pale shore" of Saint Augustine are all pedestrian places. James walked about them, around their perimeters as was his custom with European towns, then strolled through them, pausing to compare present with past, or to marvel at details missed on earlier, less leisurely visits.[9] Such places, for good or bad, were "finished," and acquiring a thin patina to time.

No such patina concealed the new roughness of urban and suburban form; most of *The American Scene* documents James's excursions in understanding it. New York was like a pin cushion, he noted after his return from New England, studded with skyscrapers "grossly tall and grossly ugly" that overpowered church-es and funneled winter gales along streets jammed with electric cars. The trolleys, "cars of Juggernaut in their power to squash," terrified him, and in a moment of desperation he determined that they were "all there measurably *is* of the Ameri-can scene."[10] They prevented his crossing streets, surrounded monuments, de-stroyed any hope of quiet, and at times kept him from entering his hotel. New Yorkers skirted death at their fenders and fought for life at their doors as franti-cally as they did at elevators. Wherever he stopped, James was jostled and shoved or warned by gongs and by shouts to get out of the way. In New York he took refuge in Central Park, in Boston he rediscovered the twisting, peaceful residen-tial streets of Beacon Hill, and in Baltimore he explored the hall and court of Johns Hopkins University, searching constantly for pedestrian islands in vehicu-lar space. Bridges, especially New York's bridges, "the horizontal sheaths of pis-tons working at high pressure day and night, and subject, one apprehends with perhaps inconsistent gloom, to certain, to fantastic, to merciless multiplication," are his symbol of urban form and existence.[11] He could not walk in the way of pis-tons and view space at his leisure.

So James began touring on wheels. Much of New York he saw from inside trol-leys, and elsewhere he traveled by train and by automobile, at first fascinated by "the great loops thrown out by the lasso of observation from the wonder-working motor car." But soon he tired of his moving vantage point, and delighted in dis-covering a small village while his automobile underwent repair. Most of the time, however, he traveled by train, "the heavy, dominant American train" which he said made the countryside exist for it and whose great terminals made the cities' only portals. Railroad schedules determined his itinerary; one told him when he must desert Salem, and another allowed him only fifteen minutes in which to glimpse Savannah. In the small hours of the morning, during a two-hour layover in a deserted Charleston station, James set out to examine the workings and meaning of the great junction. He turned back, convinced in the gloom lit only by signal lamps and flaring fireboxes that the wisest course was "to stand huddled just where one was."[12] Later, as he raced south in a well-upholstered Pullman, he was suddenly aware that he could not recall when Florida began, so uniform was his high-speed view. Caught between two kinds of space, and between two ways of seeing, James determined to go home.

Unlike earlier observers of the United States, James was forced to travel at high speed, and to interpret space ordered about lines of transportation. Astute as they are, the commentaries of his predecessors were of little use. Timothy

Dwight's four-volume *Travels in New England and New York*, for example, is superb topographical analysis. At the beginning of the nineteenth century, however, its author moved almost as slowly as seventeenth-century English travelers, walking his horse while he scrutinized his environs, and discovering local history each evening at an inn. Dwight paused to inquire into crops and wildflowers, to examine soils and industries and ferries, and to question fellow travelers about the road ahead. He gazed from hilltops on towns below, criticizing their dwellings and street arrangements, and enjoying countless moments of "profound contemplation and playfulness of mind."[13] Fifty years later, Frederick Law Olmsted explored the slave states; like Dwight, he rode horseback, following back roads and trails into the center of the South, talking with storekeepers, teamsters, and children, and puzzling over structures and fields.[14] For Dwight and Olmsted, travel was a succession of minor discoveries and observations for deciphering vague maps, and of stopping again and again to examine clusters of dwellings or a new mill or a rundown farm. Olmsted's volumes, like Dwight's, chronicle self-paced travel in knowable space, the essence of the Central Park experience Olmsted devised for New York City. Travel according to his own terms, in space immediately intelligible, was denied to James; later writers such as Post, Dreiser, and Stewart adopted automobile speed and perspective as unthinkingly as Dwight and Olmsted had adopted theirs.[15] *The American Scene*, then, perceives pedestrian and vehicular space through a paradigm congruent with neither.

Pedestrian Space

Smallness was both absolute and subjective in the typical medieval *landschaft*. The twelve- to fifteen-square-mile area was home to perhaps three hundred people satisfying almost all of their own wants. Every rod of ground was fully recognized as vitally important. Meadows, arable fields, and pastures were precisely divided and bounded by paths and balks, and everyone spoke a vocabulary of landmarks.[16] Space was symbol. A family's dwelling bespoke economic and social status as clearly as its fields expressed skill at husbandry; every spot was invested with memory or some other significance—the copse where someone saw the Devil, the corner where the cart collapsed, the hill struck twice by lightning long ago. To move about the *landschaft* was to move within symbol, to be always certain of past and present circumstance. The laborer, woodcutter, baker, and husbandman understood each other's responsibilities and associated each responsibility with a specific place. By place, men understood social position and spatial location: the woodcutter's place was hewing timber in the woodlots, not directing apprentices at the bakeshop. Cycles of birth, marriage, and death, of sowing and reaping, of building and rebuilding all found expression in space.

Individuals were subordinate to the group. Nowhere was the subordination more clearly objectified than in the common fields. Here fields of wheat, rye, or barley were plowed at the same time, planted to the same crop, harvested at the same moment, and opened at the same time to all livestock. Each householder owned one or more "strips" in the fields and was entitled to their produce, but he accepted the will of the majority concerning their care.[17] Common-field farming was never innovative. Most husbandmen distrusted new seeds and plowing techniques, and forbade would-be innovators from experiments that might destroy the harvest of all. The most respected husbandman was he who best kept the corporate tradition, not he who hoped to fence off his strip from his neighbors' and selfishly experiment.

Inside the ring of fields stood a cluster of houses and perhaps a church, all focused about a roland, that aged tree, hewn shaft, or market cross that objectified the idea of *axis mundi*, the armature about which *landschaft* life revolved. But each dwelling was of extraordinary importance too, for it alone confirmed status and rights. No one might reside in the *landschaft* unless he was a householder or under a householder's oversight, and no one might possess strips in the common fields unless he also owned all or part of a dwelling.[18] House-building and admission to the *landschaft* were strictly controlled. If the fields could be expanded and greater harvests obtained from them, new houses might be authorized for the younger sons who would not inherit their parents' dwellings and strips. Forests and swamps made any expansion difficult while slow illnesses kept populations almost static. New houses and new householders were few, every *landschaft* grew slowly. Like the gypsies, discharged soldiers, and other suspect vagrants of European folktales, drifters were rarely invited to settle, but instead urged onward, away from the place of settled men.

Houses were hardly private. In fact, privacy was scarcely understood as a concept, let alone a right. Dwellings were crowded with extended families, and rooms served many purposes. Gossip made much indoor activity known outside as well, and priests and elders were empowered and expected to enter dwellings unannounced if they suspected wrongdoing. Almost every resident of a *landschaft* accepted community values and standards as his own, however, and while domestic mischief and sin were common, major offenses were few. For such misdeeds the *landschaft* had a terrible punishment. It banished the offender and broke down his dwelling, erasing all memory of him and his crime.[19] Broken men were few in medieval Europe, but they testified to the power of group values. Position in society and in space was the essence of individual identity, and banishment and house-breaking destroyed identity completely.

Chaos surrounded every *landschaft*, whether or not the forest actually enclosed the fields and dwellings. Away from the *landschaft* individual identity diminished, and men were thought to succumb to the lure of the wild. Woodcutters, huntsmen, and others who moved between ordered space and pathless wilderness were slightly distrusted by their stay-at-home neighbors. Many folktales begin at a woodcutter's hut, already removed from strict community control on the edge of a forest. Children and young women enter the forest and discover good or evil according to their character.[20] Helpful, obedient, self-sacrificing children, those who have internalized group values, discover piles of silver, magic herbs, or other treasure after triumphing over witches and robbers, and return to the *landschaft* wiser and richer. Selfish, misbehaved children and beautiful but self-centered maidens, find ashes, dragons, and sex fiends, and are punished or destroyed, or else join the evildoers of the wild. In the folk imagination, the *landschaft* is more than the objective correlative of order and safety. It is a continuous reinforcement of character, and its inhabitants desert it with trepidation.

Sometimes, of course, chaos intruded upon order. Any field left untilled grew up at once in weeds and brush, and wolves and wild boars sometimes foraged among sheepfolds and fields, occasionally slaying a husbandman trudging home after dusk. But it was the human evil of the forest that people feared most, the eldritch robbers and traders and wanderers who practiced goety, rape, and theft, and who infected children with new ideas. Out of the forest came every evil, from sorcerers to plague, against all of which the *landschaft* was almost powerless. The order of the *landschaft* was never secure, no matter how strictly enforced from within. External disruption was always imminent.

The *landschaft* was, therefore, the spatial expression of identity, order, and value, a kind of collective self-portrait of small-group life, and a great instrument of social control. Like an island, the *landschaft* was a defined place, across which a privileged traveler walked without hesitation and without danger.

Roadsides

The integrity of the *landschaft* was broken by princes and kings intent on consolidating their rule. From the fifteenth century onward, at first hesitantly and then decisively, they made forests and other wastes safe for travel. Pacification was far more advanced in some realms than in others by the end of the sixteenth century, but everywhere alert men sensed new possibilities for adventure and profit.[21] The new concern for roads, and for exploration, developed as slowly as political unity and long-distance overland commerce, but eventually it entered the popular imagination as *strassenromantik.* The romance of the road found expression in ballads and tales and, most importantly, in wandering. The newly safe roads which passed from *landschaft* into forest promised excitement and fortune. Folktale after folktale begins with a ploughman or tradesman accosted from the highway by a traveler and enticed into adventure. The highway clearly expressed an authority beyond that of *landschaft* elders. It announced the rule of kings and promised royal protection from danger. Highway robbery was infinitely more than theft by violence—it was an affront to royal power and a disruption of the new order of the road. By any name, *camino real, richtstrasse,* or king's highway, the long-distance road was a new sort of space.

Unlike the path between fields or woodlots or dwellings, which belonged to its abutters and was limited in use, the road belonged to the wayfarer and to the king. Each *landschaft* along the highway was commanded to maintain its share of roadbed in order that armies and couriers—and merchants—might not be delayed.[22] Self-sufficiency vanished as capitals and large towns drained surrounding regions of talent and produced and flooded local markets with fashionable goods. Roads became ever more important to the places they linked, for along them flowed wealth and ideas greater than those of any one *landschaft.* Local values contested with those of the road; the husbandman prized honesty, but the peddler prized sharpness. As roads grew safer and more passable, carts and wagons and finally coaches replaced pack horses, and the flow of wealth and information increased further still. No longer was the traveler an oddity to be welcomed or turned away; he became an expected figure in—but not of—every *landschaft.*

Professional travelers—merchants, peddlers, carters—had only the road as home, although they often claimed residence in some place along it. They had a new view of the countryside, for they saw only what was visible from the road and they used only what was immediately accessible from it. Increasingly, the roadside was adapted to their needs. First came inns and stables for pack animals, then corrals for driven herds, and then bridges and toll houses and eventually directional signs and mileposts. It was at this time that *village* replaced *vill* because fields and woodlots and responsibility for them interested travelers far less than a good inn and perhaps a blacksmith shop. Eventually, the *landschaft* surrendered its identity to the highway, and was known not as a unique place but as one of many settlements along a well-known road.

For the sedentary inhabitants of the *landschaft*, the wayfarer was personified *other*, against which they evaluated themselves. In the days when roads were so few that traveling was almost unknown—and German folktales collected at mid-nineteenth century mention such times—adolescents had only their parents and adult neighbors as models. The absence of different values and exotic behavior made internalization of *landschaft* mores simple. Only when travelers provided new standards to any youth astute enough to linger about the inn or stable yard after nightfall did socialization break down. Travelers were anonymous, and their larger experience was approached with a mixture of distrust and deference by adults and adolescents alike. The road introduced the kind of marked change in interpersonal relationships which one usually associated only with city life.[23] Strangers met knowing they might not meet again, judged each other as types according to dress and occupation, and talked of matters of importance only among themselves. It was a rare carter who was deeply interested in the state of the crops, and a rarer husbandman who cared about the weak bridge thirty miles to the east. But traveler and native alike were interested in conversation, the traveler to pass his evening and the native to learn something of the larger world. While at first no one noticed, the web of corporate ties connoted by *landschaft* and *vill* was torn.

Until the nineteenth century, *landschaft* and road coexisted, but ever more fitfully. It was not that travelers were murdered in inns or daughters ran off with teamsters, but that the life of the highway was becoming removed from the life of the *landschaft*. Road signs made asking directions unnecessary, and the improved maps, road surfaces, and police ordered by stronger governments permitted travel even after nightfall.[24] The turnpike avoided some settlements in its quest for directness, and mailcoach passengers scarcely glimpsed villages far from the route.[25] Railroads only sharpened the dichotomy between traveler and inhabitant already implicit in turnpike design. Trains followed their rights-of-way too quickly for casual communication between passengers and spectators, and forced riders to look sideways at a silent blur. Soon space was ordered about the track; towns focused on stations, water towers, and grain elevators which blocked passengers' views of the towns. As factories and warehouses moved next to the rails, trains ran for miles through a tunnel of structure. The railroad traveler was denied a long-distance view and the opportunity of stopping to analyze nearby space. The inhabitants of trackside areas could in turn only gape at the faces staring behind the glass.

Increasingly, travelers suffered from a curious *anomie* described as early as 1798 in Coleridge's "The Rime of the Ancient Mariner":

Like one, that on a lonesome road
Doth walk in fear and dread
And having once turned round walks on,
And turns no more his head;
Because he knows, a frightful fiend
Doth close behind him tread.[26]

The fiend snuffling at the heels of every traveler is the fiend of homelessness, of lack of place in the world of men. Most travelers fix their heart and eye on a refuge somewhere ahead—and ignore the roadside world in their haste to arrive.

What they see from the road is landscape, the not-quite-understood complex of dwellings, fields, factories, and other artifacts of human work placed among natural landforms and vegetation or in totally modified space. If the view is chiefly natural or rural it qualifies as scenery, and the traveler choosing a "scenic route" knows he should appreciate it, if only as a relief from man-made complexity. But all too often, man-made, not natural shapes and spaces dominate his route, and the hurrying observer is stunned by elaborate man-made forms having no immediately apparent use or arrangement or uniqueness. He looks away and thinks on that place which he does understand but which no word, not *community*, not *neighborhood*, and certainly not *landschaft* accurately identifies. There is only his personal being congruent with man-made form.

James defined the power of high-speed, long-distance travel in the title of his book; *The American Scene* is concerned not with scenery but with scenes, places transformed by man. James could not avoid such places because the highway and railroad constantly directed him to them. While he was honest enough to look at them, to describe them, and to try to appreciate them, he judged them by older, far different places, the backward vestiges of America's *landschaft* past. He longed for discrete, knowable territories where social order and man-made space coincided, where the disruption of highway was unknown.

Utopias

James was not alone in his search for knowable space. Dozens of other observers rejected America's late-nineteenth-century spatial order and retreated into historical, local-color, or nature writing.[27] The reorganization of space about railroads and motor highways sparked a flurry of futurist thought in utopian novels and polemics. Earlier Americans knew little of the genre, perhaps because experimental communities had once flourished in fact, on cheap, back-country land away from censorious eyes. By the 1870s, however, most of the communities had disbanded in the face of mechanized agriculture and manufacturing, and the dream of social harmony in ordered space reappeared in the writings of utopianists.

Most of the fictional utopias are characterized by social and economic systems derived from the theories of Herbert Spencer, Henry George, or the Bible, and are served by mechanical devices—electric motors, air-conditioning units, and even automatic bed-makers—predicted in scores of magazine articles. No master plan or consensus of opinion inspired the shaping of utopian space, however, and the American visionaries were left to their own imaginations. It is startling, then, that the same spatial features, curious as they are, occur repeatedly. The striking uniformity of utopian built environments is not a matter of vagueness; most writers were lavish with detail and many delighted in engravings of city plans and building elevations. It was a shared conviction that space was no longer knowable, and a shared dislike of several especially pernicious features in particular, that prompted nineteenth-century writers, like Thomas More four centuries earlier, to envision perfection.

A geometry of well-being informs almost all utopian space. Wildernesses, mountains, and dangerous seats usually surround the perfect place and protect it from the profane world; chaotic nature rarely intrudes. The Martian utopia described in Henry Olerich's *A Cityless and Countryless World* (1893) is seemingly without hills or valleys, its topographical monotony broken only by buildings.[28] The future United States of Edward Bellamy's *Looking Backward* (1888), Bradford Peck's *The World a Department Store* (1900), and Edgar Chambless's *Roadtown* (1910) likewise lacks great forests or swamps.[29] Most of the territories are completely gardened. There is little construction or other modification because social perfection has engendered perfection of space. Every problem confronted by industrializing America is neatly solved.

In nearly every vision the marriage of country and city is consummated without the trauma of suburbia. Olerich's Martian visitor tells his American hosts that noisy, dirty cities are "detrimental to an orderly, well-regulated society," though he is quick to add that farm life is a social and economic waste too. The orderly solution to urban congestion and rural isolation, he explains, is the "community," an arrangement of apartment houses carefully sited along the perimeter of a great agricultural enclosure. Peck and Bellamy bring rural joy to everyone by filling their airy cities with shade trees and promenades, Peck by throwing together back yards and demolishing offensive structures, Bellamy by judicious planting of open squares. Chambless's Roadtown extends indefinitely across farm land, a sinuous chain of two-story row houses, each with a flower garden in front and a vegetable patch behind, and public buildings linked by a basement railroad and rooftop walks. In each utopia, farmers share in the charms of reformed city life, and all residents enjoy gardened nature—but not wilderness.

Single-family housing, except in Roadtown, is gone, along with home cooking, parlor entertainment, and housework. Communal restaurant facilities provide choice food and conversation, and the theater is the focus of recreation, though amusement parks, athletic fields, and gymnasia are scattered everywhere. A new—or very old—sense of community finds expression in malls, shopping spaces, and public parks. Even the telephone and music phone, while providing solitary entertainment of highest quality, are poor substitutes for social visits and the theater. Free time in Cooperative City and Roadtown is never private time, but time devoted to public affairs. The loner and the hermit are unknown, and indeed have no place in the *landschaft*-like utopian world.

Work is stimulating and deeply satisfying. Everyone takes turns at the boring jobs—serving in the communal restaurants or driving sight-seeing carriages—though most arduous tasks are lightened by such machines as Olerich's electric-powered farm tractor. The drudgery of housework in particular is vanquished, partly by mechanical invention and partly by communal laundries, kitchens, and furnaces. Women are free to garden, read, visit, sew, or cook, though many choose to attend the theater with friends. Factory work is meaningful too, even if it is given over to an army of young citizens, by its association with community advancement. Repeatedly, the importance of small workshops—Chambless locates a workroom in the lower level of every Roadtown house—is emphasized as the spatial manifestation of pleasant, soul-satisfying work. There is little or no separation of work space and home in the utopias. The smokeless, noiseless electric machinery is unimportant in the utopian vision of work as creative play. It is the small-group companionship, healthful work space, and beautifully finished products that satisfy the workers, not gadgets.

The companionship of the theater and shop extends beyond each citizen's immediate acquaintance to the larger community, finding expression in monumental building. Theaters are, after all, recreational, and most writers, terrified of offending sectarian readers, substituted the lecture hall and government building for the cathedrals which might otherwise order utopian space. Citizens gather for education and decision making in the grandest structures of utopia, almost invariably sited in exquisitely planted public squares. Education is rational and politics are straightforward and honest; children enter school buildings happily and adults administrative halls without hesitation. Peck's Cooperative City, for all its squares, avenues, and diagonal streets, focuses on a great administration complex, "situated in the center of several acres of land laid out by leading landscape gardeners in the most artistic manner, setting off the magnificence of the enormous building which accommodated the numerous offices and legislative halls of the executive boards."[30] At the center of almost every utopian place, rationality and right government take physical form.

Roads, and sometimes railroads, monorails, and trolley lines, touch every part of every utopia, making the monumental center immediately accessible. Nearly every writer resolves communication difficulties first with telephones, and then with highways, boulevards, and service streets. Roadtown, of course, is a road, the epitome of Chambless's determination to order linear settlement. Beneath its gently curving superstructure run several railroads for long-distance and local transportation of freight passengers. Its continuous roof is devoted to a steam-heated, glass-enclosed promenade paralleled on each side by bicycle and roller-skating paths, benches and jogging tracks. It is on this recreational street, Chambless predicts, "that Easter hats will be shown and neighbors' crops discussed and new acquaintances made and local pride developed."[31] The less ambitious plans of other writers also stress roads both as recreational and social places, and as boundary lines. Bellamy's futuristic Boston succeeds as an artistic device because of its broad streets and grand vistas, and Peck's utopia is crisscrossed with extravagant avenues. But the roads and railroads, and even Roadtown itself, lead nowhere in particular. They are not strands in larger networks, and are rarely used for long-distance travel. Indeed travel, except for an afternoon's diversion, is uncommon, for there is no place better worth seeing. Strangers, except for the nineteenth-century narrator-visitors, are unknown, for the broad avenues terminate in dimly known regions inside the wall of mountains or seashores that ring the utopian places. For all their width and beauty, utopian roads are really paths.

Thomas More would have found himself at home, for the utopian spaces are little different from his own. In 1516 More abandoned the utopian vision which had satisfied Europeans for fifteen centuries. Unlike Heaven, his Utopia was just over the horizon, somewhere beyond the New World discoveries. Like his American counterparts four centuries later, More was troubled by social and spatial changes. Feudalism was giving way to capitalism, and the old rhythms of rural life deteriorated. Intellectual authority weakened before the onslaught of empirical science, and the unity of the church was threatened. Order lay beyond imagined equatorial regions filled with wild beasts, serpents, and savage men, in an ideal commonwealth where good sense, sound learning, and mercy find expression in space. His fifty-four city-states, "all spacious and magnificent, identical in language, traditions, customs and laws," are "similar also in layout, and everywhere, as far as the nature of the ground permits, similar even in appearance."[32] Each is intimately associated with the life of its agricultural land, for every inhabitant shares in the farm work. Utopia's capital, the walled town of Amaurotum, rivals Peck's Cooperative City as a paradigm of order. Each of its three-story row houses fronts on a broad avenue and opens in the rear on a great common garden. A market building for the storage and sale of family handicrafts and the distribution of farm produce orders each quarter of the town, and common dining and recreation halls order every block. There are no hermits or idlers; everyone finds his community life and work satisfactory. The off-island world of Abraxa, the Indies, and all of Europe is only so much chaos in comparison.[33]

More's book had several successors, but by mid-seventeenth century the genre was moribund. For two hundred years, utopian writing was displaced by travel narrative. Europeans and Americans, distracted by exotic customs and topography and lulled by faith in progress, paid little attention to the continuing disruption of small-scale community existence and space. Not until the impact of railroads was fully felt, and the meaning of trolley cars and automobiles surmised, did intellectuals discover that imageable space was fiction. Distraught and baffled by the seeming chaos of society and the built environment, most shut out the confusion by concentrating on bits and pieces of personal significance—the family or circle of friends, the isolated village or farm. Others turned to utopia and city planning, reinvoking the memory of *landschaft*.

The *landschaft* is the controlling spatial metaphor in most sixteenth- and nineteenth-century utopian writing, and in city planning literature of all ages, including our own.[34] Like the *landschaft*, the utopian place is ringed by wilderness. It is also fully realized by its inhabitants, characterized by home-based farm work and handicrafts, focused about a symbolic center, and interlaced with short-distance paths. Nineteenth-century city and regional planners, like their utopian contemporaries, fastened on the *landschaft* archetype too, finding utopia in long-lost New England villages or in Main Street towns. Their successors see it in green belt suburbs or in circular condominium complexes.[35] It is always the circular, clustered archetype that reigns as standard, be it a remembered and often romanticized *landschaft* or a perfectly contrived utopia.

Archetypes

In comparison with the archetypal standard, the man-made environment of the United States seems almost chaotic. The vehicular traveler is overwhelmed, then stupefied; views change too quickly, too dramatically for sustained high-speed study. Passengers close their eyes and doze, while drivers focus on the pavement ahead and grimly ignore the roadside scene. Meaning escapes the pedestrian too. There is so much detail and so much variation that the walking seer edits his surroundings at once; he knows his house clearly enough, and his hotel and the museum, but he does not realize the buildings between.[36] Passenger and pedestrian alike have a snapshot vision of the built environment. They see bits and pieces of greater or less importance, but continuously significant space, because they see through the prism of *landschaft*.

This is the power of the ancient archetype of *landschaft*. It controls visions of what the built environment is, and shapes dreams of what it ought to be. It prevents men from realizing—and from loving—the space they inhabit.

But its power is weakening. The present post-industrial, post-modern age is also post-*landschaft*. The old significance of defined, imageable space is lost on today's children. They can hardly conceive of a medieval *landschaft* isolated from a wider world by feared wilderness, fully realized by every inhabitant, when space is defined by the road, and linearity shapes their vision of what space ought to be. In their *strassenromantik* of exploring and wandering, and in the fantasy and science fiction they so frequently read, the road and the quest are dominant metaphors. Thoreau and Henry Beston are suspect; they stay too much at home. Hobbits, apprentice magicians, and neophyte priests, on the other hand, are respected for searching in space, for following roads however dangerous and indistinct, and for confronting Coleridge's frightful fiends.[37] It is no accident that America's young people are not disturbed by interstate highways, shopping strips, and vast suburban sprawl. They are children of landscape, and see in the built environment a symmetry, order, and beauty scarcely visible to Henry James and to their parents.

The old archetypes of *landschaft*, then, no longer wholly shape images of home, neighborhood, and utopia.[38] Its resiliency, however, should not be underestimated. Archetypes are not created by the conscious, and they cannot be destroyed by it. They can only be submerged, ignored for a limited time. To ignore the archetype of *landschaft* and unthinkingly accept the landscape of linearity is as dangerous as to see all man-made space through the prism of *landschaft*. Implicit in the concept of landscape are two archetypes, the ancient one of *landschaft* and the venerable but younger one of the road.

Environmental Change in Colonial New Mexico

ROBERT MACCAMERON

In recent years scholars of North American history have paid increasing attention to the interrelationship between human societies and their physical and natural worlds. Their studies analyze the various ways in which people interact with their surrounding environment and how their choices affect not only the human community but the larger ecosystem as well. Just as nature shapes human society, humans, in significant and far-reaching ways, shape nature.

Exemplary works by William Cronon and Richard White have focused, from an ecological perspective, upon the English frontier experience in North America. Cronon, in his seminal book, *Changes in the Land: Indians, Colonists and the Ecology of New England* (1982), demonstrates that the English colonization of New England produced a number of "fundamental reorganizations . . . in the region's plant and animal communities";[1] a result, fundamentally, of the "colonists' more exclusive sense of property and their involvement in a capitalist economy."[2] Similarly, White, in his study entitled *Land Use, Environment and Social Change: The Shaping of Island County, Washington* (1980), describes how the introduction of European technologies along with the Columbian Exchange of plants, animals, and diseases dramatically altered the operation of natural systems, producing in turn what one botanist has described as the "most cataclysmic series of events in the natural history of the area since the Ice Age."[3]

This chapter focuses upon environmental change in another area of North

America: the upper Rio Grande valley which came under Spanish rule between 1598 and 1821. This largely semiarid ecosystem encompasses today the area of north central New Mexico, from approximately Belen, just below Albuquerque in the south, to Taos in the north, and from the Sangre de Cristo and Sandia mountain ranges in the east, to the Chama, Jemez, and Rio Puerco river valleys in the west. The study examines just how Spanish culture and society brought about new relationships between human societies, including Pueblo Indians[4] and the land, and how changes in the land occurred as a result of those relationships. The English experience in North America serves as a principal point of reference.

The kind of physical and natural world in which the Spanish settled provides an important context for any discussion of environmental change. Scholars today describe north-central New Mexico as possessing essentially five different life zones: the Upper Sonoran, Transitional, Canadian, Hudsonian, and Arctic-Alpine. The Upper Sonoran zone includes mostly valleys, foothills, and plains, and extends from approximately 4,000 to 7,000 or 8,000 feet in elevation. Average annual rainfall is approximately ten inches. The Transition zone encompasses the middle slopes of the higher mountains and begins at approximately 7,000 to 8,500 feet above sea level on northeast slopes and 8,000 to 9,500 feet on southwest slopes. The Canadian zone covers most of the higher peaks of the Sangre de Cristo range and extends from approximately 8,500 to 12,000 feet depending upon cold or warm slopes. The Hudsonian and Arctic-Alpine zones are found on peaks that are around or above the timberline. Each of these zones is characterized by distinct flora and fauna, as they were at the time of Spanish arrival, although the boundaries between them are sharply marked only on steep-sided slopes. The zones themselves and their individual characteristics result from differences in altitude, temperature, precipitation, and barometric pressure. Significantly, human settlement has occurred almost exclusively in the first two zones. The Upper Sonoran has encompassed, past and present, most of the agricultural and grazing lands of the region, and here farming, except at the highest elevations, requires irrigation or some other water control system. The Transition zone is a bit wetter and therefore allows for some dry farming as well as seasonal grazing. But in fact, both zones constitute a semiarid environment in which access to water has been crucial for human survival.

Within this environmental context, Pueblo-Spanish contact produced its own particular form of the Columbian Exchange, that is, the reciprocal introduction of Old and New World plants, animals, and disease. More broadly construed, the exchange also included forms of material culture and even distinct values relating to economic, social, political, and religious organization.[5] Students of New Mexico's environmental history differ in their emphasis on the effects that this exchange has had on the land and the human societies occupying it. An anthropological view emphasizes the front end of the exchange, the Spanish introduction of new seeds, wheat, vegetables, and the tools for environmental destruction, including the iron ax and livestock, which changed the ecology of the upper Rio Grande valley forever.[6] A historical view more often stresses the rather remarkable adjustment achieved by Hispanic settlers in a very rugged frontier environment. A community-based agro-pastoral system of subsistence evolved that successfully sustained, both environmentally and culturally, many small communities for generations.[7] It was only after the arrival of the Anglo-Americans into the upper Rio Grande valley and the introduction of commercial agriculture that changes in the land occurred in any dramatic fashion.[8]

This second view, in essence, subscribes to a traditional interpretation of the relationship between economics and ecology, as outlined by Donald Worster and others, that subsistence agriculture, such as that employed by the Spanish in New Mexico, tends to preserve much of an environment's diversity and complexity, producing in turn an ecological and social stability. On the other hand, capitalism, involving specialization in production and competition in markets for profit leads to "a radical simplification of the natural ecological order, number of species found in an area, and in the intricacy of their interconnections."[9]

Interpreted in the broadest sense, this essentially linear model for environmental change in colonial New Mexico is basically correct. There is no question that human impact on the land, from Pueblo to Spanish to Anglo-American, was increasingly severe. But this model also contains certain limitations when applied to the case of colonial New Mexico. A closer look at the Spanish period reveals a far more complex process at work. In fact, the evidence indicates that a number of factors or determinants worked toward or mitigated changes in the land, producing in turn, in a largely nonlinear fashion, different kinds and rates of change.

Factors include demographic features of both Pueblo and Spanish, the Spanish introduction of grazing animals into the upper Rio Grande valley, colonial New Mexico's relative social-economic isolation, Spanish material culture, dimensions of Spanish land institutions, systems of political control, the varying nature of an inclusive frontier, changes in climate, and the vulnerability and resilience of a semiarid ecosystem. Many of these factors were paradoxical in their effect. They at once simplified the ecosystem and thereby accelerated change, and sustained ecosystem diversity and complexity, and produced little or no change.

Demography, especially numbers of people and where they relocated, is an important starting point for understanding the factors affecting environmental change. Throughout the Spanish colonial period approximately 99 percent of the population of the Province of New Mexico, including the general categories of Indians, Spanish, and *castas* (people of mixed blood), occupied about only one percent of the land. Access to water, in the form of rivers, creeks, and streams, was one obvious reason. But so also was the intermittent danger posed by the nomadic tribes—Apaches, Comanches, Navajos, and Utes—surrounding the upper Rio Grande valley. Attempts to expand beyond the Rio Grande itself out to its principal tributaries and elsewhere were often discouraged by the presence of such nomads. Need for water and the presence of unfriendly Indians acted as a centripetal force upon both Pueblo and Spanish settlements. But at the same time, as Spanish population increased, especially during the eighteenth century, overcrowding along the Rio Grande occurred and access to resources, especially water and pasturage, diminished. These factors served as a countervailing or centrifugal force to Spanish settlement.[10]

The Spanish colonial population of the upper Rio Grande valley was centered principally on three villas: Santa Fe, founded in 1610; Santa Cruz de la Canada, founded in 1695; and Albuquerque, founded in 1706. The first two are located in the area known as the Rio Arriba, or the "upper river," while Albuquerque is located in the Rio Abajo, or "lower river." La Bajada mesa, 19 miles south of Santa Fe, served as the dividing line; there the Santa Fe Plateau drops about 1500 feet. Over the course of the colonial period these three centers and their environs, consisting of many smaller communities, contained the bulk of the Spanish population. Colonists also tended to radiate out from the three villas and establish small farms and ranches (ranchos) and hamlets (plazas). Yet frequently these more isolated settlements were abandoned and sometimes returned to again as shifting migration became an important demographic feature of colonial New Mexico.

The Spanish population (including *castas*), in the upper province grew very gradually at first, set back by the Pueblo Revolt of 1680 and the economic unattractiveness of the region, and then increased more rapidly by the end of the eighteenth century. It was less than 1,000 in 1600, approximately 2,900 in 1680, 2,000 in 1700, 3,400 in 1752, 5,800 in 1776, 19,000 in 1800, and 28,000 in 1821.[11] More specifically, figures for the eighteenth and nineteenth centuries indicate shifts in population between and among the three villas: enumerations indicate that Spanish concentration was highest in Santa Fe in 1752, in Albuquerque in

1760 and 1776, and in Santa Cruz in 1790, 1800, and 1817. By 1817 Santa Cruz had a Spanish population of 12,903; Albuquerque, 8,160; and Santa Fe, 6,728.[12]

More Spanish and *castas* eventually came to live in the heart of the Rio Arriba, around the Villa of Santa Cruz, than in the vicinities of either Santa Fe or Albuquerque, reflecting both environmental realities and Spanish policies based upon them. In this protective terrain, with perennial supplies of water for irrigation, Spanish governors awarded communal land grants to groups of people, believing that they could protect themselves from attack by nomadic Indians. These grants, with their concentrated agricultural communities, served the purposes of the government by establishing controls over the citizens themselves and by creating an effective defensive frontier. In contrast, the settlement pattern in the Rio Abajo, along the broad, fertile plain of the Rio Grande, came to be characterized by a more dispersed population occupying private land grants.[13] Meanwhile, the Pueblo, dramatically reduced in numbers, clung tenaciously to some twenty-one discrete settlements, ranging from Taos in the north to Isleta in the south, and from Pecos in the east to Acoma, the sky Pueblo, in the west. Paradoxically, this reduction in Pueblo population meant that there were more human beings in the upper Rio Grande valley at the beginning of the Spanish period in 1598 than at its close in 1821.

The Spanish introduction of grazing animals, termed "ganado mayor," cattle and horses, and "ganado menor," sheep and goats, into colonial New Mexico altered the face of the land in dramatic ways. In contrast to most other areas of Spanish America where cattle constituted the principal livestock, sheep came to dominate the landscape of the upper Rio Grande valley. There were several principal reasons for this development. Immense herds of buffalo on the plains to the east provided a source for hides, jerky, salted tongues, and tallow, products of a quality equal to those from domestic cattle. The hostile Indians surrounding the province, with the exception of the Navajo, usually coveted cattle and horses because they could be driven easily away. Sheep, on the other hand, could be scattered and then recovered after a raid. Throughout the course of the colonial period, the mining settlements of Nueva Vizcaya in Chihuahua and Durango provided a strong market for New Mexico sheep. During the eighteenth and early nineteenth centuries the sheep trade became the primary export industry.[14]

Unlike the human population of the region, sheep came to be concentrated principally on the broad plains of the Rio Abajo, in and around the Villa of Albuquerque. A livestock enumeration in 1827, six years after Mexican independence, reveals nearly 250,000 sheep and goats in the province, with Albuquerque possessing 155,000, Santa Fe, 62,000, and Santa Cruz de la Canada, 23,000. At the same time, there were only 5,000 cattle, 2,150 mules, and 850 horses.[15]

These grazing animals, both sheep and cattle, had an important impact on the ecology of the area. Various descriptions of the land by early visitors provide a benchmark from which to view later changes. Two pre-settlement observers of New Mexico, Hernan Gallegos and Diego Perez de Luxan, representing the Chamuscado-Rodriquez (1581) and Espejo (1582) expeditions respectively, noted lush grasslands, untouched pastures, highly suitable for both sheep and cattle.[16] After Spanish settlement, evidence indicates that portions of these grasslands were, over time, dramatically overgrazed. While many fewer in number than sheep, cattle also effected changes in the land in several principal ways. Whereas sheep were often grazed on distant pastures and required intensive labor, cattle were turned loose, unattended, on commons close by agricultural plots for safety from hostile Indians. As a result, the commons were frequently overgrazed. Marc Simmons says that Albuquerque's east mesa, for example, was virtually denuded of grass by 1750, forcing cattle to invade the fenceless crop fields of both the Spanish and the Pueblo.[17]

Over the course of the colonial period, the Spanish Archives of New Mexico contain numerous cases of New Mexico governors warning Spanish settlers to control their livestock, particularly from damaging Pueblo fields and irrigation ditches, or face severe penalties. As the number of sheep dramatically increased over the eighteenth and early nineteenth centuries Spanish settlers frequently petitioned governors for fresh pasturage away from the core settlements of the Rio Abajo and Rio Arriba, looking particularly to the south of Albuquerque, to the Rio Puerco, and to the Chama valley. Petitioners in the late 1730s, for example, complained about the inadequacy of land and water, especially the shortage of grass, near Albuquerque.[18]

Yet the needs of the settlers and goals of the state were often in conflict. Governors, representing the interests of the defensive function of the frontier, wanted to control and regulate shifts in settlement, especially on frontiers where conflicts with Indians were likely to occur. As a result, settlers' requests for fresh Crown land were often denied, and even in cases where their petitions were approved, they frequently were forced to return to the safe harbor of the three villas to escape Indian depredations. These barriers to expansion merely exacerbated livestock pressures on already settled land. By the 1820s, New Mexico sheep growers still sought fresh grass, this time to the east, onto the plains beyond the Sandia and Manzano mountains.[19]

In contrast, the isolation of colonial New Mexico from the rest of New Spain and English North America clearly moderated the process of environmental change. In 1803 Governor Fernando de Chacon, in a report to officials in New Spain, vividly portrayed the region's general economy by noting that New Mexico's "natural decadence and backwardness is traceable to the lack of development and want of formal knowledge in agriculture, commerce and the manual arts."[20] Chacon thought that Spaniards and *castas* were little dedicated to farming and contented themselves with sowing and cultivating only what was necessary for their sustenance. While remarking on the abundance of sheep in the province, he indicated that no great number of swine existed. The continual raids of the nomadic Indians discouraged raising horses and mules. Deposits of minerals such as lead, tin, and copper were located in various parts of the province but they were virtually untapped, while large scale smelting or amalgamation operations were nonexistent. Throughout New Mexico, high-quality mica or gypsum *(yeso)* covered windows in place of glass panes. Formal apprenticeships, examinations for the office of master, and organized guilds, customary elsewhere in New Spain, did not exist. But out of necessity and the natural ingenuity of the people, according to Chacon, some trades were practiced skillfully, including weaving, shoemaking, carpentry, tailoring, smithing, and masonry. The exports of the colony, transported by the annual mule trains to Sonora, Vizcaya, Coahuila, and points south, consisted of oxen, sheep, woolen textiles, some raw cotton, hides, and piñon nuts. The total value of these products, including wine from the El Paso district, was estimated at only 140,000 pesos annually. The products brought back into the colony included linen goods, chocolate, sugar loaves, soap, rice, iron, leather goods of all sorts, pelts, paper, drugs, and some money. Because hard currency was in chronic short supply throughout the colonial period, barter was the principal mode of exchange.

Trade also existed between the Spanish, the Pueblo, and nomadic tribes. Taos became a principal center for exchange during times of peace, where, for horses, an array of metal tools, corn, and trinkets, the "uncivilized" Indians traded pelts, buffalo skins, and Indians whom they had captured from other tribes.[21] The fact that pelts were obtained either through import or at Taos indicates that little hunting of small or large game went on in the upper Rio Grande valley itself.

In essence, colonial New Mexico demonstrated a low level of technological development, little or no occupational specialization, generally self-reliant local production for local consumption, and a broad utilization of the entire environment—all characteristics of a subsistence economy. It would be a mistake to assume that such characteristics did not lead to the overuse of resources or environmental degradation, but surely the degree of change was less than what might have occurred in a society distinguished by specialization in production and labor, and by the export and import of products to and from similarly specialized producers.

As an example, if large deposits of silver had been discovered in New Mexico during the Spanish period, as they were in other semiarid regions of central and northern New Spain, the ecology of the area would have been transformed to a significantly greater degree. Silver mining required the special input of both raw materials and labor: a large supply of wood for fuel, shoring in mines, construction in buildings, and machinery; water for power and washing; and thousands of grazing animals to produce hide, leather, meat, and energy for transport and powering machines. Shifts in population occurred as well to meet the intensive labor demands in the mines, and the human wastes from such concentrations were merely dumped into the local waterways. The amalgamation of silver also required the use of such products as copper sulfate, common salt, and in large quantities, mercury. Metallic mercury, mercury vapor, and lead, the residues of this process, poisoned both plants and animals as they readily invaded the air, water, and ground.

After only a few decades, the environmental effects of this economic activity were devastating. Entire areas in and around mining communities were denuded of grass, deforested, and eroded by wind and water. Vegetation loss also reduced transpiration, leading in turn to a decline in local rainfall. As food chains were destroyed, animals and fish disappeared, and it can be assumed that the results of mercury and lead poisoning in human beings, mostly Indian labor, were frequently fatal.[22] Nothing on this ecological scale occurred in the Province of New Mexico during the Spanish period.

Yet the Spanish introduction of metal tools into the upper Rio Grande valley did allow both the Spanish and Pueblo to manipulate that environment in entirely new ways. The introduction of the ax alone enabled bench lands, bounded above and below by steeper slopes, to be cleared of dense vegetation, and woodland along the rivers and in the higher elevations to be cut. Wood became an important source for construction, tools, and fuel. The piñon, a scrub pine, was used to make plowshares and the legs of spinning wheels; the cottonwood to make wine barrels and *carreta* (cart) wheels; the oak to make stirrups and the *carreta* frame; and the Douglas fir to make the shafts of plows and provide large timbers for bridges and *vigas* (beams) in roofs.[23]

The degree to which the upper Rio Grande valley was deforested as a result can be inferred from several commentaries made at the beginning and end of the Spanish period. In 1582, Luxan wrote that "this Province [New Mexico] boasts of many pine forests . . ." and that "there are also fine wooded mountains with trees of all kinds. . . ."[24] In 1839, in one of the first Anglo accounts of the region, Josiah Gregg, explorer and trader, wrote that "on the water-courses there is little timber to be found except cottonwoods, scantily scattered along their banks. Those of the Rio del Norte [Rio Grande] are now nearly bare throughout the whole range of the settlements, and the inhabitants are forced to resort to the distant mountains for most of their fuel."[25] Another Anglo account sixteen years later corroborates this description. W. W. H. Davis observed that "wood is exceedingly scarce all over the country. The valleys are generally bare of it, and that found upon the mountains consists of a growth of scrub pine called the piñon. The country is said to have been well wooded when the Spanish first settled it, but in many parts it has been entirely cut off, and in some instances without even leaving a tree for shade."[26]

The slow development of Spanish material culture is most apparent when viewed in relation to that of English North America. As a result of the colony's isolation from European influence, the use of agricultural tools remained largely unchanged over the period of Spanish rule. The scratch plow, essentially as described in the Bible, was equipped with an iron, steel, or wooden share and cut a shallow furrow about six inches deep instead of turning the soil. It was not until the end of the Mexican period in 1846 that two-handle steel plows with a moldboard for turning the soil reached New Mexico.[27]

The ecological implication of this low-level technology is significant. Whereas in English North America the use of deep cutting plows, in the absence of contour plowing, crop rotation and manuring, caused soil erosion on a massive scale, the scratch plow of New Mexico generated comparatively little soil loss. In fact, the farmlands of the upper Rio Grande valley were auto-replenishing through the agency of silt-laden irrigation and flood waters and appeared, over time, to a number of colonial visitors as particularly abundant and fertile.

Another side of material culture was the Pueblo and Spanish use of soil as a building material. While the Spanish introduced the formed standard adobe brick to the Pueblo, the techniques and materials for construction remained essentially unchanged from what the Indians had used before.[28] Adobe construction tapped several available and replenishing natural resources: loamy soil, sand, water, and straw. Wood was used only for ceiling beams or *vigas*. For the

Pueblo, the adobe was one of the most prized Indian crafts as it represented the sacredness of the land itself. In contrast, the English use of wood for both construction and sale, while entailing a sense of craft, was predicated largely on an ethos of function and profit. As a result, Pueblo and Spanish buildings, decrepit when seen through the eyes of some nineteenth-century Anglo-American visitors, arose out of and were an integral part of the physical landscape, and unlike the acute deforestation which occurred in English North America, caused few changes in the land.

After the Pueblo Revolt the Spanish imposed a pattern of land use and settlement on the upper Rio Grande valley in marked contrast to the widely scattered large estates, worked primarily by Indians through the *encomienda*, of the early seventeenth century. On the basis of individual and communal land grants the Spanish came to live in smaller units strung out along the Rio Grande and its tributaries. In the Rio Arriba, where the colony's Spanish population became most concentrated, the communal land grant dominated. In this case individuals in a group of settlers would each receive an allotment of land for a house, a plot for irrigation, and the right to use the unallotted land on the grant in common with the other settlers for pastures, watering places, firewood, logs for building, and rock quarrying, among other activities.[29] The intensity of land use under these circumstances certainly differed by degree between and among the various grants. But the very configuration of these settlements, coupled with barriers to out-migration, helped to produce such changes in the land as overgrazing and deforestation, the subsequent loss of topsoil, and the silting of irrigation ditches and streams.

A pattern of land-holding developed in the region, both on and outside of the community grants, of long-lots or narrow strips of land emanating from the water ways. They were a necessary response to the physical environment and represented a practical and equitable method of partitioning irrigable land. Yet over generations, through the institution of partible inheritance, long-lots were divided and redivided lengthwise, leading to more densely populated communities living on increasingly smaller parcels of land.[30] As a result settlers had to move their grazing and cutting operations ever higher up the mountain sides to accommodate the loss of grass and trees on the original commons. At the same time overused irrigation plots for growing wheat and garden vegetables became increasingly less productive.

On the other hand, metes and bounds, a legal system by which property boundaries were determined, was a feature of Spanish land institutions that favored environmental conservation. Boundaries were not established with any prescribed shape in mind, nor were they necessarily contiguous with any others. Instead, they were drawn to follow the natural contours of the land and to include the most valuable resources: soil, woodlands, and access to water. It was a system highly adaptive to the local environment,[31] and predicated, at least in the case of colonial New Mexico, on the needs of subsistence agriculture.

A system of grids, based upon the rectangular survey, came to prevail in many parts of Anglo-America with far more severe environmental consequences. There little allowance was made for local topography, hydrology, or climate. Fields were arranged according to a rigid north-south, east-west alignment often resulting in enhanced soil erosion. The distribution of surface water, arable land, grass suitable for grazing, and timber for securing wood was often unequal between and among individual holdings. As a result, property owners tended to make extreme demands upon natural resources that were frequently in short supply. In contrast to the communal restraints evident in parts of colonial New Mexico, in order to maximize profit in a market economy, owners were free to exploit their local environment as they saw fit.[32]

With roots deep in Iberia, Spanish irrigation was at once vital to agricultural productivity and a source of land change and deterioration. The pre-contact Pueblo use of water control devices, the complexity of which is still debated, led to the accumulation of salts and other mineral deposits in the soil and the consequent need to seek new growing areas in the face of an expanding population and shrinking bottomland. The Spanish system, borrowed by the Pueblos, placed a more complex pattern of ditches on the face of the land with even greater natural and physical effects. In both the Rio Arriba and Rio Abajo, the Spanish system of irrigation consisted primarily of a main ditch, the *acequia madre*, receiving water from a river or stream at a higher elevation than the lands being irrigated, and then relying on gravity to carry the flow. Secondary ditches (*sangrias*) branched off the main ditch and directed water to individual fields. Gates were made of earth and boulders to regulate the flow, and flumes (hollowed-out logs), provided an elevated channel for water to cross gullies and ravines.[33]

Over the years riparian lines of trees and shrubs developed alongside well-maintained ditches, replicating the biotic environment found along unimpeded streams. In other instances, grazing animals invaded both Spanish and Pueblo fields and trampled the sandy banks along the water courses, filling them or causing breaks which allowed water to escape.[34] Abandoned ditches were often transformed into gullies or small arroyos as the result of runoff and soil erosion.

The problem of salinization that confronted pre-contact Pueblo farmers was intensified under the Spanish. Alkali compounds, consisting of various salts, are characteristic components of arid and semiarid soils; they are also highly soluble in water. When dissolved by irrigation, the water and alkali enter the soil together, then return to the surface by capillary action, much the same way that oil flows up a lamp wick. The sun then evaporates the water and leaves behind the salts to act as corrosive agents on the stems of plants.[35]

The effects of such phenomena as salinization, deforestation, overgrazing, and population increase reveal that selected areas in both the Rio Arriba and Rio Abajo were exceeding their carrying capacity by the end of the eighteenth century. While remaining relatively small in numbers and employing a subsistence-based economy, the Spanish in the upper Rio Grande valley had not yet achieved a totally sustainable society in relation to their environment.

While issues of power and control, particularly during the seventeenth century, were often contested at the local level between the Franciscan missionaries and the governor, the latter official influenced the course of environmental change more strongly. Directly responsible to the Viceroy, and after 1776, the Commandante General of the Provincias Internas, the governor appointed or approved lesser officials, enforced royal decrees, and ordered the formation of militia from settlements. More importantly from an environmental perspective, he made land grants which were intended primarily to manage the settlement of the Spanish and *casta* population within the colony. For in theory at least, people were simply not allowed to live where they liked or to move at their own discretion to another location. In the process, Spanish law strictly outlawed such hallmarks of the Anglo-American frontier as land speculation and absentee ownership.

This kind of social control had several ecological effects. In some cases, a governor's decision to award, or even revoke, a land grant led to further deterioration of already occupied land. One such instance occurred in 1735 when Governor Cruzat y Gongora nullified a number of grants made by the acting governor, Juan Paez Hurtado, in the area of the Chama valley. Facing overcrowded living conditions in the vicinity of the lower valley and Santa Cruz, settlers had sought fresh land for grazing. Gongora's decision to rescind the grants was based in part on his personal need to control and regulate the advancement of settlements, but it was also based on his belief that the upper Chama was a place where colonists and Utes in close contact might precipitate a war.[36] Here environmental concerns took a back seat to issues of colonial security.

Built into the very land grant process was a step tantamount to the environmental impact study of today. The alcalde mayor, a lower-level but important official appointed by the governor, had to determine whether the proposed land grant in his jurisdiction would adversely affect any Pueblo settlement or other third party, and whether there was sufficient water for irrigation and livestock and enough cultivable, grazing, and wood-producing land to support the proposed number of settlers.[37] While this system was hardly perfect, the land grant papers of New Mexico are replete with examples of the alcalde mayor addressing these issues in writing to the governor, and sometimes recommending to him that a grant be denied because criteria were not met. While the system was intended to preserve the economic survival of the colony and not the environment, it nonetheless had the long-term effect of conserving the land and water of colonial New Mexico.

The degree to which Pueblo and Spanish society created an inclusive frontier also directly affected issues of land use and change. While Pueblo people embraced Old World plants, animals, and material culture introduced by the Spanish, they did not accept, in any profound sense, Western belief systems, particularly those centering upon religion or property. In anthropological terms, the Pueblo became acculturated, but not assimilated into Spanish society. Although greatly reduced in numbers, they lived under Spanish rule in their own compact, autonomous communities, as they do in the modern era. Through intermarriage and other forms of contact, the Spanish borrowed culturally from the Pueblo. This occurred most frequently in outlying communities rather than the missions or villas. The long-term effect of this two-way process was to produce both a hard and soft impact upon the land. The soft impact emphasizes the influence of Pueblo beliefs and customs upon the Spanish settlers. The hard impact focuses principally upon Pueblo acculturation.

Richard Ford has offered a succinct view of the existence of one Pueblo group, the Tewa, prior to contact and the subsequent effects of the Columbian Exchange upon them and their relationship to the land. The Tewa lived on maize supplemented with squash and beans, and gathered plant products. Rabbits, hares, deer, and other game provided a source for meat. Firewood came from deadwood, and construction timbers were usually recycled from older structures. In essence, according to Ford, "this was an ecosystem that could rapidly recover when fields were abandoned or when the human population founded a new village elsewhere."[38]

With the arrival of don Juan de Oñate in 1598, the Tewa economy changed dramatically. Five Spanish contributions had particular impact: the introduction of spring wheat; kitchen gardens grown with irrigation water; orchards of peach, apricot, plum, and cherry trees; grazing animals; and the metal ax. The result for the Pueblo was a strange admixture of environmental degradation to their land and "a beneficial and more secure subsistence base."[39] Deforestation and over-grazing occurred on Pueblo land just as it did on that of the Spanish. But the new food sources only reinforced the adaptive capacity of the Tewa. Wheat, in partic-ular, while not displacing corn, came to serve as "a high yield caloric safety valve" for the Pueblo. It is likely that the Pueblo partially compensated for the local over-grazing and loss of cool-season grasses such as mutton grass, Indian rice grass, and June grass, by raising wheat varieties that matured about the same time.[40] These additions to the Pueblo economy also tempered their need to migrate as they might have in pre-contact times owing to political disputes, scarcity of wood, or unproductive fields. Even as their landscape was simplified, the Tewa expand-ed their control over the productivity of their land and created a measure of eco-nomic security.[41] At least in a material sense, the Pueblo came to more closely re-semble their European neighbors.

The inclusive nature of Pueblo-Spanish society entailed accommodative ele-ments as well as more disruptive ones. Whereas students of Pueblo-Spanish rela-tions can determine with some certainty the far-reaching ecological effects of the Spanish donations to the Pueblo in the form of plants, animals, and material cul-ture, it is much more difficult to assess how Pueblo belief systems, and their prac-tice of a particular subsistence agriculture, affected Spanish settlers and the envi-ronment of the upper Rio Grande valley. As Christopher Vecsey has noted, all Native American groups "established a religious association with nature that transcended but did not nullify their effective exploitation of the environment. They achieved an integration of subsistence and religious activities."[42] He in-cludes the Pueblo among those American Indians whose religious core was mold-ed by environmental relations: their fertilization ritual, for example, was about maintaining life in humans, plants, animals, and the world at large, benefiting, in turn, nature and culture alike.

To what degree Spanish settlers adopted these attitudes from the Pueblo is im-possible to determine, but more cultural borrowing likely went on between the two groups than in almost any other region of Spanish America. Frances Swadesh has suggested that what evolved in the Province of New Mexico after the seven-teenth century was a nondominant frontier community. Relations between set-tlers and their Indian neighbors, especially in outlying areas, were far more egalitarian than the colonial model for social relations intended.[43] The traditional Spanish barriers to social mobility, including caste, class, and ethnic identity, were largely absent. Archival records also reveal that a not insignificant number of Spanish settlers moved into Pueblo communities; both men and women married Pueblo Indians and raised their children in the spouse's pueblo. As a result, according to Swadesh, Spanish settlers oftentimes "found themselves more at odds with their own colonial authorities than with their Indian trading partners, compadres, and friends."[44] As settlers acquired knowledge of herbs and wild plant foods from the Pueblo, and as they witnessed or participated in an agricultural cycle both religious and secular in meaning, they may also have internalized a view of the land predicated on principles other than the usual European ones of strict utility. And while this process hardly mitigated the harsher effects of Spanish culture on Pueblo land use, it may have changed or softened the attitude of some Spanish toward exactly how the physical and natural world should be exploited.

Any analysis of land change must also address natural forces, especially variations in climate, and their impact upon biophysical processes. Prolonged drought, heavy rains, and extremes of heat and cold may effect environmental changes entirely independent of human agency. They may also interact with human activities and speed along the process of change. In the case of New Mexico, there are essentially three types of climate: arid, semiarid and subhumid/humid. Areas of arid climate have scrubby, heat- and drought-resistant desert plant cover; semiarid areas have a vegetation of short, bunchy grasses; and the subhumid/humid areas, in the hilly or mountainous regions, have woodland or forest cover.[45] Latitude, elevation, and location with regard to moisture-laden winds primarily determine these variations in climate and consequent vegetation. Albuquerque, at the northern point of the arid range, and Santa Fe, in a semiarid range closely bounded by the mountains, although only sixty miles apart, often exhibit marked differences in local climate. Such differences mean that it is very hard to associate general climate patterns with ecological change in any specific locale.

From a broader perspective, evidence from tree ring studies (dendroclimatology) indicates that the climate of the upper Rio Grande valley between 1598 and 1821 fluctuated fairly regularly between wet and cold and warm and dry periods. For Western North America, periods of widespread drought occurred between 1626 and 1635, and 1776 and 1785, while periods of above average moisture occurred from 1611 to 1625, 1641 to 1650, and 1741 to 1755. Chronologies from the upper Rio Grande valley (lat. 35–43; long. 106–06; based on rings from Douglas fir, ponderosa pine, and piñon pine) reveal dryer than normal periods between 1661 and 1675, 1726 and 1765, and 1796 and 1830, and wetter than normal periods between 1601 and 1645, 1706 and 1730, and 1781 and 1800.[46] These changes in climate clearly had the potential to effect changes in the land, both on their own and in concert with human activity. Drought or extended periods of deficient precipitation may produce, among other effects, a marked decline in plant cover, even in the absence of grazing animals. Particularly vulnerable to drought are perennial grasses like the gramas, which may die off and then be replaced by invaders such as sagebrush, snakewood, and rabbit brush.

But several variables bear strongly on the amount of plant life lost during any particular drought. They include the time of year in which the drought occurs, winter and spring having more adverse effects than summer and fall, and the texture of the soil. More heavily textured soils may retain more moisture, release it to plant use more slowly, and thus allow perennial grasses to survive. Likewise, periods that are colder and wetter than normal can produce environmental change. An excess of rain and snow, and their runoff, have the potential to cause soil erosion, arroyo formation, changes in stream flow, and flooding. Such effects are particularly acute when the climate changes quickly and dramatically from an excessively dry to wet period.[47]

Further inferences can be drawn from the relationship between fluctuations in climate, human activity—including that of Pueblo and Spanish—and ecological change. While both climate and humans, acting independently of one another, may simplify an ecosystem, together their effects may be far greater and more long lasting. The formation of arroyos, valley bottom gullies characterized by steeply sloping or vertical walls, offers a case in point. Both human land use and climatic changes are among the complex causes leading to their creation. On the human side, logging, fire, grazing, cultivation, and the existence of roads, trails, and irrigation ditches can lead to the local removal of vegetation on valley floors and alter valley bottom soils. This increases the erodibility of valley floor material. Both an increase and decrease in humidity can contribute to arroyo formation. Increase in precipitation and decrease in temperature may increase runoff from slopes, while a decrease in precipitation and increase in temperature may reduce the vegetation cover over drainage basins. Both phenomena increase the velocity of flows through valley bottoms. Individually, but especially together, human and climatic factors lead to localized erosion and arroyo initiation.[48]

Finally, this discussion of environmental factors must be viewed in the context of the inherent vulnerability or resilience of a semiarid ecosystem. Ecologists no

longer assume that the degree of diversity and complexity of an ecosystem deter-
mines stability or change. Other factors such as elasticity, ability to recover from
damage, and inertia, and ability to resist displacement are now considered more
important. Thus desert or semidesert regions may be more resilient and less vulnerable to change than other, more complex ecosystems such as woodlands or
rain forests. The flora and fauna of arid regions evolved in an environment where
the normal pattern is more or less random alteration of short favorable periods
and long stress periods. Ecologists posit that plants and animals have preadapted resilience. Applying this idea to the upper Rio Grande valley leads to several
conclusions. On the one hand, grassland and woodland, for example, might well
have survived in some instances the invasion of sheep, cattle, and ax. In fact, some
argue that limited or optimal grazing may even enhance the growth of certain
grasslands.[49] In contrast to the total absence of grazing or excessive grazing, light
grazing favors production of some grasses by restoring nutrients through feces
and urine. Even if damaged, grasslands have the potential often to undergo a regeneration of growth. The carrying capacity of any specific ecosystem might be
sustained on account of plant and animal resilience, or even through the
"benefits" of limited human economic activities upon them. Yet it is clear that human agency, both Pueblo and Spanish, in this area contributed to fundamental
changes in the land: the carrying capacity of selected areas was exceeded as the
land proved to be more fragile than resilient. It can even be argued that a process
of desertification began in selected areas where grazing was particularly heavy.
That Anglo-Americans, upon their arrival after 1821, exacerbated those changes
manifoldly does not negate this fact.

It is tempting to assess the long-run effects of the Spanish colonial presence in
the upper Rio Grande valley on environmental change as constituting a middle
ground between those changes which occurred first under the Pueblo and then
later under the Anglo-Americans. The interaction of various factors, in effect,
produced a rate and kind of change certainly exceeding that experienced under
the Pueblo but falling far short of that produced by Anglos. The acceleration in
deforestation and overgrazing in New Mexico after the Anglo-American occupation in 1846, and especially after the introduction of the railroad in 1880, demonstrates that this assessment is essentially correct. But the kind of change that occurred between 1598 and 1821 is more complex than merely a linear one. Change
was also cyclical and layered or superimposed in nature. The rate of change was
at once fast, slow, intermittent and inexorable, and only rarely constant.

Several primary factors account for these variations in change. People did not settle evenly or randomly on the land. Instead they concentrated themselves in some areas, settled sparsely in others, on some occasions moved for environmental or other reasons, and on other occasions abandoned their land and moved back to where they had come from. Nor were these same people engaged in uniform economic activities having equal impact upon the land. Natural forces acted both alone and in concert with human activities on the land in an unpredictable and capricious manner. For instance, the frequently changing course of the Rio Grande, resulting from its own natural meander, precipitated the cyclic destruction and regeneration of the woodland or bosquet habitat of the river.[50] Overgrazing and drought together, followed by above normal precipitation, enhanced the opportunity for arroyo formation. Finally, the resilience or fragility of the land itself depended upon variations in the adaptability of local flora and fauna, and the degree to which both humans and natural forces could disturb or modify them. How changes occurred in the land was integrally tied to all of these factors.

Change was most linear in the heavier population centers of Santa Fe, Santa Cruz de la Canada, and Albuquerque. Descriptions of these areas from the sixteenth century *entradas,* when juxtaposed to Anglo-American accounts of the early to middle nineteenth century, indicate the sort of dramatic shift akin to descriptions of environmental change in colonial New England and the Pacific Northwest. The archival record for Santa Fe over the colonial period notes steady loss of natural resources in the form of water, grass, and wood to the point that a governor in the late eighteenth century recommended that the capital be moved to the confluence of the Santa Fe and Rio Grande Rivers.[51] Similarly, the Albuquerque area underwent a gradual yet inexorable process of desertification that resulted from overgrazing by sheep.

In contrast, cyclical change took place most frequently when settlers, usually on account of land pressure in and around the three villas, occupied land at the edge of the New Mexico frontier. Settlements in the north around Taos, in the northwest on the Chama River valley, in the west out to the Puerco River, in the south beyond Isleta, and in the east to the Pecos River, throughout most of the colonial period, were often in flux, sometimes abandoned outright on account of Indian hostilities, and sometimes even resettled after a several-year hiatus. As a result, changes in the land that would have occurred from permanent settlement might have commenced only to be interrupted by abandonment, and then recommenced. Deforestation and overgrazing in these areas rarely took place in a sustained way.

In this context, the year 1790 is notable in colonial New Mexico history in that it marked the beginning of relatively peaceful times between the Spanish, the Pueblo, and the nomadic tribes, especially the Apache and Comanche. The outward expansion that followed came to resemble more closely the Anglo-American frontier of an ever-receding line. It also meant that the cyclical nature of ecological change came largely to an end, as places such as the Puerco River valley began to fill with settlers so steadily that by the middle to late nineteenth century it had become as severely overgrazed and eroded as any place in New Mexico.

Environmental change might also be characterized as layered or superimposed, as in a kind of palimpsest. This occurred most often where Pueblo communities adopted the plants, animals, and material culture of the Spanish, moving from a system of extensive to intensive agriculture. In effect, Pueblo land use changed by degree rather than by radical transformation. The growing of wheat required the expansion of Spanish-style irrigation, with concomitant environmental effects; and as the Pueblo came to rely on new domesticated plants and animals, their hunting and gathering of wild plants and animals declined. So the pre-contact Pueblo practices of land use and their effects on land change intensified as Spanish agricultural practices became superimposed on Pueblo practices through the process of acculturation. The dramatic decline in Pueblo population prevented their communities from exceeding carrying capacity as, in selected cases, they were in danger of on the eve of contact.

Environmental change in colonial New Mexico was sui generis, predicated on circumstances and conditions reflective of the particular climate, geography, and cultures of the region. The dialectical relationship between different kinds of factors produced varied kinds and rates of change unlike that experienced on any other North American frontier. This study shows the limitation of relying on modes of production or economic systems as a single or even primary explanation for change. They are clearly more useful as a means of analysis from a macro perspective such as representing, in broad strokes, the environmental change which occurred under preindustrial Pueblo and Spanish society through Anglo occupation. A look at the micro level reveals the difficulty in assigning any such clear causality for change. Perhaps the challenge before us is to mediate between the two approaches.

GREEN POLITICS

Although there is no "Green" party in the United States to advance an environmentalist platform, as is the pattern of partisan organizations in several western European countries, the environment has become a powerful factor in our elections. Its influence has become increasingly evident since World War II, initially exploding during controversies over the preservation of scenic landmarks such as Echo Park and in the subsequent spawning of special interest groups dedicated to the creation of wilderness legislation, including the Wilderness and Endangered Species Acts, laws that sparked additional legislative, judicial, and cultural debate. All this occurred without the presence of a single driving political force.

An explanation for this, Samuel P. Hays asserts in "From Conservation to Environment," lies in the "massive changes" he believes transformed the American polity in the years following the war, one element of which combined an assumption that "natural environments" were an "integral part" of the rising standard of living with a perception that individual physical fitness was a social good; both demanded resolute protection. One of the ramifications of this is developed in Richard Vietor's examination of "The Evolution of Public Environmental Policy," a case study of the escalating demand to achieve "absolute air quality goals" through amendments to the Clean Air Act. This political insistence on achieving a perhaps unattainable scientific benchmark is symptomatic of a critical cultural shift that Hays also observed: "unrestrained economic growth and unfettered resource exploitation" were no longer considered the "essence of societal well-being." Just who would help articulate the nation's "wellness" remains a contentious issue, as Robert Gottlieb makes clear in "Reconstructing Environmentalism." Gottlieb not only challenges Hays' depiction of the political sources of and alterations in environmentalist beliefs, but in doing so diversifies the environmental movement's agenda and membership to include issues relevant to women, minorities, and workers. Gottlieb's interpretive thrust has a political point, reflecting a desire to develop a movement that is more broadly democratic and inclusive, an environmentalism as concerned with human and natural landscapes as it is with equity and social justice. Such a movement, he believes, could transform the human condition.

From Conservation to Environment

Environmental Politics in the United States Since World War II

SAMUEL P. HAYS

The historical significance of the rise of environmental affairs in the United States in recent decades lies in the changes which have taken place in American society since World War II. Important antecedents of those changes, to be sure, can be identified in earlier years as "background" conditions on the order of historical forerunners. But the intensity and force, and most of the substantive direction of the new environmental social and political phenomenon can be understood only through the massive changes which occurred after the end of the war—and not just in the United States but throughout advanced industrial societies.

Such is the argument of this essay. I will identify a variety of ways in which one can distinguish the old from the new, the pre- from the postwar, and sequential changes within the decades of the Environmental Era themselves. My argument will emphasize change rather than continuity. In historical analyses we are constantly forced to cope with the problem of sorting out the strands of continuous evolution from the discontinuities which mark new directions. When we are close to a broad social and political change, displaying elements of what we call social movements, we often depart from that task by a temptation to ferret out "roots" in order to give historical meaning and significance to them. So it is with the "environmental movement." Here I prefer a larger view, shaped by the overarching historical problem of identifying patterns of continuity and change. Where do environmental affairs fit in those larger patterns of evolution in twentieth-century American society and politics? In my view that "fit" lies in an emphasis on the massive changes in America after World War II and on the war itself as a historical dividing point.

The Conservation and Environmental Impulses

Prior to World War II, when the term "environment" was hardly used, the dominant theme in conservation emphasized physical resources, their more efficient use and development. The range of emphasis evolved from water and forests in the late nineteenth and early twentieth centuries, to grass and soils and game in the 1930s. In all these fields of endeavor there was a common concern for the loss of physical productivity represented by waste. The threat to the future which that "misuse" implied could be corrected through "sound" or efficient management. Hence in each field there arose a management system which emphasized a balancing of immediate in favor of more long-run production, the co-ordination of factors of production under central management schemes for the greatest efficiency. All this is a chapter in the history of production rather than of consumption, and of the way in which managers organized production rather than the way in which consumers evolved ideas and action amid the general public.

Enough has already been written about the evolution of multiple-purpose river development and sustained-yield forestry to establish their role in this context of efficient management for commodity production.[1] But perhaps a few more words could he added for those resources which came to public attention after World War I. Amid the concern about soil erosion, from both rain and wind, the major stress lay in warnings about the loss of agricultural productivity. What had taken years to build up over geologic time now was threatened with destruction by short-term practices. The soil conservation program inaugurated in 1933 gave rise to a full-scale attack on erosion problems which was carried out amid almost inspired religious fervor.[2] In the Taylor Grazing Act of 1934 the nation's grazing lands in the West were singled out as a special case of deteriorating productivity; it set in motion a long-term drive to reduce stocking levels and thereby permit recovery of the range.[3] Also during the 1930s, scientific game management came into its own with the Pittman-Robertson Act of 1936 which provided funds.[4] This involved concepts much akin to those in forestry, in which production and consumption of game would be balanced in such a fashion so as not to outrun food resources and hence sustain a continuous yield.

Perhaps the most significant vantage point from which to observe the common processes at work in these varied resource affairs was the degree to which resource managers thought of themselves as engaged in a common venture. It was not difficult to bring into the overall concept of "natural resources" the management of forests and waters, of soils and grazing lands, and of game. State departments of "natural resources" emerged, such as in Michigan, Wisconsin, and Minnesota, and some university departments of forestry became departments of natural resources—all this as the new emphases on soils and game were added to the older ones on forests and waters.[5] By the time of World War II a complex of professionals had come into being, with a strong focus on management as their common task, on the organization of applied knowledge about physical resources so as to sustain output for given investments of input under centralized management direction. This entailed a common conception of "conservation" and a common focus on "renewable resources," often within the rubric of advocating "wise use" under the direction of professional experts.[6]

During these years another and altogether different strand of activity also drew upon the term "conservation" to clash with the thrust of efficient commodity management. Today we frequently label it with the term "preservation" as we seek to distinguish between the themes of efficient development symbolized by Gifford Pinchot and natural environment management symbolized by John Muir. Those concerned with national parks and the later wilderness activities often used the term "conservation" to describe what they were about. In the Sierra Club the "conservation committees" took up the organization's political action in contrast with its outings. And those who formed the National Parks Association and later the Wilderness Society could readily think of themselves as conservationists, struggling to define the term quite differently than did those in the realm of efficient management. Even after the advent of the term "environment" these groups continued to identify themselves as "conservationists" such as in the League of Conservation Voters, especially when they wished to draw together the themes of natural environment lands and environmental protection. The National Parks Association sought to have the best of both the old and the new when it renamed its publication, *The National Parks and Conservation Magazine: The Environmental Journal.*[7]

Prior to World War II the natural environment movement made some significant gains. One thinks especially of the way in which Pinchot was blocked from absorbing the national parks under his direction in the first decade of the century and then, over his objections, advocates of natural environment values succeeded in establishing the National Park Service in 1916. Then there was the ensuing struggle of several decades in which an aggressive Park Service was able to engage the Forest Service in a contest for control of land and on many occasions won. One of the best described of these events concerns the establishment of the Olympic National Park in 1937, a former national monument under Forest Service jurisdiction until Franklin D. Roosevelt transferred all the monuments to the Park Service in June of 1933; in 1937 it was expanded by the addition of considerable acreage from the surrounding national forest.[8] Despite all this, however, the theme of management efficiency in physical resource development dominated the scene prior to World War II and natural environment programs continued to play a subordinate role.

After the war a massive turnabout of historical forces took place. The complex of specialized fields of efficient management of physical resources increasingly came under attack amid a new "environmental" thrust. It contained varied components. One was the further elaboration of the outdoor recreation and natural environment movements of prewar as reflected in the Wilderness Act of 1964, the Wild and Scenic Rivers Act of 1968, and the National Trails Act of the same year, and further legislation and administrative action on through the 1970s. But there were other strands even less rooted in the past. The most extensive was the concern for environmental pollution, or "environmental protection" as it came to be called in technical and managerial circles. While smoldering in varied and diverse ways in this or that setting from many years before, this concern burst forth to national prominence in the mid-1960s and especially in air and water pollution. And there was the decentralist thrust, the search for technologies of smaller and more human scale which complement rather than dwarf the more immediate human setting. One can find decentralist ideologies and even affirmations of smaller-scale technologies in earlier years, such as that inspired by Ralph Borsodi not long before World War II.[9] But the intensity and direction of the drive of the 1970s was of a vastly different order. The search for a "sense of place," for a context that is more manageable intellectually and emotionally amid the escalating pace of size and scale had not made its mark in earlier years as it did in the 1970s to shape broad patterns of human thought and action.

One of the most striking differences between these postwar environmental activities, in contrast with the earlier conservation affairs, was their social roots. Earlier one can find little in the way of broad popular support for the substantive objectives of conservation, little "movement" organization, and scanty evidence of broadly shared conservation values. The drive came from the top down, from

technical and managerial leaders. In the 1910s one can detect a more extensive social base for soil conservation, and especially for new game management programs. But, in sharp contrast, the Environmental Era displayed demands from
the grass roots, demands that are well charted by the innumerable citizen organizations and studies of public attitudes. One of the major themes of these later
years, in fact, was the tension that evolved between the environmental public and
the environmental managers, as impulses arising from the public clashed with
impulses arising from management. This was not a new stage of public activity
per se, but of new values as well. The widespread expression of social values in environmental action marks off the environmental era from the conservation years.

It is useful to think about this as the interaction between two sets of historical
forces, one older that was associated with large-scale management and technology, and the other newer that reflected new types of public values and demands.[10]
The term "environment" in contrast with the earlier term "conservation" reflects
more precisely the innovations in values. The technologies with which those values clashed in the postwar years, however, were closely aligned in spirit and historical roots with earlier conservation tendencies, with new stages in the evolution from the earlier spirit of scientific management of which conservation had
been an integral part. A significant element of the historical analysis, therefore, is
to identify the points of tension in the Environmental Era between the new stages
of conservation as efficient management, as it became more highly elaborated,
and the newly evolving environmental concerns which displayed an altogether
different thrust. Conflicts between older "conservation" and newer "environment" help to identify the nature of the change.

One set of episodes in this tension concerned the rejection of multiple-purpose river structures in favor of free flowing rivers; here was a direct case of irreconcilable objectives, one stemming from the conservation era, and another inherent in the new environmental era. There were cases galore. But perhaps the
most dramatic one, which pinpoints the watershed between the old and the new,
involved Hell's Canyon on the Snake River in Idaho.[11] For many years that dispute
had taken the old and honorable shape of public versus private power. Should
there be one high dam, constructed with federal funds by the Bureau of Reclamation, or three lower dams to be built by the Idaho Power Company? These were
the issues of the 1930s, the Truman years and the Eisenhower administrations.
But when the Supreme Court reviewed a ruling of the Federal Power Commission
on the issue in 1968, it pointed out in a decision written by Justice Douglas that
another option had not been considered—no dam at all. Perhaps the river was

more valuable as an undeveloped, free flowing stream. The decision was unexpected both to the immediate parties to the dispute, and also to "conservationists" in Idaho and the Pacific Northwest. In fact, those conservationists had to be persuaded to become environmentalists. But turn about they did. The decision seemed to focus a perspective which had long lain dormant, implicit in the circumstances but not yet articulated, and reflected a rather profound transformation in values which had already taken place.

There were other realms of difference between the old and the new. There was, for example, the changing public conception of the role and meaning of forests.[12] The U.S. Forest Service, and the entire community of professional foresters, continued to elaborate the details of scientific management of wood production; it took the form of increasing input for higher yields, and came to emphasize especially even-aged management. But an increasing number of Americans thought of forests as environments for home, work, and play, as an environmental rather than as a commodity resource, and hence to be protected from incompatible crop-oriented strategies. Many of them bought woodlands for their environmental rather than their wood production potential. But the forestry profession did not seem to be able to accept the new values. The Forest Service was never able to "get on top" of the wilderness movement to incorporate it in "leading edge" fashion into its own strategies. As the movement evolved from stage to stage the Service seemed to be trapped by its own internal value commitments and hence relegated to playing a rear-guard role to protect wood production.[13] Many a study conducted by the Forest Service experiment stations and other forest professionals made clear that the great majority of small woodland owners thought of their holdings as environments for wildlife and their own recreational and residential activities; yet the service forester program conducted by the Forest Service continued to emphasize wood production rather than environmental amenities as the goal of woodland management. The diverging trends became sharper with the steadily accumulating environmental interest in amenity goals in harvesting strategies and the expanding ecological emphases on more varied plant and animal life within the forest.[14]

There were also divergent tendencies arising from the soil conservation arena. In the early 1950s, the opposition of farmers to the high-dam strategies of the U.S. Army Corps of Engineers led to a new program under the jurisdiction of the Soil Conservation Service, known as PL 566, which emphasized the construction of smaller headwater dams to "hold the water where it falls." This put the SCS in the business of rural land and water development, and it quickly took up the challenge of planning a host of such "multiple-use" projects which combined small flood control reservoirs with flat-water recreation and channelization with wetland drainage.[15] By the time this program came into operation, however, in the 1960s, a considerable interest had arisen in the natural habitats of headwater streams, for example for trout fishing, and wetlands for both fish and wildlife. A head-on collision on this score turned an agency which had long been thought of as riding the lead wave of conservation affairs into one which appeared to environmentalists to be no better than the Corps—development minded and at serious odds with newer natural environment objectives.[16]

There was one notable exception to these almost irreconcilable tensions between the old and the new in which a far smoother transition occurred—the realm of wildlife. In this case the old emphasis on game was faced with a new one on nature observation or what came to be called a "nongame" or "appreciative" use of wildlife.[17] Between these two impulses there were many potential arenas for deep controversy. But there was also common ground in their joint interest in wildlife habitat. The same forest which served as a place for hunting also served as a place for nature observation. In fact, as these different users began to be identified and counted it was found that even on lands acquired exclusively for game management the great majority of users were nongame observers. As a result of this shared interest in wildlife habitat it was relatively easy for many "game managers" to shift in their self-conceptions to become "wildlife managers." Many a state agency changed its name from "game" to "wildlife" and an earlier document, "American Game Policy, 1930," which guided the profession for many years, became "The North American Wildlife Policy, 1973."[18]

If we examine the values and ideas, then, the activities and programs, the directions of impulses in the political arena, we can observe a marked transition from the pre–World War II conservation themes of efficient management of physical resources, to the post–World War environmental themes of environmental amenities, environmental protection, and human scale technology. Something new was happening in American society, arising out of the social changes and transformation in human values in the postwar years. These were associated more with the advanced consumer society of those years than with the industrial manufacturing society of the late nineteenth and the first half of the twentieth centuries. Let me now root these environmental values in these social and value changes.

The Roots of New Environmental Values

The most immediate image of the "environmental movement" consists of its "protests," its objections to the extent and manner of development and the shape of technology. From the media evidence one has a sense of environmentalists blocking "needed" energy projects, dams, highways, and industrial plants, and of complaints of the environmental harm generated by pollution. Environmental action seems to be negative, a protest affair. This impression is also heavily shaped by the "environmental impact" mode of analysis which identifies the "adverse effects" of development and presumably seeks to avoid or mitigate them. The question is one of how development can proceed with the "least" adverse effect to the "environment." From this context of thinking about environmental affairs one is tempted to formulate an environmental history based upon the way in which technology and development have created "problems" for society to be followed by ways in which action has been taken to cope with those problems.

This is superficial analysis. For environmental impulses are rooted in deep-seated changes in recent America which should be understood primarily in terms of new positive directions. We are at a stage in history when new values and new ways of looking at ourselves have emerged to give rise to new preferences. These are characteristic of advanced industrial societies throughout the world, not just in the United States. They reflect two major and widespread social changes. One is associated with the search for standards of living beyond necessities and conveniences to include amenities made possible by considerable increases in personal and social "real income." The other arises from advancing levels of education which have generated values associated with personal creativity and self-development, involvement with natural environments, physical and mental fitness and wellness, and political autonomy and efficacy. Environmental values and objectives are an integral part of these changes.

Extensive study of attitudes and values by public opinion analysts and sociologists chart these larger changes in social values in considerable detail.[19] Some have brought them together in comprehensive accounts. They can be best observed in the market analyses which have been sponsored by the American business community since the 1920s which gave rise to the initial interest in attitude surveys. Such analyses have identified value changes in almost every sub-group in the American population, from different ages to ethnic and religious variations, to regional differences and rural-urban distinctions. Two of the most comprehensive and long-term studies are now in progress, financed by American business corporations, one the Values and Lifestyles Study (VALS) conducted by

Arnold Johnson at Stanford Research Institute and the other, emphasizing content analysis of newspapers, being undertaken by John Naisbett, associated with the firm of Yankelovitch, Skelly and White in Washington, D.C.[20]

From these more general surveys, from studies specifically of environmental values, from analyses of recreational and leisure preferences undertaken by leisure research specialists, from surveys of the values expressed by those who purchase natural environment lands, and from the content of environmental action in innumerable grassroots citizen cases one can identify the "environmental impulse" not as reactive but formative.[21] It reflects a desire for a better "quality of life" which is another phase of the continual search by the American people throughout their history for a higher standard of living. Environmental values are widespread in American society, extending throughout income and occupational levels, areas of the nation and racial groups, somewhat stronger in the middle sectors and a bit weaker in the very high and very low groupings.[22] There are identifiable "leading sectors" of change with which they are associated as well as "lagging sectors." They tend to be stronger with younger people and increasing levels of education and move into the larger society from those centers of innovation. They are also more associated with particular geographical regions such as New England, the Upper Lakes states, the Upper Rocky Mountain region, and the Far West while the South, the Plains states, and the lower Rockies constitute "lagging" regions.[23] Hence one can argue that environmental values have expanded steadily in American society, associated with demographic sectors which are growing rather than with those which are more stable or declining.

Within this general context one can identify several distinctive sets of environmental tendencies. One was the way in which an increasing portion of the American people came to value natural environments as an integral part of their rising standard of living. They sought out many types of such places to experience, to explore, enjoy, and protect: high mountains and forests, wetlands, ocean shores, swamplands, wild and scenic rivers, deserts, pine barrens, remnants of the original prairies, places of relatively clean air and water, more limited "natural areas."[24] Interest in such places was not a throwback to the primitive, but an integral part of the modern standard of living as people sought to add new "amenity" and "aesthetic" goals and desires to their earlier preoccupation with necessities and conveniences. These new consumer wants were closely associated with many others of a similar kind such as in the creative arts, recreation and leisure in general, crafts, indoor and household decoration, hi-fi sets, the care of yards and gardens as living space and amenity components of necessities and conveniences. Americans experienced natural environments both emotionally and intellectually, sought them out for direct personal experience in recreation, studied them as objects of scientific and intellectual interest, and desired to have them within their community, their region, and their nation as symbols of a society with a high degree of civic consciousness and pride.[25]

A new view of health constituted an equally significant innovation in environmental values, health less as freedom from illness and more as physical and mental fitness, of feeling well, of optimal capability for exercising one's physical and mental powers.[26] The control of infectious diseases by antibiotics brought to the fore new types of health problems associated with slow, cumulative changes in physical condition, symbolized most strikingly by cancer, but by the 1980s ranging into many other conditions such as genetic and reproductive problems, degenerative changes such as heart disease and deteriorating immune systems. All this put more emphasis on the nonbacterial environmental causes of illness but, more importantly, brought into health matters an emphasis on the positive conditions of wellness and fitness. There was an increasing tendency to adopt personal habits that promoted rather than threatened health, to engage in physical exercise, to quit smoking, to eat more nutritiously, and to reduce environmental threats in the air and water that might also weaken one's wellness. Some results of this concern were the rapid increase in the business of health food stores, reaching 1.5 billion in 1979,[27] the success of the Rodale enterprises and their varied publications such as *Prevention and Organic Gardening,* and the increasing emphasis on preventive medicine.[28]

These new aesthetic and health values constituted much of the roots of environmental concern. They came into play in personal life and led to new types of consumption in the private market, but they also led to demands for public action both to enhance opportunities, such as to make natural environments more available and to ward off threats to values. The threats constituted some of the most celebrated environmental battles: power and petrochemical plant siting, hardrock mining and strip mining, chemicals in the workplace and in underground drinking water supplies, energy transmission lines, and pipelines.[29] Many a local community found itself faced with a threat imposed from the outside and sought to protect itself through "environmental action." But the incidence and intensity of reaction against these threats arose at a particular time in history because of the underlying changes in values and aspirations. People had new preferences and new personal and family values which they did not have before. Prior to World War II, the countryside, that area between the nation's cities and its wildlands, had been an area of rapid decline, a land much of which "nobody wanted," but in the years after the war it became increasingly occupied and hence defended.[30] Here was a major battleground for the contending environmental and developmental antagonists. Because of these new values developmental activities which earlier might have been accepted were now considered to be on balance more harmful than beneficial.

Still another concern began to play a more significant role in environmental affairs in the 1970s—an assertion of the desirability of more personal family and community autonomy in the face of the larger institutional world of corporate industry and government, an affirmation of smaller in the face of larger contexts of organization and power. This constituted a "self-help" movement. It was reflected in numerous publications about the possibilities of self-reliance in production of food and clothing, design and construction of homes, recreation and leisure, recycling of wastes and materials, and use of energy through such decentralized forms as wind and solar. These tendencies were far more widespread than institutional and thought leaders of the nation recognized, since their world of perception and management was far removed from community and grassroots ideas and action. The debate between "soft" and "hard" energy paths seemed to focus much of the controversy over the possibilities of decentralization.[31] But it should also be stressed that the American economy, while tending toward more centralized control and management, also generated products which made individual choices toward decentralized living more possible and hence stimulated this phase of environmental affairs. While radical change had produced large-scale systems of management it had also reinvigorated the more traditional Yankee tinkerer who now found a significant niche in the new environmental scheme of things.

Several significant historical tendencies are integral parts of these changes. One involves consumption and the role of environmental values as part of evolving consumer values.[32] At one time, perhaps as late as 1900, the primary focus in consumption was on necessities. By the 1920s a new stage had emerged which emphasized conveniences in which the emerging consumer durables, such as the automobile and household appliances, were the most visible elements. This change meant that a larger portion of personal income, and hence of social income, and production facilities were now being devoted to a new type of demand and supply. By the late 1940s a new stage in the history of consumption had come into view. Many began to find that both their necessities and conveniences had been met and an increasing share of their income could be devoted to amenities.

The shorter work week and increasing availability of vacations provided opportunities for more leisure and recreation. Hence personal and family time and income could be spent on amenities. Economists were inclined to describe this as "discretionary income." The implications of this observation about the larger context of environmental values is that it is a part of the history of consumption rather than of production. That in itself involves a departure from traditional emphases in historical analysis.

Another way of looking at these historical changes is to observe the shift in focus in daily living from a preoccupation with work in earlier years to a greater role for home, family, and leisure in the postwar period. Public opinion surveys indicate a persistent shift in which of these activities respondents felt were more important, a steady decline in a dominant emphasis on work and a steady rise in those activities associated with home, family, and leisure. One of the most significant aspects of this shift was a divorce in the physical location of work and home. For most people in the rapidly developing manufacturing cities of the nineteenth century the location of home was dictated by the location of work. But the widespread use of the automobile, beginning in the 1920s, enabled an increasing number of people, factory workers as well as white collar workers, to live in one place and to work in another. The environmental context of home, therefore, came to be an increasingly separate and distinctive focus for their choices. Much of the environmental movement arose from this physical separation of the environments of home and work.

One can identify in all this a historical shift in the wider realm of politics as well. Prior to World War II the most persistent larger context of national political debate involved the balance among sectors of production. From the late nineteenth century, on the evolution of organized extra-party political activity, in the form of "interest groups," was overwhelmingly devoted to occupational affairs, and the persistent policy issues involved the balance of the shares of production which were to be received by business, agriculture, and labor, and sub-sectors within them. Against this array of political forces consumer objectives were woefully weak. But the evolution of new types of consumption in recreation, leisure, and amenities generated a quite different setting. By providing new focal points of organized activity in common leisure and recreational interest groups, and by emphasizing community organization to protect community environmental values against threats from external developmental pressures, consumer impulses went through a degree of mobilization and activity which they had not previously enjoyed. In many an instance they were able to confront developmentalists with considerable success. Hence environmental action reflects the emergence in American politics of a new effectiveness for consumer action not known in the years before the war.

One of the distinctive aspects of the history of consumption is the degree to which what once were luxuries, enjoyed by only a few, over the years became enjoyed by many—articles of mass consumption. In the censuses of the last half of the nineteenth century several occupations were identified as the "luxury trades," producing items such as watches and books which later became widely consumed. Many such items went through a similar process, arising initially as enjoyed only by a relative few and then later becoming far more widely diffused. These included such consumer items as the wringer washing machine and the gas stove, the carpet sweeper, indoor plumbing, and the automobile. And so it was with environmental amenities. What only a few could enjoy in the nineteenth century came to be mass activities in the mid-twentieth, as many purchased homes with a higher level of amenities around them and could participate in outdoor recreation beyond the city. Amid the tendency for the more affluent to seek out and acquire as private property the more valued natural amenity sites, the public lands came to be places where the opportunity for such activities remained far more accessible to a wide segment of the social order.

A major element of the older, pre–World War II "conservation movement," efficiency in the use of resources, also became revived in the 1970s around the concern for energy supply. It led to a restatement of rather traditional options, as to whether or not natural resources were limited, and hence one had to emphasize efficiency and frugality, or whether or not they were unlimited and could be developed with unabated vigor. Environmentalists stressed the former. It was especially clear that the "natural environments" of air, water, and land were finite, and that increasing demand for these amid a fixed supply led to considerable inflation in price for those that were bought and sold in the private market. Pressures of growing demand on limited supply of material resources appeared to most people initially in the form of inflation; this trend of affairs in energy was the major cause of inflation in the entire economy. The great energy debates of the 1970s gave special focus to a wide range of issues pertaining to the "limits to growth."[33] Environmentalists stressed the possibilities of "conservation supplies" through greater energy productivity and while energy producing companies objected to this as a major policy alternative, industrial consumers of energy joined with household consumers in taking up efficiency as the major alternative. In the short run the "least cost" option in energy supply in the private market enabled the nation greatly to reduce its energy use and carried out the environmental option.[34]

In accounting for the historical timing of the environmental movement one should emphasize changes in the "threats" as well as in the values. Much of the shape and timing of environmental debate arose from changes in the magnitude and form of these threats from modern technology. That technology was applied in increasing scale and scope, from enormous draglines in strip mining, to 1000-megawatt electric generating plants and "energy parks," to superports and large-scale petrochemical plants to 765-kilovolt energy transmission lines. And there was the vast increase in the use and release into the environment of chemicals, relatively contained and generating a chemical "sea around us" which many people considered to be a long-run hazard that was out of control. The view of these technological changes as threats seemed to come primarily from their size and scale, the enormity of their range of impact, in contrast with the more human scale of daily affairs. New technologies appeared to constitute radical influences, disruptive of settled community and personal life, of a scope that was often beyond comprehension and promoted and carried through by influences "out there" from the wider corporate and governmental world. All this brought to environmental issues the problem of "control," of how one could shape more limited personal and community circumstance in the face of large-scale and radical change impinging from afar upon daily life.[35]

Stages in the Evolution of Environmental Action

Emerging environmental values did not make themselves felt all in the same way or at the same time. Within the context of our concern here for patterns of historical change, therefore, it might be well to secure some sense of stages of development within the post–World War II years. The most prevalent notion is to identify Earth Day in 1970 as the dividing line. There are other candidate events, such as the publication of Rachel Carson's *Silent Spring* in 1962, and the Santa Barbara oil blowout in 1969.[36] But in any event definition of change in these matters seems to be inadequate. Earth Day was as much a result as a cause. It came after a decade or more of underlying evolution in attitudes and action without which it would not have been possible. Many environmental organizations established earlier experienced considerable growth in membership during the 1960s, reflecting an expanding concern.[37] The regulatory mechanisms and issues in such fields as air and water pollution were shaped then; for example the Clean Air Act of 1967 established the character of the air quality program more than did that of 1970. General public awareness and interest were expressed extensively in a variety of public forums and in the mass media. Evolving public values could be observed in the growth of the outdoor recreation movement which reached back into the 1950s and the search for amenities in quieter and more natural settings, in the increasing number of people who engaged in hiking and camping or purchased recreational lands and homes on the seashore, by lakes, and in woodlands.

This is not to say that the entire scope of environmental concerns emerged fully in the 1960s. It did not. But one can observe a gradual evolution rather than a sudden outburst at the turn of the decade, a cumulative social and political change that came to be expressed vigorously even long before Earth Day.[38]

We might identify three distinct stages of evolution. Each stage brought a new set of issues to the fore without eliminating the previous ones, in a set of historical layers. Old issues persisted to be joined by new ones, creating over the years an increasingly complex and varied world of environmental controversy and debate. The initial complex of issues which arrived on the scene of national politics emphasized natural environment values in such matters as outdoor recreation, wildlands, and open space. These shaped debate between 1957 and 1965 and constituted the initial thrust of environmental action. After World War II the American people, with increased income and leisure time, sought out the nation's forests and parks, its wildlife refuges, its state and federal public lands, for recreation and enjoyment. Recognition of this growing interest and the demands upon public policy which it generated led Congress in 1958 to establish the National Outdoor Recreation Review Commission which completed its report in 1962.[39] Its recommendations heavily influenced public policy during the Johnson administration, leading directly to the Land and Water Conservation Fund of 1964 which established, for the first time, a continuous source of revenue for acquisition of state and federal outdoor recreation lands. It accelerated the drive for the National Wilderness Act of 1964 and the Wild and Scenic Rivers and National Trails Acts of 1968.

These laws reflected in only a limited way a much more widespread interest in natural environment affairs which affected local, state and federal policy. During the 1950s many in urban areas had developed a concern for urban overdevelopment and the need for open space in their communities. This usually did not receive national recognition because it took place on a more local level. But demands for national assistance for acquisition of urban open space led to legislation in 1960 which provided federal funds. The concern for open space extended to regional as well as community projects, involving a host of natural environment areas ranging from pine barrens to wetlands to swamps to creeks and streams to remnants of the original prairies. Throughout the 1960s there were attempts to add to the national park system which gave rise to new parks such as Canyonlands in Utah, new national lakeshores and seashores, and new national recreation areas.

These matters set the dominant tone of the initial phase of environmental concern until the mid-1960s. They did not decline in importance, but continued to shape administrative and legislative action as specific proposals for wilderness, scenic rivers or other natural areas emerged to be hotly debated. Such general measures as the Eastern Wilderness Act of 1974, the Federal Land Planning and Management Act of 1976, and the Alaska National Interest Lands Act of 1980 testified to the perennial public concern for natural environment areas. So also did the persistent evolution of indigenous western wilderness groups in almost every state and the formation of a western umbrella organization, the Wilderness Alliance, headquartered in Denver, in 1978.[40] One might argue that these were the most enduring and fundamental environmental issues throughout the two decades. While other citizen concerns might ebb and flow, interest in natural environment areas persisted steadily. That interest was the dominant reason for membership growth in the largest environmental organizations. The Nature Conservancy, a private group which emphasized acquisition of natural environment lands, grew in activity in the latter years of the 1970s and reached 100,000 members in 1981; this only further emphasized the persistent and enduring public concern for natural environment areas as an integral and important element of American life.[41]

Amid this initial stage of environmental politics there evolved a new and different concern for the adverse impact of industrial development with a special focus on air and water pollution. This had long evolved slowly on a local and piecemeal basis, but emerged with national force only in the mid-1960s. In the early part of the decade air and water pollution began to take on significance as national issues and by 1965 they had become highly visible. The first national public opinion poll on such questions was taken in that year, and the President's annual message in 1965 reflected, for the first time, a full-fledged concern for pollution problems. Throughout the rest of the decade and on into the 1970s these issues evolved continually. Federal legislation to stimulate remedial action was shaped over the course of these seven years from 1965 to 1972, a distinct period which constituted the second phase in the evolution of environmental politics, taking its place alongside the previously developing concern for natural environment areas.

The legislative results were manifold. Air pollution was the subject of new laws in 1967 and 1970; water pollution in 1965, 1970, and 1972.[42] The evolving concern about pesticides led to revision of the existing law in the Pesticides Act of 1972.[43] The growing public interest in natural environment values in the coastal zone, and threats to them by dredging and filling, industrial siting, and offshore oil development first made its mark on Congress in 1965 and over the next few years shaped the course of legislation which finally emerged in the Coastal Zone Management Act of 1972. Earth Day in the spring of 1970 lay in the middle of this phase of historical development, both a result of the previous half-decade of activity and concern and a new influence to accelerate action. The outline of these various phases of environmental activity, however, can be observed only by evidence and actions far beyond the events of Earth Day. Such more broad-based evidence identifies the years 1965 to 1972 as a well-defined phase of historical development in terms of issues, emphasizing the reaction against the adverse effects of industrial growth as distinct from the earlier emergence of natural environment issues.

Yet this new phase was shaped heavily by the previous period in that it gave primary emphasis to the harmful impact of pollution on ecological systems rather than on human health—a concern which was to come later. In the years between 1965 and 1972 the interest in "ecology" came to the fore to indicate the intense public interest in potential harm to the natural environment and in protection against disruptive threats. The impacts of highway construction, electric power plants, and industrial siting on wildlife, on aquatic ecosystems, and on natural environments in general played a major role in the evolution of this concern. One of the key elements of evolving public policy was the enhanced role of the U.S. Fish and Wildlife Service in modifying decisions by developmental agencies to reduce their harmful actions.[44] The effects of pesticides were thought of then in terms of their impact on wildlife and ecological food chains, rather than on human health. The major concern for the adverse effect of nuclear energy generation in the late 1960s involved its potential disruption of aquatic ecosystems from thermal pollution rather than the effect of radiation on people. The rapidly growing ecological concern was an extension of the natural environment interests of the years 1957 to 1965 into the problem of the adverse impacts of industrial growth.[45]

Beginning in the early 1970s still a third phase of environmental politics arose which brought three other sets of issues into public debate: toxic chemicals, energy, and the possibilities of social, economic, and political decentralization. These did not obliterate earlier issues, but as some natural environment matters and concern over the adverse effects of industrialization shifted from legislative to administrative politics, and thus became less visible to the general public, these new issues emerged often to dominate the scene. They were influenced heavily by the seemingly endless series of toxic chemical episodes, from PBBs in Michigan to kepone in Virginia to PCBs on the Hudson River, to the discovery of abandoned chemical dumps at Love Canal and near Louisville, Kentucky.[46] These events, however, were only the more sensational aspects of a more deep-seated new twist in public concern for human health.[47] Interest in personal health and especially in preventive health action took a major leap forward in the 1970s. It seemed to focus especially on such matters as cancer and environmental pollutants responsible for a variety of health problems, on food and diet on the one hand and exercise on the other. From these interests arose a central concern for toxic threats in the workplace, in the air and water, and in food and personal habits that came to shape some of the overriding issues of the 1970s on the environmental front. It shifted the earlier emphasis on the ecological effects of toxic pollutants to one more on human health effects. Thus, while proceedings against DDT in the late 1960s had emphasized adverse ecological impacts, similar proceedings in the 1970s focused primarily on human health.

The energy crisis of the winter of 1973–74 brought a new issue to the fore. Not that energy matters had gone unnoticed earlier, but their salience had been far more limited. After that winter they became more central. They shaped environmental politics in at least two ways. First, energy problems brought material shortages more forcefully into the realm of substantive environmental concerns and emphasized more strongly the problem of limits which these shortages imposed upon material growth.[48] The physical shortages of energy sources such as oil in the United States, the impact of shortages on rising prices, the continued emphasis on the need for energy conservation all helped to etch into the experience and thinking of Americans the "limits" to which human appetite for consumption could go. Second, the intense demand for development of new energy sources increased significantly the political influence of developmental advocates in governmental, corporate, and technical institutions which had long chafed under both natural environment and pollution control programs. This greatly overweighed the balance of political forces so that environmental leaders had far greater difficulty in being heard. In the face of energy issues environmental leaders formulated their own energy proposals which they sought to inject into the debates, but not yet with overriding success amid an overwhelming emphasis on traditional approaches to increasing energy supply.

Lifestyle issues also injected a new dimension into environmental affairs during the course of the 1970s.[49] They became especially visible in the energy debates, as the contrast emerged between highly centralized technologies on the one hand, and decentralized systems on the other. Behind these debates lay the evolution of new ideas about organizing one's daily life, one's home, community and leisure activities and even work—all of which had grown out of the changing lifestyles of younger Americans. It placed considerable emphasis on more personal, family, and community autonomy in the face of the forces of larger social, economic, and political organization. The impact and role of this change was not always clear, but it emerged forcefully in the energy debate as decentralized solar systems and conservation seemed to be appropriate to decisions made personally and locally—on a more human scale—contrasting markedly with high-technology systems which leaders of technical, corporate, and governmental institutions seemed to prefer. Issues pertaining to the centralization of political control played an increasing role in environmental politics as the 1970s came to a close.

To define stages in the evolution of environmental affairs in this manner helps to interweave those affairs with broader patterns of social change. One should be wary, perhaps, of the temptation to argue that by 1980 a "full-scale" set of environmental issues had emerged, bit by bit to form a coherent whole. For there were many different strands which at times went off in different directions. Those whose environmental experience was confined to the urban context did not always share the perspective and interest in issues of those who were preoccupied with the wildlands. Yet it was rather striking the degree to which working relationships had developed amid the varied strands.[50] What was especially noticeable was the degree to which the challenge posed by the Reagan administration tended to mobilize latent values and strengthen cooperative tendencies.[51] From the beginning of that administration, the new governmental leaders made clear their conviction that the "environmental movement" had spent itself, was no longer viable, and could readily be dismissed and ignored. During the campaign the Reagan entourage had refused often to meet with citizen environmental groups and in late November it made clear that it would not even accept the views of its own "transition team" which was made up of former Republican administration environmentalists who were thought to be far too extreme.[52] Hence environmentalists of all these varied hues faced a hostile government that was not prone to be evasive or deceptive about that hostility. Its anti-environmental views were expressed with enormous vigor and clarity.

We can well look upon that challenge as a historical experiment which tested the extent and permanence of the changes in social values which lay at the root of environmental interest. By its opposition the Reagan administration could be thought of as challenging citizen environmental activity to prove itself. And the response, in turn, indicated a degree of depth and persistence which makes clear that environmental affairs stem from the extensive and deep-seated changes we have been describing. Most striking perhaps have been the public opinion polls during 1981 pertaining to revision of the Clean Air Act. On two occasions, in April and in September the Harris poll found that some 80 percent of the American people favor at least maintaining that Act or making it stricter, levels of positive environmental opinion on air quality higher than for polls in the 1960s or 1970s.[53] One can also cite the rapid increases in membership which have occurred in many environmental organizations, most notably the Sierra Club, as well as financial contributions to them.[54] And the initial forays into electoral politics which environmentalists have recently undertaken seem to have tapped activist predispositions mobilized by the fear of the new administration.[55]

We might take this response to the Reagan administration challenge, therefore, as evidence of the degree to which we can assess the environmental activities of the past three decades as associated with fundamental and persistent change, not a temporary display of sentiment, which causes environmental values to be injected into public affairs continuously and even more vigorously in the face of political adversity. The most striking aspect of this for the historian lies in the way in which it identifies more sharply the social roots of environmental values, perception, and action. Something is there, in a broad segment of the American people which shapes the course of public policy in these decades after World War II that was far different from the case earlier. One observes not rise and fall, but persistent evolution, changes rooted in personal circumstance which added up to broad social changes out of which "movements" and political action arise and are sustained.[56] Environmental affairs take on meaning as integral parts of a "new society" that is an integral element of the advanced consumer and industrial order of the last half of the twentieth century.

The Environmental Economy and Environmental Ideology

There remain two larger modes of analysis which help to define the historical role of environmental affairs—one economic and the other ideological. In neither case can one associate environmental politics with either the pre–World War II economy or its ideology. In both cases we must look to innovations rooted in postwar changes.[57]

Environmental impulses served as a major influence in shaping the newer, more "modern," economy. They brought to the fore new demand factors which in turn generated new types of production to fill them; they placed increasing pressure on greater technological efficiency in production to reduce harmful residuals and resource waste. In many aspects of the economy one can distinguish between older and newer forms of demand and supply, institutions and modes of economic analysis. The transition represents a shift from the older manufacturing to the newer advanced consumer economy. In this transition environmental influences were an integral part of the emerging economy that was struggling for a larger role in America amid more established economic institutions. From this context of analysis we can establish further elements of the role of environmental affairs in long-run social change.

In public debate there was a tendency to set off the "economy" versus the "environment" as if the latter constituted a restraint on the former.[58] But environmental affairs were a part of the economy, that part which constituted new types of consumer demand, giving rise in turn to new modes of production to supply that demand, some in the private market and some in the public. The ensuing controversies were between older and newer types of demand, and the allocation of resources as between older and newer types of production as patterns of demand changed. It was difficult for the older manufacturing economy, with its emphasis on consumer necessities and conveniences and physical commodities to fill them, to accept the legitimacy of the newer economy which gave rise to newer consumer needs and types of production.[59] The tension between old and new was reminiscent of the similar tension in the nineteenth century between the older agricultural and the newer manufacturing economies.

Much of the American economy had moved beyond necessities and conveniences to encompass amenities. It is difficult to identify this change if one begins the analysis with the traditional focus on modes of production; it is more easily identified if one starts with changing patterns of consumption. The former approach lumps together many and varied changes as one "service economy," that beyond raw material extraction and manufacturing. The latter identifies varied new sectors associated with consumption such as the "recreation economy," the "leisure economy," "the health economy," the "creative arts economy," and the "environmental economy" each of which identifies a new direction of economic change. Much of this involves discretionary income, the allocation of expenditures not just to amenities but also to reshaped and restyled necessities and conveniences themselves to make them aesthetically more appealing, to add to them elements other than traditional characteristics of "utility."

The most serious question of resource allocation raised by the environmental economy lay in the appropriate balance which should be struck between natural and developed environments. The new environmental consumer society called for more of the former. This gave rise to massive debates over such issues as wilderness and other natural environment proposals. It was difficult, if not impossible, for those associated with the older developmental economy to accept the notion that natural environments should play a major role in modern economic affairs. Hence they tended to argue that in this matter a proper "balance" should involve only minimal allocation of air, land, and water to natural environments. Often they maintained that such allocations should end. They might approve some role for natural environments in the modern economy but only if they were on sites which developers themselves did not want.[60] Hence, mining companies argued that wherever minerals were to be found they should be developed irrespective of their implication for the degradation of natural environments.

The environmental impulse also had major implications for the technology of production, serving as a force toward more rapid modernization of plants and equipment.[61] In any given segment of industry plants constituted a variable spectrum ranging from the most obsolete to the most modern. In the normal course of private market choices the more obsolete were discarded and the more modern added, giving rise to a general tendency for the entire industry toward modernization. But environmentalists felt that the pace of change was too slow. They were especially interested in the environmental efficiency of production, the degree to which it reduced the output of residuals per factor input; they believed that plants which were more obsolete in material product output were also more obsolete in environmental output. Hence they urged that the most modern plants, the "average of the best," should serve as examples against which the rest of the industry should be judged. In focusing on these "best technologies" as models for achievement by all, environmentalists served as a force for technological innovation.[62]

It was often difficult, however, for industry to move at the pace which environmentalists desired. Many corporate leaders were from sales and marketing origins, rather than engineering and production, and tended not to press continually for cost-reducing technologies but to maintain cost-increasing ones so long as they were profitable.[63] At the same time the corporate response to regulatory requirements often led to superficial changes to reduce the immediate burden of governmental decisions such as legal action or limited "add on" pollution control technology rather than to re-examine production technologies in order to seek combined efficiencies in both product and environmental output.[64] Those who took a leading role in that direction, such as Joseph Ling of the 3-M corporation, were often thought of as "eccentrics" by their fellow executives. The internal politics of trade associations which spoke for business in the larger political scene often required that their public positions not be too "advanced," since their members included both the more obsolete and the more modern firms; in water pollution control, for example, they argued that the median firm rather than the most efficient 10 percent should serve as the model for the rest of the industry to follow. Corporate leaders often argued that regulation was a roadblock to greater production efficiencies, but when such a proposition was subject to empirical examination it was found, on the contrary, that if regulation was sufficiently firm it gave rise to more serious examination of manufacturing processes and resulted in innovation.

We might also profitably identify the environmental impulse more precisely in terms of its ideological component. What is the place of environmental ideas amid the political ideologies inherited from the recent past? These customarily divide political forces between the "liberal" and the "conservative." The corporate business community and critics of growth in government are thought of as "conservatives," while more subordinate sectors of society who look to government to aid them are thought of as "liberal." While these ideological patterns have roots deeper in history than the 1930s, they were given a new twist during the New Deal when controversies over public spending for social programs such as welfare and social security were added to those of earlier vintage which involved disputes among business, labor, and agriculture over the distribution of the fruits of production.

Environmental issues and environmental ideas are difficult to classify in this way. If one raised the question as to whether or not environmentalists favored public or private enterprise in principle, one would have to observe that while they called for greater governmental initiatives in behalf of their objectives such as in public land management or environmental controls on private production, they were as skeptical of public as they were of private enterprise. The Tennessee Valley Authority, the major example of public ownership of the means of industrial production in recent times, was roundly condemned when its actions with respect to air pollution, dam building, and coal and uranium mining were environmentally detrimental and was applauded when it took up innovative energy measures during the Carter administration.[65] Was it associated with ideological traditions sustained by the politics of the industrial working class? Certainly not with socialist ideologies, and with more reformist movements only partially. For while worker movements grew out of the struggle among producers for varied shares of the profits of production, environmental values were associated more with consumption which tended to draw lines of demarcation between environmentalists and producers as a whole. Only when it came to environmental health, which brought occupational and community health concerns together, did workers and environmentalists find common ground.[66]

Environmental values and ideas tended not to fit into traditional political ideologies, but to cut across them.[67] They tended to define corporate leaders as radicals, as responsible for massive, rapid, and deep-seated transformations in modern society that threatened to destroy prized natural environments, that uprooted stable ways of life, and generated pervasive and persistent chemical threats. Corporate leaders were ever demanding that people change their lives markedly in order to accommodate developmental objectives, and to accept the risks of their proposals for rapid and far-reaching change. In response to these demands, environmentalists sought to slow up the pace of innovation, to restrain it. Hence they were conservative. It would not be accurate to describe them as one industry leader did as "stone-age neanderthals," for environmentalists shared, with approval, the material benefits of modern production. But they were willing to argue that the pace of change in America in the 1960s and 1970s was far too rapid and should be slowed down so as not to destroy values important to a society of modern patterns of life.[68]

They were also often fiscal conservatives when the use of public funds was an important instrument of material development and engaged in many a political struggle to cut back public spending. The 1960s and 1970s were decades of rapid economic "growth" in which jobs and product increased dramatically and public programs with public funding played a major role in it: construction of dams and highways, rebuilding on flood plains after floods, channelization of streams and rivers, development of barrier islands, a host of "rural development" programs which had become extended from the "depressed" area of the Appalachians to the entire nation.[69] All these tended to encourage more rapid economic development. The most widely known cases of environmental action on this score pertained to funds for construction of public works in rivers and harbors under the auspices of the U.S. Army Corps of Engineers.[70] It was no wonder that in fashioning coalitions to scale back such expenditures environmentalists joined with the National Taxpayers Union and other "fiscal conservatives" in Congress who tended to give ideological support to reduced public spending.

At the same time, in social values environmentalists could be thought of as innovative rather than conservative. Their views about natural environments and human health were associated with newer rather than older ideas about human wants and needs; they had a larger association with other innovations in values such as the more autonomous role of women, more cosmopolitan rather than traditional ways of life, and "freer" ways of thinking that were associated with social modernization. Such value changes had taken place at a number of times in the nation's past and these historians understand by sorting out newer values from older, distinguishing those people who espoused the newer with enthusiasm from those who drew back in defense and fear against cultural change. In the mid-nineteenth century, for example, the Republican party had been associated with innovations in cultural values and the Democratic party with a defense of older ones. But in the mid-twentieth century, the party roles were reversed, as the Democrats seemed to harbor cultural innovation and the Republicans spoke out in defense of older values. These patterns of cultural change tended to define what was "conservative" and what was "liberal" in terms different from the issues of economic controversy. And so it was with environmental values in which environmentalists both expressed the defense of daily life from technological radicalism and espoused innovations in cultural values.[71]

Within the context of these more "modern" and more innovative values, however, there was in environmental affairs a deeply conservative streak in a different sense that went far beyond the role of corporations and their defense to the larger ideology of conservatism—a search for wider human meaning. Environmentalists tended to work out their values amid a "sense of place" that provided roots to life's meaning much in the same way as "local" community values long had displayed. It was their involvement with the natural environments of given places that had engaged emotions and minds. It was the threat to that "place" of home, work, and play from large-scale developments, from air and water pollution, and from toxic chemical contamination which aroused them to action. Environmentalists sought roots in the less developed and more natural world, and rapid change threatened those roots with impairment and destruction. Insofar as one could describe conservatism as more generally a search for roots, for stability and order amid the larger world of rapid change, then environmentalists shared that impulse.[72]

This essay has constituted an attempt to place the environmental affairs of the past three decades in the perspective of historical evolution. I have sought not just to search for antecedents which would serve to link the more recent and the more remote pasts through some similarity of human activity. Instead I have sought to determine the degree to which a relatively full range of characteristics of environmental affairs, from values to political controversy to economic change and political ideology, constitute merely an elaboration of earlier tendencies or something that was relatively new, a departure from the past. I have argued that these cannot be understood adequately unless they are associated with the newer society, the newer economy, and the newer politics of the decades after World War II. Moreover, they can be understood only as an evolving phenomenon within those postwar decades, amid the patterns of change in the advanced consumer society as it steadily took shape. American society today is far different than it was in the 1930s. It can best be understood not as an implication of the New Deal years, but as a product of vast social and economic transformations which took place after World War II which brought many new values and impulses to the American political scene. And so it is with environmental affairs. While displaying some roots in earlier times they were shaped primarily by the rapidly changing society which came into being after the war which, in so many ways, constitutes a watershed in American history.

The Evolution of Public Environmental Policy
The Case of "No-Significant Deterioration"

RICHARD H. K. VIETOR

The technological complexities of our industrial society have come to permeate nearly every facet of public policy during the past twenty years. The development of environmental regulatory policies particularly reflects this concrescence of technology, law, and the industrial economy. A case in point is the evolution of "no-significant deterioration" as a controversial premise of federal air pollution control. Significant deterioration refers to the degree to which air, cleaner than the national standards mandated by the 1970 Clean Air Act, shall be further degraded by industrial pollutants. This matter became a serious problem of public policy in May, 1972, when Federal Court Judge Pratt granted a Sierra Club petition enjoining William Ruckelshaus, Administrator of the Environmental Protection Agency, from approving state control programs which did not provide controls for the maintenance of existing clean air.[1] This decision, affirmed by the Supreme Court in 1973, forced the EPA to promulgate regulations to define and restrict the significant deterioration of clean air. Those regulations satisfied neither industrialists nor environmentalists and have induced a plethora of court challenges and initiatives in the Congress to amend the Clean Air Act. Whether or not the federal government should adopt and enforce a policy of "no-significant deterioration" (NSD) has become a controversial issue. Implicit in this issue are broader questions involving conflicting environmental values, the federalistic structure of administrative law, pluralist politics of environmental and industrial interest groups, and the limits of technology to cope with economic growth. This essay traces the process by which NSD policy evolved in order to examine the interaction of political and economic forces which has determined the course of public environmental policy.

Industrial opponents of a "no-significant deterioration" policy do not acknowledge that the concept has any substantive roots in legislative intent or administrative policy prior to 1972. According to Carl Bagge, president of the National Coal Association, the entire question is a rhetorical construct, created by rabid preservationists and amplified into reality "because a federal judge once decided two words in the preamble of the Clean Air Act prohibit the 'significant deterioration' of air that is cleaner than the most rigid federal standards now in effect."[2] But in fact, the NSD concept originated, albeit in an ephemeral and ill-defined form, as an element of environmental policy in the mid-1960s. Early concern with the problem of air pollution, since 1958, focused most exclusively on its ill-effects on human health, particularly in urban-industrial trial areas. The national conservation organizations were not especially involved in the conferences and legislative battles which set the course for air pollution control regulations. Such groups as the Sierra Club and the National Audubon Society were devoting their energies to nonurban policy, including the Wilderness Act (1964), the Highway Beautification Act (1965), and the preservation of Dinosaur National Monument and North Cascades National Park. This gap between urban and nonurban environmental problems led to the temporary obscurity of the problem of maintaining clean air by the more critical need to enhance filthy, unhealthy air.

The first National Conference on Air Pollution convened in 1958 and met again in 1962 and 1966. The entire focus of these conferences was the serious health peril which industrial pollutants increasingly posed for urban populations. While local smoke control campaigns of earlier years had improved visibility, dangerous sulfur oxides, particulates, and hydrocarbon compounds were polluting ambient air in increasingly heavy volumes. The most vocal participants in these early conferences were traditional health organizations, spearheaded by the American Cancer Society and the National Tuberculosis and Respiratory Disease Association.[3] The discussions concerned methods by which government and business could cooperate to reduce these airborne health hazards. By 1963, when the first Clean Air Act was legislated, these health organizations were more broadly supported by the American Public Health Association, the Association of State and Territorial Health Officers, the American Municipal Association, and the National Association of Counties.[4] In the 1963 air pollution hearings, important testimony supporting federal regulation was given by Mayors Joseph Barr of Pittsburgh and Richard Daley of Chicago.[5] Their support was crucial, since the regulatory policies under consideration were directed at the serious pollution hazards in their industrial cities. The resulting 1963 Clean Air Act was at best ineffectual, largely because local authorities, backed by industrial interests, had sought federal support without sacrificing local regulatory autonomy. The law merely provided for intervention by the Department of Health, Education, and Welfare in instances of severe inter-urban pollution.

During the 1960s, however, the focus of environmental activism began to shift to the national conservation organizations. Their legislative successes in preserving natural areas, their rapidly growing memberships, and their own broadening of environmental concerns effectuated this development. The Sierra Club alone expanded its membership from 5,000 after World War II to 141,000 by 1973.[6] This expanded constituent base provided involvement in the whole range of environmental problems plaguing industrial society. As environmental concern grew more sophisticated, the linkages between urban industrial problems and the preservation of the countryside became more evident. This contextual shift was just underway by 1967, when Congress once again addressed the problem of air pollution.

A great many industrialists and several environmental groups participated in the public debate which resulted in the 1967 Air Quality Act. The debate centered on federal v. state control authority and emissions v. ambient air standards. Senator Edmund Muskie, chairman of the Subcommittee on Air and Water Pollution, was very much a central figure in the controversy. The Johnson Administration had been convinced by Vernon MacKenzie, HEW's air pollution chief, to support national emission standards for stationary industrial pollution sources.[7] However, Muskie believed that emission standards would be "minimal rather than uniform and hence would not eliminate economic inequities of plant compliance with varying local rules." He preferred ambient air standards (pollutant levels in the air at large) which would give "equal priority to critical areas and areas where no problem presently existed."[8] Industrial interests opposed any national standards, but were most adamantly against the prospect of emission standards.[9] Environmental groups, by then including the Izaak Walton League and the National Wildlife Federation, favored emission controls. Here is where the NSD issue was first drawn, although it is unlikely that any of the protagonists understood the significance of ambient v. emission standards. Senator Muskie certainly misunderstood the implications of his proposals. It is evident from his logic that Muskie was concerned that the standards not induce pollution of "areas where no problem presently existed." But ambient standards were to have precisely that effect. By establishing national levels of pollution which were to be tolerated (or achieved), the door was opened to allowing clean air to deteriorate down to those levels.

The problems of setting air quality standards for various pollutants and developing criteria on which to base those standards were new and exceedingly complex. Concern for maintaining existing clean air was thus lost in the shuffle as Congress legislated the Air Quality Act of 1967. The Act authorized HEW's National Air Pollution Control Administration (NAPCA) to develop health criteria for each pollutant—the maximum concentrations of each pollutant which would not hazard human health. The states were mandated to establish standards of air quality based on those criteria, and to eventually develop implementation plans to achieve those standards. The NAPCA was to develop criteria and guidelines for state planning. A stated purpose of the Air Quality Act was "to protect and enhance the quality of the Nation's air resources so as to promote the public health and welfare and the productive capacity of its population."[10] This phrase, "protect and enhance," was interpreted by Judge Pratt in *Sierra Club v. Ruckelshaus* to mean that Congress not only intended to clean up ("enhance") polluted air, but also to maintain ("protect") the existing quality of clean air.[11] The Senate Report accompanying the 1967 Act further stated that "The Air Quality Act . . . serves notice . . . that there will be no haven for polluters anywhere in the country."[12] Thus the early legislative history of the significant deterioration concept reflects an ambiguity, suggesting that maintenance of very clean air was certainly a goal of the lawmakers, but one which they were incapable of, or unwilling to define clearly.

Industrial energy developers had a vague sense of the NSD concept as early as 1967. Shortly after the enactment of the Air Quality Act, the National Coal Policy Conference, an inter-industry trade association representing coal, railroad, electric utility, and equipment manufacturer interests, published *A Guide to the Air Quality Act of 1967*. This was a handbook for industrialists, explaining the implications of the Act, and recommending appropriate policy inputs. The NCPC manual emphasized the importance of "including in a control region only those areas absolutely necessary. . . ." Doing so would "permit a wider choice in location of new power plants outside of the area where the pollution problem is in fact serious."[13] However, even were this policy pursued, warned the NCPC, "any State . . . may set more stringent standards for stationary sources in order to achieve a higher level of ambient air quality. . . ."[14] Evidently, the NCPC recognized the significance of industrial development in clean air regions, but did not perceive restrictions of such development in the Act itself.

In 1968, the National Air Pollution Control Administration set about implementing the mandates of the Act. While establishing health criteria for sulfur oxide and particulate matter concentrations, the NAPCA also established an in-house task force to develop guidelines for the formation of air quality standards and implementation plans by the states. In an early discussion paper (May, 1968), Sidney Edleman, chief of the NAPCA's Environmental Health Branch, addressed the problem of significant deterioration. Based on his interpretation of the "protect and enhance" language of the Air Quality Act, Edleman proposed that "Not only is air quality to be 'enhanced,' but it is to be *protected*, obviously against a loss of quality or degradation."[15] Accordingly, he felt that "standard setting activity by the State is called for even if the level of air quality in the region is better than that which the criteria would require. . . ."[16] For precedent, Edleman recalled that the Federal Water Pollution Control Administration had adopted a no-degradation policy after the enactment of the 1965 Water Quality Act. That policy had stated that "In order to 'enhance the quality of the water,' standards shall include a provision to assure that present water quality will not be degraded."[17] Thus, Edleman recommended the following:

> To the extent that it [air quality] is better than the criteria with respect to other pollutants, the protection of the public welfare would call for the maintenance of such better air quality to the extent possible. In my view, an adaptation of the FWPCA policy which would protect the existing air quality in a region which is better than that called for by the air quality criteria should be established to serve the purposes of the Clean Air Act.[18]

Significantly, however, this FWPCA control policy had been qualified by the disclaimer that clean water qualities might be lowered where "such change is justifiable as a result of necessary economic or social development. . . ."[19] Just such a qualification was vigorously supported after 1972 by industrial energy developers for the no-significant deterioration air pollution policy. This early recommendation is important to the present debate over NSD policy. Opponents of the policy argue that the entire issue resulted from the Sierra Club's manipulation of extraneous verbiage in the preamble of the 1970 Clean Air Act. But Edleman's paper suggests to the contrary, that the NSD concept was a fundamental aspect of administrative implementation of air pollution control from the start. Moreover, it was logically preceded by an even earlier water pollution control policy.

As the NAPCA guidelines developed through various draft stages, the difficulties of the no-significant deterioration concept became increasingly evident. An early October, 1968, draft emphasized, but did not go beyond, the need to "protect" the air against a loss of quality.[20] The second guidelines draft, however, included a negative NSD qualification: "If the calculated air quality is appreciably less than the proposed standard, the implementation plan regulations will be considered unduly restrictive. States will be advised and some relaxing of the regulations may be made."[21] But in the third draft, circulated in December, 1968, the Deputy Chief of Abatement Programs criticized its "inference that a policy of degradation is permissible."[22] The January, 1969, draft drew similar comments from the Bureau of Abatement and Control, which recommended that "A nondegradation statement is needed under [the] A[ir] Q[uality] standard section."[23] In the next to final draft (February, 1969), the issue was addressed, but without the "degradation" or "deterioration" language:

> Where air quality is already satisfactory . . . steps to protect and preserve it will be necessary. . . . In any event, air quality standards that would permit a significant rise in pollution levels . . . would not be consistent with the intent of the Air Quality Act.[24]

During the Spring of 1969, John T. Middleton, Commissioner of the NAPCA, was making speeches explaining the emerging air pollution control policy. In Phoenix, Middleton cautioned against "employing air quality criteria . . . as a license to pollute the air up to a given level."[25] By June, when Middleton addressed local air pollution regulators in Pennsylvania, the policy of precluding deterioration of existing air quality had fully crystallized.[26] In mid-1969, the HEW published its "Guidelines for the Development of Air Quality Standards and Implementation Plans." Section 1.51 of those guidelines clearly stated the government's policy position:

> Air quality standards which, even if fully implemented, would result in significant deterioration of air quality in any substantial portion of an air quality control region clearly would conflict with this expressed purpose of the law.[27]

When the HEW guidelines were promulgated in 1969, industrial coal interests immediately recognized the implications of NSD as an immense threat to future industrial expansion, particularly in the West. The American Mining Congress, representing hundreds of major energy developers and users, vigorously responded to the guidelines. It claimed that the NAPCA had "usurped primary responsibility . . ." for air pollution control from the states.[28] In uncharacteristically drastic action, the American Mining Congress sent a letter criticizing the

"Guidelines" to HEW Secretary Finch, and a copy to the governors of all fifty states. A central complaint in that letter focused on the NSD clause of the "Guidelines," which the AMC recognized as the first blatant expression of a policy which might seriously threaten future industrial development.[29]

By 1970, both industrial and environmentalist interests were calling for another amendment of the Air Quality Act. Industrialists felt that the health criteria were too rigorous, and that the HEW was dictating terms to the states. Environmentalists feared weak state standards, and felt that the entire implementation process had bogged down. Public concern for the environment reached a crescendo in the spring of 1970. On April 22, environmental activists and school children throughout the nation celebrated the first Earth Day, dramatizing with gas masks and rallies the seriousness of environmental degradation. The Gallup polls recorded the public's heightened priority for environmental cleanup.[30] In May, "Nader's Raiders" published *Vanishing Air,* a widely read critique of Senator Muskie's "Mr. Clean" image and the failure of previous air pollution control efforts.[31] This new context of public awareness set the stage for legislation which political scientist Charles Jones describes as "speculative augmentation."[32] Responding to public demand, politicians rewrote the Clean Air Act to achieve absolute air quality goals, regardless of the economic cost and technological feasibility of doing so.

The Nixon Administration, recognizing the opportunity to co-opt Senator Muskie's leadership in environmental matters, proposed tough amendments to the 1967 Act.[33] Senator Muskie, by then a presidential hopeful smarting from the Nixon challenge and the Nader criticism, took the bull by the horns and presented the Congress with an extraordinary new law. Both industrialists and environmentalists participated in the Senate hearings on various proposed amendments. When Muskie's bill, with bipartisan support, was reported out of the committee, it passed the Senate unanimously (73–0). Afterward, Joseph Mullan of the National Coal Association reflected industry's astonishment with the new measure: "The bill that came out of the Senate was not the bill that anybody testified to."[34] The 1970 Clean Air Act mandated national, rather than state, ambient air quality standards based on rigid HEW health criteria. The Environmental Protection Agency, now responsible for implementing the law, was to establish national primary standards based on the health criteria, and secondary standards based on "public welfare" (the ill-effects of air pollutants on natural and man-made environments). States were to develop implementation plans to achieve the new federal standards. Furthermore, the Act authorized the EPA to promulgate emission standards for new stationary pollutant sources to be built in the future.[35]

Once again obscured by the health orientation of the Act, the NSD concept was retained in Section 101(b) of the 1970 Act. While the "no-significant deterioration" phraseology was not used, Section 101 retained the "protect and enhance" language of the earlier law.[36] But the intent of that language was made clear in the House hearings on the bill, in which Robert Finch, Secretary of HEW, made a statement clarifying HEW policy:

> One of the express purposes of the Clean Air Act is "to protect and enhance the quality of the Nation's air resources." Accordingly, it has been and will continue to be our view that implementation plans that would permit significant deterioration of air quality in any area would be in conflict with this provision of the Act.[37]

Senator John Cooper (R-Ky.), seeking clarification of this point, asked, ". . . if the region or an area had a certain air quality which might be higher than other areas of the country. . . . It could not be degraded; is that correct?" HEW Under Secretary Veneman responded: "Yes. . . . we did not want deterioration of the air in those areas that may be below what the standard is at the present time."[38] The Senate Report accompanying the Clean Air Act explicitly stated that "The Secretary [HEW] should not approve any implementation plan which does not provide . . . for the continued maintenance of . . . areas where current air pollution levels are already equal to, or better than, the air quality goals. . . ."[39] The fact that the Clean Air Act did not adopt this explicit language was a major point of the EPA's appeal of the *Sierra Club v. Ruckelshaus* decision.[40]

Early in 1971, the Environmental Protection Agency proposed national standards for sulfur oxide, particulate, and nitrogen oxide pollutants.[41] Fifty-one environmentalist groups, including six national conservation organizations, submitted written comments on these proposals. Several of these groups demanded inclusion in the standards of a no-significant deterioration statement. The Natural Resources Defense Council made their case most effectively:

> The ambient air quality standards should explicitly state that because of the legislative history of the Act, in air quality regions where present levels of pollutants are lower than the national standards, the present levels should constitute the federal standards of ambient air quality for these regions.[42]

None of the industrial comments made reference to the NSD concept, no doubt reflecting their relief that it had not been explicitly expressed in the Act or in the EPA's standards proposal. However, the EPA, responding to environmentalist criticisms, included in the final standards the disclaimer that promulgation of national standards "shall not be considered in any manner to allow significant deterioration of existing air quality. . . ."[43] It was on the basis of this statement and the "protect and enhance" language in the 1970 Act, that the Supreme Court affirmed *Sierra Club v. Ruckelshaus* without an opinion (4–4).[44]

On the heels of the standards promulgation, the Environmental Protection Agency prepared a new set of implementation guidelines for the states. When the proposed guidelines were published in April, 1971, the EPA solicited industrial and environmentalist comments.[45] As in 1969, representatives of industry reacted vigorously. Nearly 400 industrial groups and firms submitted comments, virtually all of which were critical of the guidelines. Once again, industrialists felt that the EPA, by dint of the specific instructions in the guidelines, was preempting state prerogatives and exceeding the mandates of the 1970 Act. On the other hand, environmentalist groups were satisfied with most provisions of the guidelines. However, several of the groups, particularly the Natural Resources Defense Council, vehemently recommended inclusion of a NSD clause in the guidelines to prohibit further pollution of air cleaner than the national standards.[46]

By June, 1971, the EPA had responded to these comments with a revised draft of the implementation plan guidelines. That draft was never made public, but some of the revisions it proposed were explained in a briefing memorandum prepared in June by John Middleton. According to this memo, most of the changes responded to environmentalist demands. The most crucial change was the inclusion in the guidelines of a no-significant deterioration statement. The Middleton memo stated the following:

> Environmental groups, in particular, urged that EPA officials establish a non-deterioration policy applicable to clean air areas. . . . Accordingly, the regulations have been modified to include the following statement . . . "Approval of a plan shall not be considered in any manner to allow significant deterioration of existing air quality in any portion of the State."[47]

The June 28 version of the revised guidelines, which included this clause, never saw the light of day. It was not promulgated in the *Federal Register*, but rather, submitted to the Office of Management and Budget for a thoroughgoing interagency review. When the final guidelines emerged in August, 1971, and were officially promulgated in the *Federal Register*, they were substantially different from either the April or June versions. Among other things, the crucial NSD clause emerged as the following, thoroughly innocuous, disclaimer:

Nothing in this part shall be construed in any manner to encourage a state to prepare, adopt, or submit a plan which does not provide for the protection and enhancement of air quality so as to promote the public health and welfare and productive capacity. (36F.R.15487)

Two congressional oversight hearings, a few journalists, and at least one historian have tried to determine how and why the OMB managed to force the EPA to accept this crucial revision.[48] In the House Clean Air Act hearings of 1972, Congressman Paul Rogers (D-Fla.) closely questioned EPA Administrator Ruckelshaus on the subject of that June version of the "Guidelines." Rogers requested that Ruckelshaus submit to the committee "your proposed guidelines before they went to the OMB [for further revision]. . . ."[49] Accordingly, Ruckelshaus submitted a document entitled, "Plans for Implementation of National Ambient Air Quality Standards," subtitled, "Submitted to Office of Management and Budget from EPA, June, 1971." This document was not, however, the June 28 version which included the changes recommended by Middleton, but rather the original April guidelines proposal. Not only did it not include the revisions indicated in the Middleton memo of June, 1971, but it had the April version section numbers, not the section numbers of the draft the OMB reviewed and sent back to the EPA.[50] Furthermore, when Congressman Rogers asked Ruckelshaus for the OMB's revisions, Ruckelshaus responded, "I do not know that they submitted any recommendations to us."[51] However, another Middleton briefing paper, dated July 19, was circulated to top EPA administrators, stating the OMB's specific recommendations and the EPA's objections to them.[52]

In the Senate oversight hearings, Ruckelshaus testified that "OMB did not get any final crack at the regulation. OMB is nothing more than a conduit [for] other federal agencies who want to comment on any regulations. . . . I take great exception . . . that these substantial changes in any way reflect anybody's opinion except mine on what changes should be made."[53] In a 1973 interview, Edward Tuerk, a top assistant of the Administrator, corroborated this statement.[54]

In fact, the OMB interagency review did force the significant changes on the EPA, against its will. The July 19 discussion paper authored by John Middleton firmly documents this fact. Referring to the revisions demanded by the OMB review, this memorandum states:

The Department of Commerce maintains that the non-deterioration provision [sec. 420-07 (b)] is neither justifiable nor legally supportable. The provision in question states: "Approval of a plan shall not be considered in any manner to allow significant deterioration of existing air quality in any portion of the state."

This provision is a statement of principle nearly identical to a statement included in the regulations setting forth the national ambient air standards. There is nothing in EPA's regulations that would require a state to have, in its implementation plan, a strategy designed specifically to implement section 420-07(b). *The statement should be retained* [emphasis added] to keep EPA's regulations from being construed as preventing States from preserving clean air where they choose to do so.[55]

The memorandum discussed and rejected four other major revisions suggested by the OMB review. The source of each suggestion was indicated in parenthesis reflecting the leading role of the Department of Commerce and the Federal Power Commission in that review.

Thus, the Department of Commerce, through the OMB interagency review, forced the EPA to make the crucial change which eventually led the NSD issue into the courts. A strong case can be made that the National Industrial Pollution Control Council was behind the Department of Commerce's impositions on the EPA's guidelines. In 1970, President Nixon established the NIPCC as an advisory council to the Department of Commerce. This council, and its twenty-two subcouncils, was made up of corporate chief executives from every major polluting industry who met irregularly between 1970 and 1973. Air pollution was the subject of most serious concern to NIPCC members. At a NIPCC meeting in May, 1971, there was an extended discussion that "indicated widely-held and growing concern on the part of the members that the Government and the public are underestimating the extent of the disruption . . . that is being set in motion by stringent pollution control measures being applied in unrealistically constrained time frames." At that meeting, Commerce Secretary Maurice Stans urged the industrialists to work together to help revise the standards.[56] Undersecretary James Lynn told NIPCC members about the implications of the "implementation of the Clean Air Act," and urged that "industry should make economic study data on pollution impacts available to the Government whenever possible."[57] By June, 1971, the NIPCC had established systematic liaison with both the EPA and the White House, and there was a steady flow of correspondence and meetings between NIPCC members, the EPA, and the American Mining Congress. NIPCC's director, Walter Hamilton, kept a special file on the guidelines and the oversight hearings in 1972.[58]

Environmentalists were furious with the eleventh-hour interference which had resulted, according to Richard Ayers of the Natural Resources Defense Council, in "drastically weakened guidelines."[59] Having been outflanked on the NSD concept, environmentalists turned to the courts to challenge the EPA's regulations. In

May, 1972, several environmental groups, headed by the Sierra Club, petitioned the federal district court of Washington, D.C., to enjoin William Ruckelshaus from approving state control programs which did not provide for the maintenance of existing clean air. *Sierra Club v. Ruckelshaus* was the first important clash between the Environmental Protection Agency and those environmental groups which had come to be the EPA's only significant political constituency. By accepting the crucial changes in the guidelines, under pressure from the OMB and the Department of Commerce, EPA Administrator Ruckelshaus had compromised his amicable relationship with organized environmental interests. Starting with its brief in *Sierra Club v. Ruckelshaus* the EPA was forced to defend its opposition to NSD policy, an artificial position which contradicted well-established NAPCA policy since 1968. Judge Pratt held that:

> Having considered the stated purpose of the Clean Air Act of 1970, the legislative history of the Act and its predecessor, and the past and present administrative interpretation of the Acts, it is our judgment that the Clean Air Act of 1970 is based in important part on a policy of non-degradation of existing clean air and the 40 C.F.R. 51.12(b), in permitting the states to submit plans which allow pollution level of clean air to rise to the secondary standard level of pollution, is contrary to the legislative policy of the Act.[60]

This opinion was affirmed without opinion by the U.S. Court of Appeals and by the Supreme Court.[61] It is ironic that administrative pressures, initiated by corporate energy interests, forced the NSD issue into the courts. While it had remained an ill-defined element of the 1970 Act, and even as a clause of the EPA's standards document, it had not affected industry compliance with the law. But forced into the open by the courts, the NSD concept had to be defined precisely in terms of specific regulations. Those specifics were to pose serious problems for public policy with regard to future industrial development.

The Supreme Court's affirmation of *Sierra Club v. Ruckelshaus* in 1973 brought a storm of protest from the industrial community. Coal, electric utility, steel, smelting, and petroleum interests were among the most vocal critics of the Court's decision. Even the *New York Times* was dubious of this "decision of sweeping impact. . . . It almost certainly prohibits even small industry from building new plants in rural areas. . . ."[62] The National Coal Association called for immediate congressional review of the Clean Air Act. Carl Bagge, NCA president, claimed the decision "will stop the construction of any new fossil fuel power plants in most of the United States. . . . It will also wash out any prospect of producing synthetic natural gas or petroleum . . . killing our best chances of meeting the energy crisis from our domestic reserves."[63] The Edison Electric Institute and the National Rural Electric Cooperative Association likewise encouraged congressional resolution of the court decision's implications. Peabody Coal's president, Edwin Phelps, insisted that "the Clean Air Act should be amended" as quickly as possible.[64] The Senate Public Works Committee convened a special hearing on the NSD issue in July, 1973, wherein Carl Bagge spoke for industrial interests and Lawrence Moss, president of the Sierra Club, represented environmentalists.[65]

As the no-significant deterioration controversy intensified, the Environmental Protection Agency began developing NSD regulations so that state implementation plans could be revised and approved. In July, 1973, the EPA proposed four alternative plans for preventing significant deterioration.[66] The EPA held public hearings and elicited comments and alternate proposals. Both the Sierra Club and the Natural Resources Defense Council rejected the EPA proposals, and suggested their own plans. The Sierra Club offered a plan in which air quality deterioration would be defined in terms of the concentration of air pollution averaged over a one-kilometer sphere, the center of which would be an emission point. The EPA rejected this plan because of its arbitrary definition of significant deterioration, and because the Sierra Club's specific increments (of pollutants) "would impose severe growth restrictions" for which the "Sierra Club has presented no rationale. . . ."[67] Under the NRDC proposal, "the total emissions in clean areas, plus a five percent increase, would be divided by the total population in clean areas to arrive at the allowed per capita emissions."[68] EPA rejected this scheme because the location of resource processing industries should be determined by the location of natural resources, not the population served.[69]

In December, 1974, the EPA selected the least innocuous of its alternate proposals and promulgated the official regulations for the "Prevention of Significant Deterioration."[70] These regulations were to apply to new industrial emissions in air regions which are cleaner than the pollution levels defined by the secondary ambient air quality standards. Neither the Clean Air Act nor the Court had defined "significant deterioration." The EPA felt that a definition must be predicated on social, economic, and environmental factors varying from one area to another. Thus, the EPA's rules sought to shift the responsibility for defining and implementing significant deterioration to state and local governmental units. The new regulations called for the "establishment of 'classes' of different allowable incremental increases in total suspended particulates and sulfur dioxide."

Class I applies to areas in which practically any change in air quality would be considered significant;

Class II applies to areas in which deterioration normally accompanying moderate well-controlled growth would be considered significant;

Class III applies to those areas in which deterioration up to the national standards would be considered significant.[71]

Initially, all areas of the country were designated Class II. Provisions were made for redesignation by the states, pending approval by the EPA. A Class I area would be allowed incremental deterioration of about 8 percent of the secondary standard for particulate matter, and about 3 percent for sulfur oxides. Class II areas would be allowed deterioration of 16 percent of the secondary particulate standard and 25 percent of the secondary sulfur oxide standard.[72] These incremental allowances would be measured on the basis of existing pollutant concentrations as of January 1, 1975.

These NSD regulations pleased no one. Environmentalists felt that the increments for Class II and Class III certainly did allow air deterioration that would be significant.[73] Several environmental organizations immediately launched court challenges which are now pending. Industrial energy developers were utterly dismayed. Representatives of coal, electric utilities, paper, steel, chemicals, and oil industries spoke against the new regulations in another round of Senate hearings.[74] The industrial interests claimed that the significant deterioration regulations were being used as land use planning restrictions by environmentalists. As such, the limitations would circumscribe "the future development of coal gasification, coal liquefaction, mine power generation, and even clean coal facilities. . . ."[75]

By 1974, the NSD concept had become the most controversial aspect of public environmental policy. No-significant deterioration was no longer, if it had ever been, an isolated air pollution issue. NSD limitations were increasingly perceived as having broad implications for the "available technology" debate, strip mining in the west, Project Independence, and the "limits to growth" idea. After 1972, the evolution of NSD policy became fundamentally intertwined with these other major issues of public policy.

Since 1970, the definition of "Best Available Control Technology" (BACT) has been at the heart of political conflict over federal air pollution policy. The Clean Air Act called for federal emission standards for new sources of pollution that would reflect the "degree of emission limitation achievable through the application of the best system of emission reduction which . . . has been adequately demonstrated."[76] The argument focused on stack gas scrubbers, devices for removing sulfur oxides from industrial emissions. Businessmen were convinced that scrubbers were too expensive and were not technically capable of achieving the sulfur oxide standards. Environmentalists and EPA officials claimed that scrubber systems were already proving themselves in commercial use.[77] After national hearings and an exhaustive study, a government panel (SOCTAP) issued a report in mid-1973 which concluded that scrubbers were feasible.[78] Although the issue was officially settled, industrialists remained unconvinced. Most notably, American Electric Power instituted a massive publicity campaign throughout 1974, deriding scrubber technology to the American public. At this juncture, the Supreme Court affirmed *Sierra Club v. Ruckelshaus,* and the EPA set about promulgating incremental standards to achieve no significant deterioration. The new NSD standards had the effect of escalating the controversy over BACT to a heated pitch. Industrialists were certain that scrubber technology could not possibly achieve the incremental standards, and that to even try would be prohibitively expensive.[79] Environmentalists countered this claim by arguing that the NSD limits would repeat the "technology forcing" effect of the 1970 Act and would stimulate new engineering development of even more effective pollution control devices.[80] Resolution of this issue became central to the congressional agenda for amending the Clean Air Act after 1975.

Meanwhile, NSD was exacerbating political conflict over federal strip mining regulation. As the locus of coal strip mining had shifted from Appalachia to the thick seams under the western Plains, opposition to large-scale western mine development had grown among environmentalists and land owners. After 1973, the prospect of growth-limiting NSD restrictions held great promise for environmental groups promoting federal legislation that would circumscribe strip mining in the clean air areas of Wyoming, Colorado, and Montana. As might be expected, those environmental interests became vigorous proponents of a federal policy of no-significant deterioration. Conversely, the National Coal Association and the Federal Energy Agency recognized that a rigorous NSD policy would constitute a "de facto" constriction of western strip mining development.[81]

For Project Independence, the Ford Administration's solution to the "energy crisis," NSD policy posed a serious threat. According to the March, 1975, Energy Independence Environmental Impact Statement,

> The significant deterioration regulations could have a major inhibiting effect on the location of new power plants and new energy resource development projects. . . . The

regulations could preclude the siting of such facilities in areas zoned for minimal, or even moderate amounts of deterioration [i.e., Class I and Class II].[82]

To vitiate these perceived problems, the Ford Administration joined with the business community in calling on the Congress to amend the Clean Air Act so as to "remove the EPA Administrator's authority to establish standards more restrictive than primary and secondary ambient air quality standards."[83] Environmentalists also sought amendment of the Act, but to the opposite effect—that more strict NSD provisions be mandated.

At this time, a fourth public issue, the idea of "limits to growth," began to further exacerbate the NSD question. Just five months after the *Sierra Club v. Ruckelshaus* decision, the Club of Rome had published a study entitled, *The Limits to Growth.*[84] In a computer model, its authors had predicted that continued economic and population growth posed a worldwide threat to the quality of life. Moreover, the study calculated that neither technology nor discovery of new resources would mitigate the problems that would result from continued growth. The issue of limiting growth through some sort of steady state economy became a vocal public debate as other scholars, including E. F. Schumacher, Lewis Perlman, Herman Daly, and Herman Kahn, joined the fray.[85] For the most part, business interests devoted to energy and industrial expansion found limits to growth thinking abhorrent, and they believed that the NSD policy was the first explicit step toward such a goal. Keith Doig, a Shell Oil vice-president, was among those businessmen who saw in the NSD policy something more than air pollution control. "I am not at all sure," said Doig, "of the sincerity of some of the proponents [of NSD]—in fact, it appears that genuine concern for our environment is being used by others to achieve quite different objectives."[86] Mr. Doig warned Shell stockholders that "some 79 percent of all the electrical generation capacity now planned for the next decade could not be built. . . ." were the NSD limitations proposed by the EPA accepted by the Congress.[87] To Mr. Doig and many others, the NSD policy was a concrete expression of an idea that appeared to threaten one of capitalism's basic tenets.

From all of these perspectives, Congress came under tremendous pressure in 1975 to amend the Clean Air Act. That year, it held extensive hearings and mark-up sessions in both houses. Several Washington correspondents claimed that the Clean Air Act amendments generated more intense lobbying by more industrial interest groups than any other legislative issue since the Taft-Hartley Act. Coal, oil, electric utilities, railroads, banks, steel, equipment manufacturers, and dozens of national trade associations sought to convince congressmen that energy growth was essential and would be obstructed by a policy of no significant deterioration. In 1976, an eleventh-hour filibuster in the Senate killed an amendment package that both the House and the Senate had approved. It is noteworthy that Senators Frank Moss and Jake Garn conducted the filibuster. Their state, Utah, because of its proximity to several national parks, would have been most affected by the no-significant deterioration provision. Finally, in the summer of 1977, after still another year of heated political conflict, President Jimmy Carter signed into law the 1977 Clean Air Act Amendments. The most crucial amendment to the Act was the assertion of a NSD policy.

In the course of a decade, the no-significant deterioration concept had evolved from an ill-defined goal of "protecting" air quality to a central premise of federal air pollution regulatory policy. Most certainly, the new policy was a compromised version of the standards which the EPA had promulgated in 1974. The deterioration increments were expanded, the definition of mandatory Class I areas changed, provision for buffer zones abandoned, and variance conditions were established.[88] Even with those changes, however, no-significant deterioration, regardless of its potential effect on industrial growth, had been articulated by the Congress to be national policy. That policy had evolved through the 1970 contextual revolution in environmental values, the 1973 advent of energy crisis perception, and intense industrial opposition to growth-limiting policy. It remains to be seen, of course, the extent to which the government succeeds in implementing this new policy (remembering that the 1970 ambient air quality standards will still not have been achieved by 1983). Nevertheless, the evolution of the NSD concept to the level of a fully articulated national policy reflects some fundamental changes in traditional American values which have accepted and promoted unrestrained economic growth and unfettered resource exploitation. Our new national policy of no-significant deterioration reflects not merely concern with unpaid social costs, but perhaps an acknowledgment that growth is not the essence of societal well-being.

Reconstructing Environmentalism
Complex Movements, Diverse Roots

ROBERT GOTTLIEB

Where We Live, Work, and Play

There was a great deal of anticipation in the conference room at the Washington Court Hotel on Capitol Hill when Dana Alston began her address before the first National People of Color Environmental Leadership Summit in October 1991. Conference delegates included grassroots environmental activists from across the country: African-Americans from the petrochemical industry corridor in Louisiana; Latinos from urban and rural areas of the Southwest; Native American activists like the Western Shoshone protesting underground nuclear testing on their lands; organizers of multi-racial coalitions in places like Albany, New York, and San Francisco. The purpose of the Summit was to begin to define a new environmental politics capable of placing the concerns, method of organizing, and constituencies of such grassroots groups and activists at the center of the environmental discourse.

Alston, a key organizer of the Summit, symbolized the new kind of environmentalist the Summit had sought to attract. Born in Harlem, she first became active in the mid-1960s in the black student movement, addressing issues of apartheid and the Vietnam War. Pursuing an interest in the relationship between social and economic justice issues and public health concerns, Alston completed a master's degree in Occupational and Environmental Health at Columbia University. She subsequently worked at the Red Cross (where she dealt with the new emergency issues associated with toxics and nuclear power problems) and Rural America (where she organized conferences on pesticide issues). In February 1990, Alston joined the staff of the Panos Institute, an organization that dealt with the intersection of environment and development issues from the perspective of Third World needs and concerns. Alston was hired to develop a program related to the rise of domestic "people of color" organizations concerned with environmental justice. In that capacity, she became part of the planning committee organizing the People of Color Summit.

In her talk, Alston suggested to the delegates and participants that she would try to "define for ourselves the issues of the ecology and the environment, to speak these truths that we know from our lives. . . ." "For us," she declared, "the issues of the environment do not stand alone by themselves. . . . The environment, for us, is where we live, where we work, and where we play. The environment affords us the platform to address the critical issues of our time: questions of militarism and defense policy; religious freedom; cultural survival; energy-sustainable development; the future of our cities; transportation; housing; land and sovereignty rights; self-determination; employment—and we can go on and on." In pursuing such issues, Alston concluded (restating a dominant theme of the Summit): "We refuse narrow definitions."[1]

This question of definition, as Alston had stated, lies at the heart of understanding the past, present, and future of the environmental movement. Today, the environmental movement, broadly defined, contains a diverse set of organizations, ideas, and approaches: professional groups whose claims to power depend on scientific and legal expertise; environmental justice advocates concerned about equity and discrimination; traditional conservationists or protectionists whose long-established organizations have become a powerful institutional presence; local grassroots protest groups organized around a single issue; direct action groups bearing moral witness in their defense of the natural environment.

Environmental organizations range from multimillion dollar operations led by chief executive officers and staffed by experts to ad hoc neighborhood associations formed around a local environmental concern. There are groups concerned with the need for efficiency in existing economic arrangements and those that seek to remake society; groups who promote market solutions and those who want to regulate market failures; conservative environmentalists hoping to strengthen the system and radical environmentalists interested in an agenda for social change.

Given the diverse nature of contemporary environmentalism, it is striking how narrowly the movement has been historically described. In nearly all the standard environmental histories, the roots of environmentalism are presented as differing perspectives over how best to manage or preserve "Nature"; that is, "Nature" outside of the cities and the experiences of people's everyday lives.[2] The primary players in numerous historical texts—the Muirs and the Pinchots—represent those perspectives to the exclusion of other figures not seen as engaged in environmental struggles because their concerns were urban and industrial. There has been no place in this history for Alice Hamilton, who helped identify the new industrial poisons and spoke of reforming the "dangerous trades"; for empowerment advocates like Florence Kelley, who sought to reform the conditions of the urban and industrial environment in order to improve the quality of life of workers, children, women, and the poor; or for urban critics like Lewis Mumford, who spoke of the excesses of the industrial city and envisioned environmental harmony linking city and countryside at the regional scale.

In part because of these historical omissions, scholars offer sharply divergent views about the origin, evolution, and nature of contemporary environmentalism. Most common explanations place the beginning of the current environmental movement on or around Earth Day 1970. The new movement, they emphasize, came to anchor new forms of environmental policy and management based most directly on the cleanup and control of pollution rather than simply the management or protection of the natural environment. This explanation thus provides a convenient way to distinguish between an earlier conservationist epoch where battles took place over national parks, forest lands, resource development, and recreational resources, and today's environmental era where pollution and environmental hazards dominate contemporary policy agendas.

The problem with this explanation of such a historical divide in environmentalism is who is left out and what it fails to explain. Pollution issues are not just a recent concern; people have recognized and struggled about these problems for more than a century in significant and varied ways. A history which separates resource development and its regulation from the urban and industrial environment disguises a crucial link which connects both pollution and the loss of wilderness. If we see environmentalism as rooted primarily or exclusively in the struggle to reserve or manage extra-urban "Nature," we find it difficult to link the changes in material life after World War II—the rise of petrochemicals, the dawning of the nuclear age, the tendencies towards over-production and mass consumption—with the rise of new social movements focused on issues of quality of life. And by defining contemporary environmentalism primarily in reference to its mainstream, institutional forms, such historians cannot account for the spontaneity and diversity of an environmentalism rooted in communities and constituencies seeking to address issues (as Dana Alston tells us) of where and how people live, work, and play.

Situating Bob Marshall

To understand this complex movement with diverse roots, it might be best to begin with Wilderness Society founder Bob Marshall, who remains a paradoxical figure within environmentalism. A champion of the poor and powerless, deeply committed to wilderness, and equally forceful about the need to make nature a direct part of people's lives, Marshall is an especially enigmatic figure for those who have defined environmentalism in narrower terms. Yet this intense, always curious radical forester proposed a common thread for a movement split between managers and protectors of the natural environment and those defining their environmentalism on the basis of daily life experiences.

The son of the senior partner in the prestigious, Washington, D.C., law firm of Guggenheimer, Untermeyer, and Marshall, Robert Marshall grew up steeped in liberal values, including defense of civil liberties, respect for minority rights, and the fight against discrimination. Encouraged by his father, who maintained a strong interest in forest conservation, Marshall decided to launch a forestry-related career. He worked in various capacities for the Forest Service, where he began to view the forests as a necessary retreat "from the encompassing clutch of a mechanistic civilization," a place where people would be able to "enjoy the most worthwhile and perhaps the only worthwhile part of life." Marshall quickly became a strong critic of development pressures on forest lands from private logging companies which had led to a decline in productivity, increase in soil erosion, and "ruination of the forest beauty."[3]

Marshall's compassion for people and powerful desire to be in touch with wilderness eventually led him to adopt two distinctive, yet, for him, compatible positions about wilderness "protection." On the one hand, Marshall feared a loss of the wild, undeveloped forest lands in their spectacular western settings and in the less "monumental" forest areas of the East, such as the Adirondacks. In a February 1930 article for the *Scientific Monthly*, Marshall laid out this concept of wilderness as a "region which contains no permanent inhabitants, possesses no possibility of conveyance by any mechanical means, and is sufficiently spacious that a person crossing it must have the experience of sleeping out." To achieve that goal, Marshall urged a new organization be formed "of spirited people who will fight for the freedom of the wilderness," and be militant and uncompromising in their stance.[4]

At the same time, Marshall argued that wilderness belonged to all the people, not simply to an elite who wanted such areas available for their own use. Already by 1925, Marshall was writing that "people can not live generation after generation in the city without serious retrogression, physical, moral and mental, and the time will come when the most destitute of the city population will be able to get a vacation in the forest."[5] Marshall was particularly critical of the policies of the National Park Service with their expensive facilities and concessions. Though he argued against more roads and increased development in either park or forest lands, Marshall nevertheless wanted wilderness accessible to "the ordinary guy." This was a particularly appealing position to the Forest Service during the New Deal era, which convinced Marshall to head a new outdoors and recreation office. Through this office, the Forest Service hoped to contrast itself as a "blue collar" alternative to the Park Service.[6]

Despite his agency role, Marshall remained a critic of the Forest Service's pro-development positions as well as the Park Service's recreation-oriented policies which ended up destroying wilderness. The criticism of the Forest Service, spelled out in his best-known work, *The People's Forests,* was tied to Marshall's overall critique of private forestry and its role in injuring the workforce, the community, and the land itself. In this book, Marshall advocated public ownership of forest lands in order that "social welfare" be "substituted for private gain as the major objective for management." To Marshall, that meant a new labor and rural economic development strategy and careful land use planning, more research and science, and safeguarding recreational values from "commercial exploitation." His concept of linking protectionist objectives within a social policy framework was, according to one reviewer from the *Nation* magazine, the best assurance for future generations that the forests could provide "a green retreat from whatever happen to be the insoluble problems of their age."[7]

This search for a green retreat, or a "green utopia," became a continuing passion for Marshall, both in his governmental activities and advocacy work. After his return to the Forest Service in 1937 following a stint with the Bureau of Indian Affairs, Marshall laid out this combined social and environmentalist vision. It included subsidizing transportation to the public forests and operating camps at nominal cost for low income people, changing Forest Service practices that discriminated against blacks, Jews, and other minorities, and acquiring more recreational forest land near urban centers. At the same time, he sought to designate wilderness as places "in which there shall be no roads or other provision for the motorized transportation, no commercial timber cutting, and no occupancy under special use permit for hotels, stores, resorts, summer homes, organization camps, hunting and fishing lodges or similar uses. . . ."[8]

Marshall also sought to integrate some of these ideas into the approach of the Wilderness Society, an organization he helped found and finance in its first years of operation. In 1937, Marshall enlisted his close friend Catherine Bauer, a leader in the regional planning movement, to explore the issues of wilderness, public access, and social policy. In a long letter to Marshall, Bauer noted that wilderness appreciation was seen as "snobbish" but that a great many people, even the majority, could enjoy the wilderness, given a chance to experience it. Bauer suggested that "factory workers, who experience our machine civilization in its rawest and most extreme form," could most benefit from wilderness and by doing so could broaden wilderness's political base.[9]

Though Bauer's suggestions reflected Marshall's own approach, they caused concern and consternation among other key figures in the Wilderness Society, especially its executive director Robert Sterling Yard. Yard worried that the New Deal forester might interest too many "radicals" like Bauer in wanting to influence wilderness policy. Yard and others were also concerned about the red-baiting of Marshall during the late 1930s. These attacks, led by members of the House Un-American Activities Committee, sought to tar Marshall through his nonwilderness activities and financial contributions. Wilderness Society leaders like Yard feared that Marshall's activities might reflect on the organization, and remained skeptical of his advocacy of a democratic wilderness policy. In response to the "democratic wilderness" concept, key Wilderness Society figure Olaus Murie would later write that "wilderness is for those who appreciate" and that if "the multitudes" were brought into the backcountry without really understanding its "subtle values," "there would be an insistent and effective demand for more and more facilities, and we would find ourselves losing our wilderness and having these areas reduced to the commonplace."[10]

Murie's pivotal essay was written a few months after Bob Marshall unexpectedly died in his sleep during an overnight train ride from Washington, D.C., to New York. In his will, Marshall divided his $1.5 million estate into three trusts: one for social advocacy including support for promoting "an economic system in the United States based upon the theory of production for use and not for profit"; a second to promote civil liberties; and a third, for "preservation of wilderness conditions in outdoor America, including, but not limited to, the preservation of areas embracing primitive conditions of transportation, vegetation, and fauna." This last trust came to be controlled by Wilderness Society officials whose approaches were more narrowly conceived (in terms of membership and constituency) and politically limiting (in terms of resource policy) than Marshall's own inclinations. Over time, Bob Marshall's life and ideas began to undergo reinterpretation. His love for wilderness, it was eventually claimed, had really been an exclusive concern. With his death, Bob Marshall, the "people's forester" whose life's mission had sought to link social justice and protected wilderness, would become an uncertain historical figure representing environmentalism's divide between movements, constituencies, and ideas.[11]

Exploring the Dangerous Trades

At the other edge of this historical environmental divide stand the urban and industrial reformers, including advocates such as Alice Hamilton who sought to situate the concern about environmental hazards in the context of urban and industrial life. A compassionate advocate yet cautious and careful researcher, Alice Hamilton can be considered this country's first major urban/industrial environmentalist. Born in 1869 in New York City and raised in Fort Wayne, Indiana, Hamilton decided to study medicine, one of the few disciplines available to this first generation of women able to enter the universities and embark on a professional career. "I chose medicine," Hamilton would later write in her autobiography, ". . . because as a doctor I could go anywhere I pleased—to far-off lands or to city slums—and be quite sure that I could be of use anywhere."[12]

Even prior to entering medical school at the University of Michigan, Hamilton thought of combining her interest in medicine and science with humanitarian service and social reform. She found the ideal outlet when she moved into the Hull House settlement in Chicago while accepting a position as professor of pathology at the Woman's Medical School of Northwestern University. Hull House for Hamilton, as her biographer, Barbara Sicherman noted, "was an ideal place from which to observe the connections between environment and disease." There, she organized a well-baby clinic, looked into the cocaine traffic endemic in the neighborhood, took part in efforts to improve the quality of health care for the poor, and investigated a serious typhoid epidemic, among other activities. The typhoid epidemic was particularly instructive since Hamilton's investigation eventually helped reveal that a sewage outflow (an episode covered up by the Board of Health) bore direct relationship to the outbreak of the disease in specific neighborhoods. It became Hamilton's first experience with how the issues of health, the environment, and politics intersected.[13]

Hull House also became a staging ground for Hamilton's growing interest in the little understood and poorly treated area of industrial disease. At the settlement house, Hamilton heard countless stories about "industrial poisoning": carbon monoxide in the steel mills, pneumonia and rheumatism in the stockyards, "phossy jaw" from white phosphorous used in match factories. Though industrial medicine had become an accepted discipline in Europe, its detractors in the United States suggested, as Hamilton wryly noted, that "here was a subject tainted with Socialism or with feminine sentimentality for the poor."[14] For Hamilton, however, the exploration of industrial poisons joined her passion for reform with her desire to pursue a real world–based science.

In 1908, Hamilton's interest in the subject of industrial poisons was further stimulated by her appointment to the Illinois Commission on Occupational Diseases for whom she later became chief medical investigator. It was in this capacity that Hamilton began her famous investigation of the lead industries. Hamilton sought to identify which industries used lead and the kinds of health problems associated with them. In pursuing her research among the lead companies, Hamilton frequently encountered the belief (a kind of ideological rationale for lack of action), that worker unwillingness to "wash hands or scrub nails" represented the primary cause for occupational lead poisoning and that its occurrence was therefore "inevitable." Hamilton also quickly came to realize that lead hazards and health impacts were underreported by workers who concealed their illness out of fear of losing their jobs.[15]

Faced with lack of documentation and information, few resources, company resistance, and workers' fears, Hamilton's investigations, first with the Illinois Survey and subsequently with the Bureau of Labor within the Commerce Department, demonstrated an extraordinary resourcefulness in the pursuit of her "shoe-leather epidemiology." Her search for data required long hours and uneven information; a duty, she felt, "to the producer, not to the product." Hamilton recognized that her compassion as a woman for the victims of the dangerous trades gave her certain advantages in soliciting information. "It seemed natural and right that a woman should put the care of the producing workman ahead of the value of the thing he was producing," Hamilton remarked; "In a man it would have been [seen as] sentimentality or radicalism."[16]

During the next several decades, Hamilton became the premier investigator of occupational hazards in the United States. Her research (and advocacy) ranged over a number of industries and toxic substances. Her insights and investigative techniques broke new ground in the areas of worker and community health and anticipated the later interest in the occupational and environmental problems associated with such substances as heavy metals, solvents, and petroleum-based products. Forty years prior to the major environmental debates about the uses of science and technology and the nature of risk, Hamilton was already warning that workers were being used as "laboratory material" by industrial chemists who were introducing new products such as petrochemicals and petroleum distillates "about whose effect on human beings we know very little." This rush to introduce new industrial products such as solvents, she argued, represented new hazards in the workplace and the general environment.[17]

Hamilton was also convinced that "control" techniques, such as respirators or other protective devices, were far from adequate, anticipating similar debates within OSHA (the Occupational Safety and Health Administration) during the 1970s and 1980s. She focused on the impacts from low exposures from substances like lead, anticipating the "no acceptable threshold" argument about certain toxic substances. During debates over the decision by the automotive industry to introduce tetraethyl lead in gasoline during the 1920s, Hamilton became a key critic of the claim that the small amounts of lead involved were not significant. In a 1925 article, Hamilton wrote, "I am not one of those who believe that the use of this leaded gasoline can ever be made safe. No lead industry has ever, under the strictest control, lost all its dangers. Where there is lead, some case of lead poisoning sooner or later develops. . . ." The question of when environmental or public health factors needed to be considered was critical for Hamilton. "It makes me hope," Hamilton said of the tetraethyl lead controversy, "that the day is not far off when we shall take the next step and investigate a new danger in industry before it is put into use, before any fatal harm has been done to workmen . . . and the question will be treated as one belonging to the public health from the very outset, not after its importance has been demonstrated on the bodies of workmen."[18]

As her research began to receive attention, Hamilton's standing grew in an area that had largely failed to elicit interest in academic, industry, or government circles. In 1919, she was appointed assistant professor of industrial medicine at Harvard University following Harvard's decision to initiate a degree program in industrial hygiene. The appointment attracted attention since she was the first woman professor in any field in Harvard and the university did not admit women to its medical school. But Hamilton was chosen partly because there weren't any men interested in the position and the medical field and academic world in general still viewed occupational and environmental issues with little interest.[19]

By the 1920s, with the publication of her classic text, *Industrial Poisons in the United States,* her increasing prominence in issues of occupational and environmental health, and her participation in organizations like the Workers' Health Bureau, Alice Hamilton had become the country's most effective voice for exploring the environmental consequences of industrial activity. Her interest touched on issues of class, race, and gender in the workplace and the long-term hazards of the production system. A powerful environmental advocate in an era when the term had yet to be invented, Alice Hamilton must be seen as a core figure in the reconstruction of the urban and industrial roots of environmentalism.

Rachel Carson's Legacy

While Alice Hamilton provided a crucial link between work and environment, Rachel Carson, far more celebrated in the annals of environmentalism, anticipated, through her cautionary exploration of the world of synthetic pesticides,[20] a new language of environmental concern, linking urban and industrial issues with fears about the degradation of the natural environment. The publication of *Silent Spring* in 1962 and the ensuing controversy that made it an epochal event in the history of environmentalism can also be seen as helping launch a new era of environmental protest in which the idea of Nature under stress also began to be seen as a question of the quality of life.

Born in the small Western Pennsylvania town of Springdale on the Allegheny River, Rachel Carson as a young woman developed two related passions: nature writing and science research. While teaching biology courses in the evening, Carson took a job with the Bureau of U.S. Fisheries in the late 1930s, and began writing about undersea life. Though her first book, *Under the Sea Wind,* failed initially to generate interest, she continued her desire to write about the oceans and other science-based environmental topics. During the war, as editor of the Bureau of Fisheries publications, she became familiar with new research about the ocean environment which became the genesis of her first successful published book, *The Sea Around Us.*

The Sea Around Us combined Carson's knowledge of oceanography and marine biology, her concern for the harm that had been done to the sea and its life, and a readable style that made her work immediately accessible. The book was an extraordinary success, on the best-seller list for eighty-six weeks, more than two million copies sold, and translated into thirty-two languages. Carson was not as surprised as some about the book's public reception and its indication of a popular interest in science. In accepting the National Book Award for *The Sea Around Us,* Carson defined this interest in science as reflecting daily life concerns. "Many people have commented on the fact that a work of science should have a large popular sale," she commented. "But this notion, that 'science' is something that belongs in a separate compartment of its own, apart from everyday life, is one that I should like to challenge. We live in a scientific age; yet we assume that knowledge is the prerogative of only a small number of human beings, isolated and priestlike in their laboratories. This is not true. The materials of science are the materials of life itself. Science is part of the reality of living; it is the what, the how, and the why in everything in our experience. It is impossible to understand man without understanding his environment and the forces that have molded him physically and mentally."[21]

Carson's environmental pursuit of "the what, the how, and the why" in daily experience also made her a logical candidate to investigate the most striking petrochemical success story of the postwar era. By the late 1950s, when she began her work on *Silent Spring,* pesticides had already become a fixture in both agricultural production and other commercial uses, having already fully supplanted all other pest control methods. Their use was of such magnitude that significant episodes of harm to wildlife and immediate health impacts on farmworkers began to be recorded throughout the country, although scientific and technical publications either ignored or dismissed those concerns.[22]

For Carson, pesticides assumed significance as a defining environmental hazard of the post–World War II industrial order. In an interview prior to the publication of *Silent Spring,* Carson told the *Washington Post* that while pesticide impacts were not directly equivalent to nuclear fallout, which she characterized as the major environmental hazard of the period, the two were still "interrelated, combining to render our environment progressively less fit to live in."[23] In a period when the question of pollution was only just beginning to receive significant public attention, Carson's research suggested that public health and the environment, human and natural environments, were inseparable. Her insistence that "expertise" had to be democratically grounded, that pesticide impacts were a public not a technical issue decided in expert arenas often subject to industry influence, anticipated later debates about the absence of the public's role in determining risk and in making choices about hazardous technologies. For Carson, natural and human environments were under siege from a science and a technology that had "armed itself with the most modern and terrible weapons." This technology, she declared, was being turned "not just against the insects [but] against the earth" itself. Through such writing, Carson sought to not only present information but convince her audience about a new kind of danger, to create in effect a new environmental consciousness.[24]

While sharply criticized by the chemical industry and other pro-industry commentators—one favorable reviewer reported how Carson had been compared to "a priestess of nature, a bird, cat, or fish lover, and a devotee of some mystical cult having to do with the law of the Universe to which critics obviously consider themselves immune"[25]—Carson's thesis about reducing pesticide use was even deemed controversial among certain members of conservationist groups. In letters to the *Sierra Club Bulletin,* for example, several Club members employed by the ag-chem industry complained that a favorable review of *Silent Spring* in the publication did not bode well "for the future of the Sierra Club as a leading influential force in furthering objectives of conservation." The *Bulletin* editor, Bruce Kilgore, defended the review while acknowledging that "some members of the Club would disagree" since the book's subject matter was controversial within the Club.[26] Privately, Club executive director David Brower complained that some of the group's board members, including those tied to the chemical industry, remained skeptical of *Silent Spring.*[27]

The period following the release of the book was a difficult time for the shy and reserved nature writer who was also going through a debilitating bout with the cancer that caused her death eighteen months after *Silent Spring*'s publication. During this period, Carson continued to counter her critics by elaborating the key elements of her argument: that science and specialized technical knowledge had been divorced from any larger policy framework or public input; that "science" could be purchased and thus corrupted; that the rise of pesticides was indicative of "an era dominated by industry, in which the right to make money, at whatever cost to others, is seldom challenged"; and that the pesticide problem revealed how hazardous technologies could pollute both natural and human environments.[28] Carson in fact had hoped to refine her core concept of interrelated ecological systems in a book about ecology that she was never able to complete, though her thesis would ultimately be developed by others, both analysts and activists. Thus, while an earlier critic of the chemical industry, Alice Hamilton, laid the groundwork for discussing environmental themes in an urban-industrial age, Rachel Carson, with her evocative cry against the silencing of the "robins, catbirds, doves, jays, wrens, and scores of other bird voices," brought to the fore questions about the urban and industrial order that an evolving environmentalism would soon need to face.

Interest Group or Social Movement?

Bob Marshall, Alice Hamilton, Rachel Carson: each serve as signposts for a broader, more inclusive interpretation of the roots of environmentalism. This interpretation situates environmentalism as a complex of social movements that first appeared in response to the rapid urbanization, industrialization, and closing of the frontier that launched the Progressive Era in the 1890s. Pressures on human and natural environments can then be seen as connected, integral to the urban and industrial order. The social and technological changes brought about by the Depression and World War II further stimulated environmentalist views. And if we see Earth Day 1970 not simply as the beginning of a new movement but as the culmination of an era of protest and as prefiguring the different approaches within contemporary environmentalism, we can more fully explain the commonalities and differences of today's complex environmental claims.

Today, the possibilities for environmental change have become linked to the redefinition of the environmental movement and its capacity to transform society itself. Issues of policy, agenda, and constituency remain uncertain and in contention between and among environmental groups as well as between environmentalists and their opponents. Yet we are also in a period where environmental questions reflect crucial social outcomes and where new opportunities for change have emerged both within the environmental movement and throughout the society. These opportunities for change involve questions of technology and production, decision making and empowerment, social organization and cultural values. They reflect changes at the global level—most significantly the end of the Cold War. They also reflect local struggles and actions, as demonstrated by new constituencies and social claims which may influence the direction of environmentalism.

How then to define an environmental group? Does environmentalism include those social movements, democratic and populist insurgencies, who have sought a fundamental restructuring of the urban and industrial order? Should it be defined as a set of interest groups influencing policy to help rationalize that same urban and industrial order? Has it been a movement primarily about Nature or about Industry, about production or about consumption, about wilderness or about pollution, about natural environments or about human environments, or do such distinctions themselves indicate differing interpretations of what environmentalism ought to be?

In exploring the distinctive roots and contemporary forms of environmentalism, I have sought to analyze it as both interest group and social movement. The origins of environmentalism derive in part from the development of a set of social claims about Nature and urban and industrial life. During the Progressive Era, one set of groups came to be associated with the rise of resource agencies and environmental bureaucracies, adopting a perspective related to the uses of science and expertise for a more efficient and rational management of the natural environment and urban and industrial order. Similarly, concerns about protecting wilderness came to be linked to the activities of specific constituencies who saw themselves as "consumption users of wildlife" (as Thomas Kimball of the National Wildlife Federation described members of his organization) or consumers of other environmental amenities. At the same time, there were groups who saw themselves fighting for the public interest by challenging powerful special interests such as the timber companies and mining interests in their attempt to exploit the environment for private gain.[29]

These Progressive Era conservationist and protectionist groups were similar to and contrasted with urban and industrial groups of the same period who focused on the environmental conditions of daily life. Settlement House workers, regional planning advocates, and occupational health activists all shared a faith in scientific analysis and the power of rational judgment pervasive in that era. But these groups also sought to empower powerless constituencies and organize for the minimal conditions of well-being and community identity in a harsh urban and industrial age.

With the significant social and technological changes of World War II and the postwar era, environmentalism became increasingly identified with changes in the patterns of consumption associated with an expanding urban and industrial order. Scenic resources came to be valued as recreational resources, although the mobilizations in defense of wilderness of the post–World War II era, such as the campaigns against water development projects at Dinosaur National Monument in the 1950s and the Grand Canyon in the 1960s, tentatively sought to define a public interest value in protecting Nature for its own sake. The changes in consumption also helped to shape the rebellious politics of the 1960s activists with their concerns about "quality of life" and the hazards of technology. Consumption thus became at once an economic activity, with a defined set of interests associated with it, and a social condition giving rise to the new social movements of the post–World War II period.

By Earth Day 1970, some confusion about the new environmentalism had set in. This reflected the uncertain origins of the new environmentalists, which included 1960s activism, recreational/leisure time–oriented protectionist politics of the post–World War II period, and the resource management emphasis of several of the old line conservationist groups. Even the trend toward professionalization that so quickly prevailed among environmental groups during the 1970s nevertheless incorporated both adversarial and system management perspectives on how to accomplish environmental change. Environmentalists were activists and lobbyists, system opponents and system managers. But by the late 1970s, most of the professional groups had become thoroughly linked to the environmental policy system designed to manage and control rather than reduce or restructure the sources of pollution and other environmental concerns. Though at times contesting how policies were implemented, the ties that developed between professional or mainstream groups and the environmental policy system itself made mainstream environmentalism especially vulnerable to the charge of interest group politics.

This interest group association most directly separates the mainstream groups from alternative environmental groups, grassroots networks and community-based groups as well as direct action or more ideologically oriented "green" organizations. The use of the term "alternative" for environmental groups suggests different scenarios for these organizations as well as different historical antecedents. On the one hand, "alternative" implies "outside the mainstream," different from the way that environmentalism has come to be defined, especially in terms of its interest group status. It is in this context that many activists within alternative groups, such as anti-toxics organizers, don't like to call themselves "environmentalists."

At the same time, "alternative" approaches are noteworthy for their critical view of the existing urban and industrial order, similar to ways that the new social movements of the 1960s defined their activities or the urban and industrial reformers of the Progressive Era defined themselves. Such a view is "critical" in the sense that environmental problems are seen as rooted in the structures of production and consumption in the society. Such a viewpoint also extends to the critique of mainstream/"interest group" forms of environmental organization and activity. The term "alternative" in this context suggests a social change movement, as well as efforts to construct a prefigurative movement seeking new forms of community and technology. As opposed to the mainstream environmentalist association with "interest group," alternative groups can be seen as inheriting the mantle of the new social movements or at least the tradition of social claims about equity, empowerment, and daily life concerns.

Unlike the mainstream groups, the alternative groups seek to more directly address questions of gender, race, and class. Many, though not all, of the alternative groups have established a "feminist" perspective that is both more interactive and democratic. By pursuing a message of environmental justice and equity, some alternative groups, especially those in the anti-toxics area, have necessarily become focused on issues of discrimination and racism embedded in the toxics producing economy. And with the tentative efforts to link workplace and community environmental concerns, some alternative groups have begun to shift the definition of environmentalism away from the exclusive focus on consumption to the sphere of work and production. By elevating issues of work and production, the dynamics of race and gender, questions of community and empowerment, a reconstituted environmentalism has the capacity to establish a common ground between and among constituencies and issues, bridging a new politics of social and environmental change.

Who then will be able to speak for environmentalism in the 1990s and beyond? Will it be the mainstream groups with their big budgets, large staff, and interest group recognition? Can alternative groups, many of whom reject the term "environmentalist," lay claim to a tradition yet to be considered environmental? Can mainstream and alternative groups find a common language, a shared history, a common conceptual and organizational home?

The figures of Bob Marshall, Alice Hamilton, and Rachel Carson provide a clue. These compassionate, methodical, bitterly criticized figures, accused of being romantics and sentimentalists, biased researchers and pseudo-scientists, opened up new ways of understanding what it meant to be concerned about human and natural environments. They were figures who transcended the limited discourse of their era, forcing their contemporaries to realize that much more was at stake than one damaged forest or one industrial poison or one dying bird. Their language was transformative, their environmentalism expressed in both daily life and ecological dimensions.

To understand these environmental figures as connected rather than disparate parts of a tradition helps answer the question posed by Dana Alston at the People of Color Summit discussed at the beginning of this essay. How should environmentalism be defined? Alston asked. Should we keep to the narrow definitions that have provided environmental legitimacy for some groups and ideas to the exclusion of others? To learn the lessons of Bob Marshall, Alice Hamilton, and Rachel Carson and how they are linked in their concern for the world we live in helps begin that process of redefining and reconstituting environmentalism in the less narrow terms that Alston urged. It involves a redefinition that leads toward an environmentalism that is democratic and inclusive, an environmentalism of equity and social justice, an environmentalism of linked natural and human environments, an environmentalism of transformation. The complex and continuing history of this movement points the way toward these new possibilities of change.

URBAN FIELDS

The intersection between environmental and urban history has only recently been platted, or so some have argued—an argument whose source draws upon the field's historic emphasis on wilderness, a natural terrain generally thought to be the antithesis of the human-created cityscape. This sense of incompatibility represents a venerable cultural tension that has deflected our attention from acknowledging the degree to which we seek to be at home with both landscapes. The distinction between environmental and urban historiographies, in short, has been overdone.

The distinction, in any event, first was challenged in the spring 1979 issue of *Environmental Review,* which was devoted to urban issues. Since then, historians such as Joel Tarr and Martin Melosi have excavated the rich veins that bind together these two subdisciplines. The mining metaphor is particularly apt, given the subject of Tarr's "Searching for a 'Sink' for an Industrial Waste," a probe of the process whereby the Pennsylvania coke industry disposed of its toxic byproducts in the late nineteenth and early twentieth centuries. The industry initially dumped wastes into streams, befouling regional water supplies, and then later "fixed" this by venting the noxious wastes into the atmosphere. Each act damaged the health of both the human population and the environment.

Less visible and less quantifiable forms of pollution were also products of the Industrial Revolution, and chief among these was noise. An invention of the late nineteenth century, noise was not just an oppressive sound, but an unregulated din urbanites hoped to quiet through legal challenges or political legislation. In tracking these regulatory activities through the early twentieth century, Raymond W. Smilor observes that the enactment of public ordinances against industrial clamor, though often ineffective, nonetheless established an essential precedent: "environmental regulation was necessary and possible."

Yet rarely did such regulation protect the working class, the poor and disenfranchised, who continued to bear a disproportionate share of the various environmental burdens industrialism generated. That this remains true, Martin Melosi observes in "Equity, Eco-racism, and Environmental History," appears to be one source for the rise of a more "people-centered" environmental justice movement which is "shifting attention to urban blight, public health, and urban living conditions," as well as questioning "the demands for economic growth at the expense of human welfare." Predictably, as the language of environmental politics has begun to change it has already altered the scholarly agenda, a significant by-product of this blend of urban and environmental history.

Searching for a "Sink" for an Industrial Waste

JOEL A. TARR

The creation of wastes from any process, be it a natural, consumer, or production process, requires location of a place of deposit or a "sink" for disposal. Much of the history of industrial waste disposal, as well as the disposal of wastes from other sources such as an urban population, involves the search for a "sink" in which wastes could be disposed of in the cheapest and most convenient manner possible.[1] Often, however, such sinks proved only temporary, and substances placed in them created severe pollution, interacted with other substances to produce serious nuisance and health consequences, or leaked out or migrated into other media. Society reacted against such pollution problems in various ways, depending upon the problem's severity, the existence of technological means to reduce or control the pollution, the prevailing legal and legislative norms relating to the environment, and the existence and availability of other possible sinks. Oftentimes action or policy intended to remove a polluting waste from one environmental sink resulted in its deposition in another sink.

One major industrial process, the making of fuel to be utilized in the smelting of iron, produced a series of pollution problems and environmental damages. In the United States iron makers first used charcoal as a metallurgical fuel in their furnaces but began switching to mineral fuels in the 1830s and 1840s. Anthracite coal was used initially, but in the post–Civil War decades iron and steel manufacturers increasingly replaced anthracite with coke—the solid residue of almost pure carbon left when heat is used to drive out the volatile matter from bituminous coal. By 1911, coke fueled the smelting of 98 percent of the nation's pig iron. From about the 1850s through World War I, the technology used to produce coke was the beehive oven, but after that date there was a rapid transition to the by-product coke oven, the technology in use today.

After considering the environmental damages of the iron-making fuels in use before the substitution of coke, the article will focus on the different coke-making processes and especially their effects in the Pittsburgh region. It will examine how the transition from the beehive to the by-product oven altered the location of much of the nation's coke-making facilities from rural to urban sites and shifted the pollution burden from the air to the water and then back to the air. In this context, it will explore the various pollution control strategies followed by concerned groups at a time when industrial waste disposal was of limited public concern.

Iron-Making Fuels Before Coke

Mineral fuel, in either the form of anthracite or bituminous coal, did not become important in the making of iron and steel in the United States until the mid-nineteenth century. Before this time, the fuel most commonly used was charcoal, an ideal furnace fuel because it was relatively free from sulphur or phosphorus impurities and because its ash furnished part of the flux required to smelt the ore. Charcoal manufacture usually took place on what was called an iron plantation, a large forested tract of land that also contained an iron furnace, a charcoal "pit" or "hearth," casting beds, and workers' housing. On the plantation, bundles of cord wood in six- to ten-foot lengths were stacked in a cone with a base of about twenty-five feet in diameter, covered with damp leaves and turf, and burned for between three and ten days. No attempt was made to condense any of the by-product wood chemicals vented from the stack during the charcoaling process.[2]

Charcoal production had major environmental effects both because of the amount of timber consumed and because it involved the slow burning of wood with no emissions control. Geographer Michael Williams estimates that it took 150 acres of woodland to produce 1,000 tons of pig iron, the annual output of an average iron furnace. Plantations ranged in size from the 3,000 acres of woodland usually required for a profitable iron plantation to over 10,000 acres.[3] In 1862, a year of low production, 25,000 acres were cleared, while in the peak year of charcoal-iron production, 1890, 94,000 acres (147 square miles) were cleared. While timber production for iron production was only 1.3 percent of the land cleared for agriculture, it was heavily concentrated in certain states such as Ohio, Pennsylvania, and West Virginia.[4]

In addition to forest depletion, the charing process produced severe local air pollution effects. While the timber was being chared, the smoking piles of wood, covered with wet leaves, gave off a dark, heavy smoke with a disagreeable odor that gathered over the stream valleys where the operations of the iron plantations were often located. Many Pennsylvania boroughs passed ordinances prohibiting charcoal manufacture within town limits. The iron furnaces themselves were also heavy air polluters, as the opening at the top of the furnace permitted carbon monoxide, heat, and smoke to escape and created great clouds of smoke and fumes.[5]

While technological improvements that improved productivity were gradually made in both charcoal-making and iron furnaces during the course of the nineteenth century, growing demand for iron depleted conveniently located timber supplies.[6] Beginning in the 1830s, however, the pressure on fuel supply was relieved by the increasing substitution of anthracite coal for charcoal. Anthracite possessed almost pure carbon content and burned with a small blue flame that produced intense heat and little smoke. A ton of anthracite was equal in heat energy to 200 bushels of charcoal, giving it a strong cost advantage. The primary deposits of anthracite were in the rugged, mountainous areas of northeastern Pennsylvania making access difficult, but in the 1830s the construction of three major coal canals sharply reduced the cost of transportation and permitted large-scale production.[7] Iron makers had seldom used anthracite in their blast furnaces before the 1830s because the fuel was difficult to ignite. In the middle of that decade, however, an American and a Welshman both discovered that the "hot blast," an innovation developed in Great Britain in 1828 for use in charcoal furnaces, facilitated the use of anthracite as a fuel. Anthracite had greater strength than charcoal and could be burned in larger furnaces with higher thermal efficiency. Furnaces utilizing the mineral fuel sprang up along the established coal trade routes close to urban areas, with the Lehigh Valley of Pennsylvania becoming the center of the anthracite iron industry. By 1853 there were 121 anthracite blast furnaces in eastern Pennsylvania, the nation's center of iron manufacture.[8]

From the 1850s through the 1870s, anthracite served as the dominant fuel in the making of iron.[9] The shift to anthracite from charcoal shifted the environmental burden of providing fuel for iron-making from the forests to the coal-bearing lands of eastern Pennsylvania.[10] But while anthracite retained its position as a desirable fuel in manufacturing and in domestic heating until the middle of the twentieth century, its importance in the iron-making was relatively short-lived. In 1852 the railroad crossed the Appalachian Mountains, opening the rich bituminous coal fields of Western Pennsylvania to extensive exploitation and development. In the process, some of the most fertile valleys in Western Pennsylvania were exposed to widespread environmental damages from the mining of coal, the making of coke, and the smelting of iron.

Bituminous Coal and Coke

The bituminous coal located in the Connellsville fields of Western Pennsylvania (part of the Pittsburgh seam) was the world's richest coking coal—it was a critical factor making the Pittsburgh region the nation's iron and steel center in the late nineteenth century. "Coking," noted John Fulton in 1895, "is the art of preparing from bituminous or other coal a fuel adapted for metallurgical and other special purposes."[11] It involves using heat to expel volatile matter such as water, sulphur, and hydrocarbons from the coal, leaving a solid residue consisting primarily of fixed carbon, ash, and the nonvolatilizable sulphur. The best coking coal was a coal with few impurities such as sulphur or phosphorus, a sufficient proportion of volatile or gaseous matter to supply the necessary heat in coking, and one that would produce a coke with a porous structure and strong physical strength.[12]

Coke is low in cost compared to anthracite and ideal in structure for blast furnace use. It possesses a porous structure that permits it to be burned at a rapid rate and also ample strength at high temperature to carry the burden of the blast furnace charge. These advantages of cost and quality led to coke's rapid displacement of anthracite and of charcoal in the iron-making process. In 1854, for instance, charcoal iron composed about 47.5 percent of the nation's total pig iron production, anthracite pig iron 45 percent, and bituminous pig iron 7.5 percent. By 1880, however, the percent of pig iron made with bituminous coal and coke had risen to 45 percent while mixed anthracite and coke had dropped to 42 percent and charcoal to only 13 percent. In 1911, bituminous coal and coke provided the fuel for 98 percent of the pig iron manufactured in the nation and charcoal had practically disappeared from the picture.[13]

Coke was initially produced following a technique borrowed from that used in charcoal preparation: bituminous coal would be placed in piles or rows about fourteen feet wide on level ground surfaced with coal dust; wood was interspersed to ignite the mass. In the most effective process, the coal was burned slowly in a moist, smoldering heat, driving off the sulphur and the hydrocarbon gases and leaving a silvery white coke high in carbon. When the burning of the gaseous matter had ceased, the heap was smothered with a coating of dust or duff, followed by application of a small quantity of water. No attempt was made to capture any of the escaping gases. The time necessary for coking a heap was usually between 5 and 8 days, with a yield of approximately 59 percent of coke.[14]

Most of the coke produced in the nation from about 1850 to 1920 was manufactured by the so-called beehive coke oven. These ovens were first constructed in the 1850s in Western Pennsylvania and by the late nineteenth century most coke makers had adopted the technology. The oven consisted of a circular, dome-like chamber, lined with fire brick and with a tiled floor. Although there were improvements in oven design over the years, changes were relatively minor.[15] The ovens were first preheated by a wood and coal fire and then charged with coal from a car (a "Larry") running on tracks above the oven. The coke was leveled by hand with a scraper and the front opening bricked up with clay, with a two- or three-inch opening at the top. The coking process proceeded from the top downward, with the burning of the volatile by-products escaping from the coal producing the required heat. The coking process was complete when all the volatile material was burned. The brick-work was removed from the door, the coke was cooled by a water spray from a hand-held hose, and then removed from the oven by hand or by mechanical means.[16]

On average, it took about 1.5 tons of high quality bituminous coal "coked" over a period of forty-eight hours to produce a ton of coke, the composition of which was between 85 and 90 percent carbon.[17]

In 1855 there were 106 beehive ovens in the nation; by 1880 the total had increased to over 12,000 ovens in 186 plants; and by 1909 there were almost 104,000 ovens in 579 plants, the maximum number ever to exist. Initially almost all of the beehive ovens were located in the Connellsville Coke Region and as late as 1918, over half the country's ovens were still in this area.[18] Beehive ovens were normally arranged in banks of single or double rows, often built into a hillside in order to conserve heat. Normally, they were constructed in banks of about 300, considered the optimum scale for efficient operation. Numerous small operations were scattered over the countryside in coking areas. Some installations, however, were much larger, with 600 to 900 ovens. The Jones & Laughlin Iron and Steel Company beehive coking plant in the Hazelwood area of Pittsburgh was the world's largest, with 1,510 ovens. The J & L ovens were also unusual in that they were located close to the blast furnaces of an integrated mill, whereas most beehives were sited in proximity to coal deposits in order to reduce transportation costs.[19]

The uncontrolled emissions vented from the beehive ovens had a devastating impact on the nearby environment. The various hydrocarbons, fumes, and ash released by the coal distillation process killed and stunted trees and crops within the locale of the ovens and often left a layer of coal dust, ash, and particles on the surrounding fields. Writing in 1900, a Pennsylvania state botanist observed that "the most conspicuous feature in coke oven surroundings is the general wretchedness of everything of the nature of shrub or tree, either individual or collective." In those areas where there were large numbers of ovens scattered over the countryside, he wrote, "the district becomes almost continually one of a highly vitiated air, seldom without the overhanging clouds of smoke. . . ."[20] "Cloudy by day and fire by night" was the phrase often used to characterize the coke region.[21] In addition to the damage from the fumes, solid wastes from the coking process, such as coal wastes and ashes, were often dumped into nearby creeks or valleys, damming the flow and causing flooding and further damage and undoubtedly damaging stream ecology.[22] In several cases in the late nineteenth and early twentieth centuries, Pennsylvania courts held that beehive coke plants, like other manufacturing firms, were liable for damages caused by their operations.[23] In these cases, the courts made a clear differentiation between responsibility for injuries from coking operations and those resulting (as they held in the Sanderson cases of the 1880s) from the "natural and necessary result of the development of his own land by the owner."[24] Since the manufacture of coke was not the "natural and necessary use" of property, even on coal lands, individuals whose property was injured by such operations were entitled to compensation.[25] But while judges might award damages, they seldom would issue injunctions closing down a firm's operations. Rather they usually performed a rough type of cost-benefit analysis that Christine Rosen has called "private" or "social cost" balancing.[26] In such a situation, judges weighed the benefits the victims would receive from an abatement of pollution with the economic injury either the firm (in private balancing) or society (with social cost balancing) would suffer if the court required the polluter to reduce its pollution.[27] Invariably, in late-nineteenth-century Pennsylvania, the courts found the projected costs to society of an injunction to be too high even though they awarded damages for property damages to individuals. These awards, however, were usually relatively small since most coking operations were in rural areas, and the cost of damages to surrounding farmers were minor compared to the potential cost of pollution-control equipment such as tall stacks.

Because of the heavy pollution produced by beehive ovens, cities were reluctant to allow them to operate within their boundaries. In 1869, for instance, a Pittsburgh ordinance forbade the construction of coke ovens within the city limits and assessed a penalty of $100 per day for operating an oven. The ordinance, however, was probably not enforced, since the 1884 Sanborn Maps for Pittsburgh show well over a hundred coke ovens within the city boundaries. In 1892, the city revised its policy and enacted a new ordinance permitting coke oven construction if the ovens were supplied with "smoke control devices" approved by the chief of the Department of Public Safety.[28] The Jones & Laughlin Steel beehive oven plant in Hazelwood (constructed between 1899 and 1907), for instance, was fitted with tall stacks to burn and disperse the oven gases and smoke away from nearby residential areas.[29]

While these stacks undoubtedly reduced the emissions flow compared to the average beehive oven, the pollution was still substantial. In 1914, the Mellon Institute "Smoke Investigation of Pittsburgh" reported that coke ovens located in the Pittsburgh district (but not specifically naming Jones & Laughlin), smoked "almost constantly, although the smoke . . . is no denser than 60 percent black," a figure probably derived from the smoke measurement device known as the Ringlemann Chart.[30] Although by today's standards this would be considered very substantial air pollution, the level was not unusual for industrial emissions in Pittsburgh in the first decade of the twentieth century. The city passed several smoke control ordinances in this period, but they either exempted iron and steel mills or were unenforceable. Court records show no lawsuits that specifically referred to the urban coke operations.[31]

Another drawback of the beehive process was the waste of the valuable coal by-products vented during coking. For instance, in 1910, the maximum year for beehive production, 53 million tons of coal were processed into coke. One study suggests that this amount of coal had the potential of producing approximately 530 billion cubic feet of gas, 400 million gallons of coal tar, nearly 150 million gallons of light oils, and 600,000 tons of ammonium sulfate.[32] A few attempts were made to construct beehive ovens in which the by-products could be captured, but engineers had little success except in regard to waste heat.[33] Even if the by-products were recovered, aside from the coal gas, few markets existed in the United States in the late nineteenth century for other coal distillation by-products.[34]

In addition to wasting valuable by-products, the beehive oven possessed other disadvantages as a production technology. Most important was that it required very high quality bituminous coal in order to produce good coke and it was a difficult technology with which to achieve scale economies. The beehive's principal advantage over alternatives was its low cost and ease of construction and operation. As long as high quality bituminous supplies existed and markets for coal distillation by-products were limited, there was little incentive to alter the technology.

The primary alternative coke-making technology to the beehive oven at the turn of the century was the retort or by-product oven, developed by Belgian, French, and German engineers because of the absence of good coking coal in Europe. The by-product or retort oven was a narrow, slot oven constructed in batteries in which coking chambers alternated with heating chambers. Coal was charged through openings in the top of the oven and the coke pushed out by a power-driven ram at the end of the combustion process, to be quenched outside the oven. The gas evolving from the coal supplied the heat required for distillation and was also used for other purposes throughout the mill.[35]

The by-product oven had the advantage of capturing the volatile elements freed by the coking process. The yield of by-products was determined by the quality and quantity of coke desired.[36] The freed gases were collected by a system of pipes at the top of the oven and then cooled, resulting in condensation of tar and water vapor into a liquid of about 70 percent tar and 30 percent of ammonia in water. The ammonia underwent further separation at a still and was eventually transformed into various ammonium compounds. In addition to the ammonia compounds, other by-products from the coking process included benzene, toluene, naphtha, and xylene.[37] Because it salvaged useful materials that might otherwise be wasted, one engineer observed that the by-product oven was "a part of the movement for the conservation of our natural resources," thus placing this development in industrial technology as part of the Conservation Movement of the early twentieth century.[38]

The by-product oven had several other advantages over the beehive oven in addition to capturing the coal distillation products: it could produce high-grade coke using blends of various qualities of coal; it produced a higher coke yield per ton of coal (average of 70 percent compared to 64.5 percent); and, it provided for more rapid coking. But the technology also had several important disadvantages, most critical of which was its high capital and operating costs. Markets for by-products could, in principle, make up some of this differential but the United States was largely without a well-developed chemical industry capable of utilizing coal distillation products such as tar, ammonia, and benzene. Coal tar products produced in Germany dominated the American markets. These factors, plus the availability of large supplies of superior coking coal, retarded the technology's widespread adoption in this country.[39]

Coal, iron, and steel firms did construct several by-product ovens at the beginning of the century, transferring and adopting different versions of the European technology, but more rapid development did not occur until after 1914. One important factor stimulating change was the discovery by U.S. Steel that coke oven gas and tar could produce important fuel efficiencies in the integrated steel mills.[40] A war-induced cutoff of organic chemical supplies from Germany provided an especially strong impetus for rapid development. Faced by the loss of synthetic dyes, drugs, and solvents, as well as benzene, toluene, and phenols for the production of explosives, the United States had to rapidly create a chemical industry largely based on aromatic compounds.[41]

Many by-product plants were constructed to take advantage of the new markets. They were most frequently located at integrated steel mills, where the coke could be used in the blast furnaces and the gas and tar could be used as fuel to provide heat in the iron and steel works. In addition, a number of cities including St. Paul, St. Louis, Baltimore, and Jersey City, began obtaining their municipal gas supplies from by-product ovens.[42] By 1929, in what was a "revolutionary" industrial change, the new technology was supplying 75 percent of the coke manufactured in the United States.[43] Between 1909 and 1940, the number of beehive ovens in the country shrank from nearly 104,000 to about 15,000, located in 75 different plants and consuming 4,802,996 tons of coal. By the same year, the number of by-product ovens had increased to 12,734 in 89 plants, consuming 76,582,780 tons of coal.[44]

The By-Product Coke Oven and Drinking Water Supplies

By-product coke ovens were customarily located on the banks of rivers or lakes, usually in close proximity to the blast furnaces of an integrated steel mill. The water bodies provided them with the water necessary for their operations, with cheap barge transportation for the coal, and with a place to dispose of their liquid wastes. These locations were usually in or near urban areas, in contrast to the mostly rural locations of the beehive ovens. Although the by-product oven produced fewer air emissions than did the beehive oven, it still had a heavy pollution flow both to the air and to the water. Its wastewater stream, for instance, contained heavy concentrations of ammonia, cyanide, and phenolics, as well as various acid, base, and neutral hydrocarbons.[45] While all of these substances could damage stream life and the quality of water supplies, the phenol-containing effluent stream discharged by the ammonia stills had the most noticeable impact, creating severe taste and odor problems in drinking water.[46] The volumes discharged were extremely large; one chemical engineer estimated in 1923 that by-product coke plants discharged approximately 38,000,000 tons of still wastes in that year.[47]

While phenols would produce obnoxious tastes and odors in any water supply, the effect on drinking water was especially severe if the receiving waters were chlorinated. During the late nineteenth and early twentieth centuries, a number of American cities with centralized water systems attempted to protect their supplies from infectious wastes from sewage by adopting filtration technology or chlorinating their supplies. The first successful use of chlorination occurred in 1908, when Jersey City proved the effectiveness of the technique in protecting its water supply from pathogens. A number of other municipalities, also faced by the threat of sewage pollution of their water supplies, adopted the technology.[48] As chlorination spread to various industrialized areas, however, waterworks managers observed that the water supply often developed strong medicinal tastes and odors. This was initially blamed on the chlorine alone but in 1918 chemists from the Milwaukee Department of Health and the city waterworks determined that the tastes and odors were due to interaction between the chlorine and the coal-tar wastes.[49]

Cities in the Ohio River Basin were particularly subject to these problems because of their heavy use of chlorine to protect their water supplies (drawn from sewage-polluted rivers) and the concentration of by-product coke ovens in the region. In the mid-1920s the U.S. Public Health Service (USPHS) identified 25 cities in the Ohio River Valley where the interaction of chlorine with phenol wastes made the water almost undrinkable.[50] Other cities in different watersheds, such as Chicago, Cleveland, Detroit, Milwaukee, Rochester, and Troy, also experienced similar problems.[51] In this case, two new technologies had interacted chemically to produce harmful results for drinking water quality.

Aside from the taste and odor nuisances, public health officials worried about the health effects of the phenols in water supply. They had four principal concerns: they could injure the health of persons drinking the polluted water; the offensive tastes and odors might discourage individuals from drinking sufficient water for good health or perhaps cause them to drink biologically contaminated water; and, because phenols reacted with chlorine, waterworks managers might reduce the amount of chlorine treatment, thereby increasing the risk of exposure to infectious disease.[52] A typhoid outbreak in 1925 in Ironton, Ohio (18 cases, 3 deaths), attributed to the fact that "choloro-phenol" tastes in the municipal water supply were so offensive that local residents were forced to drink unprotected water, highlighted these fears.[53]

The focus on the health risks presented by the phenol wastes rather than upon their injuries to property represented a shift in concern in regard to pollution from coke manufacture. The emissions from beehive ovens were viewed as producing primarily private property damages that could be compensated for monetarily but the threat to water supply threatened the public good. Public authorities had several courses of action, the most available of which, given the absence of specifically applicable state statutory industrial waste regulations, was to use the courts to secure relief under the common law.[54] Here state authority in regard to the police power was a well-established basis for court action to protect the public health.

The most important court action occurred in McKeesport, Pennsylvania, an industrial city of approximately 50,000 people located at the junction of the Youghiogheny and Monongahela Rivers about fifteen miles above Pittsburgh. From 1881 until 1916, McKeesport obtained its water supply from the Youghiogheny River, but increased pollution by mine acid drainage and other industrial wastes caused the city to shift to the Monongahela River in the latter year. Although the Monongahela was also polluted by industrial wastes, its waters required less treatment than that from the Youghiogheny, significantly lowering the city's treatment costs.[55]

Approximately eighteen months after McKeesport had moved to the Monongahela for its supplies, the Carnegie Steel Corporation (subsidiary of the U.S. Steel Corporation) brought on line the world's largest by-product coke operation, located at Clairton, Pennsylvania. This plant was sited on the banks of the Monongahela River about two miles above the McKeesport water intake and was meant to provide coke for the U.S. Steel integrated steel mills in the Monongahela Valley. The plant initially had 768 ovens, each of which had a capacity of 13.2 tons of coal per charge and produced about nine tons of coke. In addition, the plant produced large quantities of coal by-products such as gas, tar, benzene, and ammonium sulphate.[56]

The Carnegie by-product works at Clairton discharged the liquid wastes from their plant through a pipe into Peters Creek, a stream which ran through their property and into the Monongahela River. The city of McKeesport charged that in the river the wastes "permeated" the water and caused "foul" odors and "offensive and nauseating taste" in its water supply. In September of 1918, McKeesport sued in the county courts for an injunction to prevent the Carnegie Steel Corporation from discharging these polluting wastes, charging that the plant's discharges rendered its water "unwholesome, unpalatable and unfit for drinking purposes and domestic uses, as well as injurious to health."[57] The courts, following the well-established precedent that the police power could be used to protect the public health, granted the injunction, observing that the city's "duty to supply its inhabitants reasonably pure and palatable water cannot be questioned; nor can the right of its inhabitants to such quality be injuriously affected by foul and noxious fluids." If it wanted to continue operations, Carnegie Steel would have to find a different means to dispose of the wastes from its by-product coke plant.[58]

The Clairton decision reflected the fact that, by 1918, legal precedent existed under the police power to prevent the pollution of water supplies by industrial wastes if they clearly injured the public health. In the case of by-product coke oven wastes, cause and effect were usually clearly definable and the courts were willing to act. In regard to other industrial wastes, however, cause and effect were less obvious and therefore the courts and municipal and state health authorities were much more reluctant to use the police power to take action.[59]

Another strategy to deal with industrial pollution was to push for new legislation to enlarge the power of State Health Departments over industrial wastes. In most states control over water pollution originally rested in the Departments of Health because of its relation to infectious disease and, aside from Massachusetts, little attention was paid to industrial pollution in the first decades of the twentieth century. In the 1920s, however, such pollution increasingly became a matter of concern, and several states enacted legislation to attempt to regulate its discharge.[60] In Pennsylvania, for instance, the legislature established a Sanitary Water Board in 1923 to determine the proper use of streams in the Commonwealth and to control pollution.[61] In 1924, the Deputy State Attorney General ruled that the Sanitary Water Board had the power to prevent the pollution of public water supplies by industrial wastes (such as phenols) that could cause the public to avoid drinking the water because of tastes and odors.[62] In Ohio, in 1925, the legislature approved a bill that gave the State Health Department the authority to zone sources of water supplies and to require new industries, or industries developing new processes, to file plans for satisfactory treatment plants. The State Director of Health called it, "the most advanced piece of public health legislation bearing on the subject of stream pollution prevention in the country."[63]

In both Pennsylvania and Ohio, the directors of the relevant administrative bodies argued that cooperation rather than legal action was the proper strategy to follow in regard to industrial pollution, an approach to regulation that was typical of the 1920s.[64] Underlying this strategy in the case of industrial pollution was a belief that "no existing satisfactory and economically possible means of treatment existed for the wastes of many important industries," and that legal action could drive industry out of the state rather than produce environmental improvement.[65] State authorities believed that firms themselves wished to find a means to deal with pollution and would be willing to cooperate in seeking means of remedial action.

Pennsylvania pioneered in the area of cooperative agreements and during the 1920s made compacts with industries such as pulp and paper and tanning to achieve voluntary pollution reduction. This strategy was also used in regard to phenol pollution, and by 1928 the Sanitary Water Board had reached agreements with all Pennsylvania by-product coke operators to prevent taste and odor-producing substances from reaching state waters.[66] In Ohio, similar agreements between the state and industry were reached during the decade.[67] In spite of these agreements, state authorities worried about the limitations on their ability to secure action by firms located on interstate streams such as the Ohio River. Without such agreements, it would be extremely difficult to secure voluntary compliance by intrastate firms because of their fear of being at a disadvantage in competitive markets.[68]

In a 1923 letter to Surgeon General Homer S. Cummings, Dr. John E. Monger, Director of the Ohio Board of Health, argued that the problem of interstate pollution required federal leadership since "it is apparent that any particular State is powerless to secure a remedy."[69] The federal government, however, had very limited responsibilities in regard to water pollution. In 1912 the U.S. Congress had assigned the Public Health Service (USPHS) the task of "investigating" the relationship between health and water pollution. The USPHS was concerned with industrial wastes primarily because of their relation to the quality of drinking water supplies. Beginning in 1913, it launched a series of stream pollution investigations centering on the Ohio River that were directed from what later became its Center for Pollution Studies in Cincinnati.[70] In February, 1923, after having received considerable complaints about phenol pollution of water supplies, the USPHS launched a national survey of the problem.[71]

Dr. Monger of the Ohio State Board of Health urged the USPHS to take a leading role in addressing the interstate problem, but the Surgeon General, Homer S. Cummings, was extremely cautious about overstepping his authority.[72] It was only after considerable urging that he agreed to take the minimal step of sponsoring a national conference on the phenol question. This conference, held in Washington on May 18, 1923, was attended by representatives from fifteen State Health Departments as well as federal agencies such as the USPHS, the Bureau of Mines, and the Bureau of Standards. Although the federal role remained limited, the meeting served important organizational and informational purposes.[73]

A second national conference, also called by the Surgeon General, was held in Washington in January, 1924, and led to the formal organization of the Ohio River states. Three months later, in April, 1924, health department officials from the Ohio River states met in Pittsburgh with representatives of the basin's by-product coke firms to see if they could reach agreement to control phenol wastes in the Ohio River. The tone of the meeting was cooperative rather than confrontational. By this time, many of the firms had already signed agreements with the Ohio and Pennsylvania Departments of Health to control their phenol wastes, and now the remainder agreed to take similar action.[74]

At the Pittsburgh meeting the Health Department representatives from Ohio, Pennsylvania and West Virginia also signed the Ohio River Interstate Stream Conservation Agreement. In this compact, the signatories agreed upon uniform policy in regard to the protection of water supplies from phenols and other tarry acid wastes. The Chief Engineers of the three Departments of Health constituted a "Board of Public Health Engineers of the Ohio River Basin," with the responsibility for protecting the watershed from industrial pollution.[75] The prominent role of the engineers rather than the public health physicians (Directors of State Public Health Departments were required to be physicians in almost all states) in this case reflects both the rise to prominence of engineers in public health activities and the extent to which industrial pollution was viewed as a technical problem.[76]

By 1929, 17 of the 19 Basin firms had installed phenol elimination devices, sharply reducing the most severe taste and odor problems. The procedure utilized by fourteen of the plants having 88 percent of the nation's coke-making capacity was to use the phenolic wastewater to quench the glowing coke, vaporizing the wastewater into the atmosphere rather than disposing of it in waterways.[77] The Carnegie Steel Corporation had adopted quenching at its Clairton works in 1918 after the McKeesport injunction, and it was to become the model for most of the other by-product plants, actively pushed by Judge E. H. Gary, Chairman of the Board of Directors of the U.S. Steel Corporation. Having themselves installed a quenching system, U.S. Steel was anxious to persuade its competitors to adopt a technology at least as costly as the one they were utilizing. After a visit to the

Clairton works to examine their quenching operation, Monger wrote to coke plant managers urging them to adopt the technique. "We trust," he said in a thinly veiled warning, "that voluntary action will be taken by various coke producing companies in Ohio without formal order by this department."[78]

Quenching with the phenolic wastewater was an expensive environmental control technology, especially for companies not used to expenditures on pollution control. The capital costs of installing such a system at one plant of the Youngstown Sheet and Tube Company with a 5,000-ton daily capacity in the mid-1920s was $250,000, with much higher costs at larger plants. The companies disliked quenching, not only because of its capital costs, but also because it greatly increased the rate of equipment and structural corrosion. U.S. Steel's maintenance costs at its Clairton works (30,000-ton capacity) in the mid-1920s were over $60,000 per year.[79] The phenolic wastewater used in the quenching system also imparted an odor to the coke, limiting its use for home heating.[80] The coke manufacturers, however, had little choice, since alternative approaches such as phenol recovery or chemical treatment were either uncertain or prohibitively expensive.[81] In the absence of other more cost-effective recovery methods, and under pressure from the threat of lawsuits, the companies voluntarily (and reluctantly) adopted the quenching technique.

The importance of the threat of legal action in achieving "cooperation" from industry was illustrated in the case of phenol pollution in the Chicago area. Chicago, as well as other smaller cities in Illinois and Indiana, drew their water supply from Lake Michigan. In 1927, phenolic wastes from coke ovens in the Calumet region of Indiana, located on the Lake, so badly polluted the water supply that it was undrinkable for a week. As in the Ohio River Basin, conferences were held between industry representatives and the state health authorities. On June 13, 1928, representatives of the Health Departments of states bordering the Great Lakes (Illinois, Michigan, Wisconsin, Indiana, Minnesota, Ohio, Pennsylvania, and New York) signed the Great Lakes Water Covenant, an agreement similar to that in the Ohio River Basin in regard to reducing phenol pollution.

At the time of the signing of the agreement, the involved industries gave the health authorities "verbal assurances" that they would control their phenol wastes within a two-year period. In spite of the "assurances," however, no action was taken, and during the following year a number of episodes of tastes and odors in the water supply related to phenol wastes occurred. The Chicago Health authorities "realized that unless pressure was exerted by the Department of Health in an unmistakable way, the citizens of Chicago might be forced to drink unpalatable water for many years to come," and the Department threatened to obtain a court injunction shutting down the firms. Faced by the threat of legal action, the Indiana coke companies capitulated and installed quenching technology.[82]

Coke quenching with the phenol wastewater did not actually eliminate the waste stream but rather shifted the burden of the pollution from the water to the air. While engineers originally thought that quenching destroyed the pollutants in the wastewater, it soon became apparent that the volatile components, such as phenol, cyanide, and ammonia, were actually being steam distilled and discharged to the atmosphere. Some engineers believed this an undesirable solution. At the 1927 convention of the American Water Works Association, for instance, an engineer and chemist for the Rochester Gas & Electric Corporation, a firm itself having problems with phenol wastes from its manufactured gas retorts, commented that quenching was "one of the most objectionable things that could be inflicted on any community because we simply delay the time of getting them [the phenols] into the water supply."[83] But this was a minority view, and most engineers and public health figures accepted quenching as the only option that came close to meeting both environmental and economic constraints.[84] In the absence of air pollution statutes, and with only a very limited understanding of the effects of airborne industrial pollutants, the air became a preferable sink to the water.

This examination of the connection between the production of metallurgical fuels for the iron-making process, especially that of coke, and the creation of environmental damage and pollution, illuminates several aspects of the relationship between industrial technology, the environment, and public policy in the period before stringent local, state, or federal environmental legislation.

As technologies, both the beehive and the by-product coke ovens were major polluters, but their effects on the environment were substantially different. Beehive coke oven operators, as has been noted, made only very limited attempts to capture emissions, most of which went into the atmosphere, fouling the air and destroying nearby vegetation. Since these ovens were largely located in rural rather than urban areas (as were charcoal plantations), a somewhat smaller population was affected, but because the ovens were usually scattered over the landscape, they impacted a larger geographical area. Air quality, for instance, in the Connellsville coke area, with the largest concentration of beehive coke ovens in the world, was notoriously bad. A city such as Pittsburgh could either ban the ovens outright or attempt to impose pollution-control requirements, although enforcement was uncertain at best. The only recourse urban and rural dwellers often had was the common law—an option that was both expensive and cumbersome and brought limited damages.

By-product coke ovens were most frequently located near heavily populated metropolitan areas where the potential existed for pollution impacts on a larger population. However, in contrast with the beehive ovens, the by-product ovens were viewed as an improvement in regard to air emissions. The Mellon Smoke Investigation of Pittsburgh, the first comprehensive study of the impact of smoke pollution on a major American city, for instance, found it encouraging in 1914 that by-product coke ovens, "which give no [visible] smoke," would soon replace the city's "antiquated" beehive ovens.[85] Unanticipated, however, was the phenol waste stream's severe impact on chlorinated drinking water supplies. Thus, two new technologies, one of which had produced significant public health improvements, and the other which both improved productivity and reduced air emissions, had interacted to produce a health- and nuisance-related environmental problem.

The quenching technology utilized to keep the phenol wastewater out of waterways and thus to protect drinking water quality itself ultimately proved problematical from an environmental and health perspective. Quenching shifted the pollution burden from the water to the air, but it was not until the 1960s, at a time of growing environmental consciousness, that the technique received a serious challenge.[86] Until this decade, air pollution (because of uncertainties about health effects) clearly held a lower position on the environmental policy agenda than did water pollution.[87]

Finally, we come to the policy process itself. The Ohio River Basin agreement between coke producers and public health authorities that removed the phenolic wastes from water supplies was the result of a cooperative strategy (backed by a threat of court action) rather than a policy of command and control. Such an approach became the norm in Pennsylvania and many other industrial states in regard to industrial pollution for most of the period from the 1920s through the 1960s.[88] State government officials, as well as industrial managers, were concerned over possible federal inroads into their authority and uniform standard setting from a distance. In 1926, for instance, when Congress was considering national legislation controlling stream pollution, the Ohio River Basin Association passed a resolution opposing federal control.[89] The leadership of the USPHS, which had strong ties with the state health organizations, also held this position. Agreements such as the compact over phenols, said Dr. Wade H. Frost, Director of the Cincinnati Research Station of the Public Health Service, gave "infinitely more promise of success than the enactment of federal laws to regulate the pollution of interstate streams."[90] This skepticism about federal command and control regulation and a preference for interstate agencies and state standard setting remained the preferred line of attack on industrial pollution for Ohio River states through the 1970s.[91]

The case of waste disposal from coke manufacturing highlights the extent to which, in the past, because of economic and technological factors, we have often shifted our pollution burdens from one media and one sink to another rather than adopting a more holistic approach to the environment. Lack of knowledge and expensive control costs clearly limited choices at the time of the innovations but progress in pollution control was often slowed by company resistance to large expenditures for control technology. As the case illustrates, society attempted a range of options in attempting to cope with industrial pollution, including, in this example, local and county ordinances, state legislation, regional compacts, and eventual federal standards. The movement from one governmental level to another reflects the difficulties of regulating interstate pollution, public frustration over the slowness of state regulation and industry response, and a shift from concern over property damages to the environmental and health effects of pollution.

Today our society confronts the necessity of paying for past waste disposal practices that both affect environmental quality and pose potential health dangers. Many current environmental problems involve cross-media effects, such as groundwater pollution by runoff from hazardous waste dumps or sanitary landfills, that reflect past disposal choices. The history behooves us, therefore, to sharpen our sensitivity to the interconnected nature of the environment and the implications of these connections for environmental policy.

Personal Boundaries in the Urban Environment

The Legal Attack on Noise, 1865–1930

RAYMOND W. SMILOR

After the Civil War, Americans found themselves living in surroundings that were drastically different from their notions of what constituted an ideal way of life. Cities had always been a part of the American experience. However, the swift and unrelenting process of urbanization produced what amounted to an entirely new environment. The urban environment seemed hostile, frustrating, and bewildering. Adjusting to that new environment required a dramatic shift in perception. More importantly, it demanded a reevaluation of the individual's relationship to society. Living at the doorstep of a foundry or next to a stranger with widely different tastes and habits presented enormous practical problems for the individual. The same quarrel arose innumerable times—the issue of impinging rights. As a result, people needed to redefine their personal boundaries in city life.

In the legal attack on din, Americans demonstrated the great difficulties involved in adjusting to the urban environment. Noise was an evasive pollutant. But its very evasiveness allowed for the expression of deeper concerns than simply the desire for quiet. The urban environment called basic rights into question. The exploitation of natural resources and the fact of overcrowding collided with human needs. That collision led to a reassessment of individual rights. How much freedom could an individual exercise? How far could one person carry on an activity before he interfered with the rights of another? How much responsibility did a person have toward his community? Where did one person's rights end and another's begin? In the confines of urban life these were difficult and puzzling questions that gained urgency as congestion increased.

Noise represented an infringement of an individual's rights. One writer helplessly asked, "How soon shall we learn that one has no more right to throw noises than they have to throw stones into a house?"[1] A person had a right to his property, to his health, and to the comfort of his home. Racket violated these rights: "People dare not enter a man's house or peep into it, yet he has no way of preventing them from filling his house and his office with nerve-racking noise."[2]

Din especially threatened the right to privacy. For immigrants who came seeking the spaciousness of the farm and for native-born Americans who believed in the Daniel Boone ideal that if you could see your neighbor's chimney smoke then you were too close to him, the lack of privacy resulting from urbanization was a jolting experience. As the possibilities for solitude diminished, the desire for privacy increased. More than any other environmental pollutant, noise was an invasion of privacy. One could close his windows to keep out smoke and odors or turn off the tap if the water was foul; he had some control over these intrusions. But an individual had no control over din. E. L. Godkin declared that racket "invades the house like a troop of savages on a raid, and respects neither age nor sex."[3] Others found it hard "to think of noise as protective, as sheltering, not invading and not upsetting the processes of the mind."[4] No matter how quiet one attempted to be, there seemed no escape. As a practical, everyday result of noise, sleep and rest in one's own home were often impossible. Again, the issue of rights was at stake: "Conditions favorable to rest and sleep are fundamental rights of individuals everywhere, if there are any fundamental rights. They have a direct relation to sanity, happiness, and life."[5]

With their rights at stake, Americans turned to the law for solutions. In the common law, a man suffering from a serious noise disturbance had legal recourse through the courts in upholding his right of personal comfort and the undisturbed use of his property. By 1900, this alone was not enough to deal with the mounting urban noise problem. Responding to public pressure, cities began to pass ordinances to eliminate din. In so doing, they reflected a growing belief that a person was capable of offending an entire community by his noise disturbance. To preserve the public welfare, people began to accept, even demand, that law limit the individual citizen in pursuit of his own desires.

For Americans first confronting urban din, noise was simply disagreeable sound. While that was a convenient definition for the layman, it was almost useless in a court of law. The concept of "disagreeable" was relative to the judgment of the listener. One man's noise was another man's music. This subjective quality made noise more difficult to deal with than other environmental pollutants. One could see smoke or point to garbage or taste foul water, but how did one make a sound tangible? Determining exactly what noise was, and continues to be, is a perplexing legal question. Until the development of the audiometer and decibel in the 1920s, courts and municipalities had no scientifically accurate instruments to measure sound. Individual judges had to decide where to draw the fine line that separated noise from sound. Even with the audiometer and decibel, determining when sound becomes noise has not been easy because courts must take several factors into consideration. Decibel level alone is insufficient. Consequently, the law dealing with noise pollution developed on a case by case basis as the courts tried to provide relief for discomfort or damage resulting from din.

Individuals first attacked noise as a nuisance. A nuisance is "a wrong done to one by unlawfully disturbing him in the enjoyment of his property or in the exercise of a common right."[6] It may be anything that endangers life or health, offends the senses, or violates the laws of decency. The term encompasses almost all situations or activities that cause discomfort, dissatisfaction, or annoyance. It eludes exact definition because the amount of annoyance or inconvenience constituting a nuisance is largely a matter of degree.

At common law, noise was not a nuisance per se. But it had the potential of becoming a legal nuisance:

It is now well settled that *noise* alone, unaccompanied with smoke noxious vapors or noisome smells, may create a nuisance and be the subject of an action at law for damages, in equity for an injunction, or of an indictment as a public offense.[7]

In determining when noise created a nuisance, courts followed a haphazard and sometimes tortuous path. Noise constituted a nuisance if it materially interfered with the ordinary comfort of life or impaired the reasonable enjoyment of a person's house:

If unusual and disturbing noises are made, and particularly if they are regularly and persistently made, and if they are of a character to affect the comfort of a man's household, or the peace and health of his family, and to destroy the comfortable enjoyment of his home, a court of equity will stretch out its strong arm to prevent the continuance of such injurious acts.[8]

What exactly were "unusual and disturbing noises"? The long list of ifs indicated the difficulty of proving noise a nuisance. The test as to whether a noise constituted a nuisance manifested an inherent problem in deciding to file suit. The plaintiff had to prove actual pecuniary loss or actual physical discomfort to "the average person of ordinary sensibilities" in order to show that a specific noise was a nuisance.[9] Of course, in every case there was disagreement on who qualified as a "person of ordinary sensibilities." The defendant would argue that the plaintiff was a person of extraordinary sensibilities and therefore not qualified to judge the intensity of the sound. Still, there was some consensus on who did not qualify. Highly nervous and oversensitive people did not receive exceptional immunity or protection, nor did the sick and afflicted, those most in need of quiet surroundings.[10] The practical result of such a test for noise was that "No definite rule can be given that is applicable to every case, as each case must necessarily stand by itself, and be determined by a jury with reference to the circumstances peculiar to itself."[11] Several circumstances entered the picture. Before a complaining party could receive relief at law or in equity for a noise nuisance, courts first had to consider the character, volume, time, place, duration, and locality of the sound as well as the nature of the trade or use of the property and the resulting effects.[12]

In weighing these many factors, courts attempted to reconcile the conflicting interests of two property owners—one who thought that his ownership entitled him to use his property in whatever way he wished and the neighbor who believed that his ownership entitled him to enjoy his property without disturbance. In the process, courts displayed two primary concerns. On the one hand, they looked at the harm that the annoyance caused the plaintiff. On the other hand, they considered the reasonableness of the defendant's conduct and the harm that the discontinuance of the activity would cause him.[13]

However, these concerns did not exist in a vacuum. During the Industrial Revolution in the United States, the values of Americans dictated that production must continue. Environmental quality was secondary to economic gain whether by a business or an individual. This attitude continued to hamper the anti-noise campaign. Racket represented someone's work, and everyone hesitated to interfere with that right. "Money is something *real*," one critic emphasized, "And it is difficult for anyone to prove that he is losing more money by a given noise than someone else is making from it."[14] If the individual could not put his protest on an economic basis, his ingrained reverence for property rights and the work ethic impeded his actions:

> If we are to suppress much of the clang and clatter that we dislike, we shall certainly interfere with some people's efforts to make money. Considered from our present ethi-

cal point of view, this is a terrible thing to do unless the provocation is very great indeed.[15]

Courts tended to agree. Consequently, "trifling or occasional" noises that were dependent on the ordinary use of property or the pursuit of an ordinary trade did not constitute a nuisance. The law expected individuals to endure some discomforts for the protection of industry: ". . . people must yield somewhat of their rights . . . and some of their quiet and repose, so desirable in dwellings, must be given up, if they live in towns or in public places, that business may go on."[16] At the same time, courts tried to acknowledge the individual's right to enjoy his home. A person could not establish his trade in the vicinity of dwellings so as to seriously disturb the quiet and repose of those residing or doing business there. A lawful trade that produced noise became a nuisance only when it was so near to houses and at such unusual and unreasonable hours that it disturbed the other dwellings in the area.

Few of the complaints concerning industrial noises involved the large plants of nationally important heavy industries as was common in the smoke and gas cases. Chemical fumes and smoke annoyed people at greater distances than noise. Furthermore, larger plants operated in neighborhoods fully industrialized or purposely placed in comparative isolation where they gained a type of immunity from prosecution. More often, noise complaints involved enterprises of moderate size established within or near a residential section. Occasionally, residential building spread into a partially industrialized area. If an industry operated in an appropriate locality, that is, a mainly industrialized area, and functioned prudently during working hours, then residents in the area had to endure the noise that was inseparable from the enterprise. Although the householder could not expect the same peace and quiet as in a strictly residential district, he could still insist that the owner conduct the business in a reasonable manner with due regard to his rights as one who lived in a manufacturing district.[17]

While courts nodded to the individual, they expected citizens to put up with annoyance since "No one can move into a quarter given over to foundries and boiler shops and demand the quiet of a farm."[18] The American experience sanctioned not just the use but the exploitation of natural resources. Legal conditions further encouraged that exploitation. Courts hesitated to interfere with production. The *Boston Advertiser* declared, "The law is tender to a steam engine or a boilermaker. . . . the sum of the matter is that in the interest of trade or manufactures you may vibrate or deafen your neighbor with few restrictions."[19] A barometer of the intensity of the problem and the desire to redefine personal boundaries in city life was the multitude of suits that arose to stop racket.[20] Despite the difficulties in proving noise a nuisance, people attacked particular sounds in order to gain some relief. Because of the pounding and hammering, grounds for litigation were most conspicuous for noise produced in the fabrication of metals. Suits involved iron mills, machine shops, forges, and foundries as well as jewelers and silversmiths. Courts held that under certain circumstances noise from skating rinks, roller coasters, circuses, electric light companies, and steam engines constituted a nuisance. Loud and profane talk as well as boisterous conduct provided grounds for suit.

Courts generally agreed that noises that were not a nuisance in the daytime could be one at night. Factory whistles and sirens, while not nuisances per se, became nuisances when their sound interrupted sleep thus injuring health and affecting the comfort of one's home. For people who had to live near a factory, the whistle nuisance could cause "a marked depreciation of real estate and often a retardation of promising growth on account of this health destroying and intolerable noise."[21]

Barking dogs, bleating calves, crowing roosters, stamping and kicking horses frequently led to litigation. In close quarters, pets could be particularly annoying: "The noise of certain domesticated birds are as bad as the 'cruel clarions' and ought to be pursued by penal laws. Macaw and manslaughter, crime and canary bird, should go together in a code, and be tried by a jury of annoyance."[22]

The din arising from many activities was not a nuisance per se but could become one from the manner in which a person conducted it. The music of a brass band or hand organ over a long period of time, for example, constituted a nuisance. The question of what is music further complicated cases involving singing and musical instruments: "Justice is deaf, as well as blind, when it comes to distinguishing between the music which is a 'concourse of sweet sounds' and music which is not."[23] Again a noise nuisance became a question of degree. If one's singing or musical instrument was a willful and malicious disturbance or destroyed the comfort of another's residence, then it could be a nuisance. An apartment dweller noted the hazard of thin walls and the lack of soundproof ceilings: "If ever I am driven to suicide by noise, it will be after an hour of 'Silver Threads Among the Gold' upon an organ at one end of the block, together with 'Hear me! Norma!' upon an organ at the other end."[24]

Even church bells came under attack. In the close confines of urban life, pealing bells could be more a nuisance than a call to worship where their ringing was unreasonable and unusual. Bells no longer served the same purposes as they did in the countryside. Amid the divergent population of the city, they could be offensive and again brought into focus the need to arrive at a new set of personal boundaries:

> For why should a Quaker be wakened by a Roman Catholic bell; or a Presbyterian by an Episcopal bell or a Methodist by a Baptist bell? If church-bells could be so constructed that they would be guaranteed to waken only the members of the church in which they are hung, they could be tolerated, but so long as they continue to arouse believers in opposing faiths, our non-sectarian laws ought to be strong enough to silence them.[25]

Still, a certain amount of give and take had to exist among the residents of a city. And courts expected people to tolerate some of their neighbor's racket that "accompanies and is incident to the reasonable recreation of a crowded population."[26]

Given the various factors that entered into each case, it was impossible to arrive at a guiding principle to judge noise nuisance suits. However, there was one area in which a definitive ruling emerged. In *Richards v. Washington Terminal,* which remains the rule of law today, the United States Supreme Court held that property owners along a railroad right-of-way must bear without redress the noise, vibration, and smoke incident to the running of the trains.[27] Thus, where law sanctions the activity of the government or government-sponsored entity, and where the facility creating the noise is properly designed and operated, private property owners have no recourse in law even though the noise may cause a decline in the value of affected property.[28]

Court cases demonstrated that if a nuisance-noise injured or threatened with injury the rights of a private property owner, the owner could sue for damages or sue to enjoin commencement or maintenance of the nuisance. Private nuisance suits, however, were not an effective way of preventing urban noise pollution. Following the rule of live and let live, private citizens hesitated to bring suit against a noise-maker, especially if he was a neighbor. Indeed, hesitancy in filing a grievance was characteristic of the vast majority of Americans. Private litigation could also be expensive and time-consuming. Many people simply could not afford to follow the general law of nuisance through the courts even if they wished to seek legal redress. A more difficult problem was the elusiveness of city noise. Amid the everyday din, identifying the source of the racket was often no easy task. Most importantly, courts demurred to restrain noises that business, government, or government-sponsored entities created. While a private suit might solve one person's noise problem, it did little to abate the larger problem of urban noise pollution.[29]

Pressure for more effective control of urban din began to develop in the 1890s. Adjusting to a new environment required a new public policy that extended municipal powers to deal more adequately with noise pollution. Americans came to recognize that noise was not only a private annoyance, not just a menace to an individual's comfort and safety, but also a public nuisance and a threat to the whole community's well-being. As a result, they stressed their community rights in restricting noisy intruders. Where a reasonable number of people requested the removal of noises "the burden of proof should lie with the opponents; and the fact that the latter do not mind the noises objected to should not count too strongly for them."[30] In response to the growing desire for quieter surroundings, cities began to combat urban noise after the turn of the century by passing ordinances outlawing "excessive" or "unnecessary" noise.

Facing the same difficulty in defining noise, anti-din reformers left it to the courts to decide what was "excessive" or "unnecessary." They placed their faith in the power of law. As they came to regard racket as a public nuisance, reformers focused on "the crime of noise" and tirelessly sought legislation. They believed they could legislate the noise pollution problem away.

Anti-din advocates attacked noise on the local level for two reasons. At common law, the character, volume, time, place, duration, and locality entered into consideration in determining whether a particular noise constituted a nuisance. These factors varied from one community to the next. Therefore, an ordinance, or local law, was the proper and most effective form of legislation for a public authority to handle a local community problem. The community possessed the power to enact noise-abatement ordinances under the police power that the state delegated to it.[31]

More importantly, at the local level community action was most effective. Reformers could mobilize public opinion to support legislation because people responded to noise disturbance in their own neighborhoods. Citizens supported the legal attack on noise not as members of a special class or particular interest, but as listeners concerned about the safety and quality of their surroundings.

By 1913, cities across the country had adopted anti-noise ordinances. Setting boundaries on personal conduct, these local laws fell into five categories: occupational noises, noises by motor vehicles, animals and steam whistles, and zones of quiet.[32]

Occupational noises were those that individuals made in carrying out their jobs. Laws in this category limited the activity of peddlers, vendors, and hawkers as well as auctioneers and some store owners. They prevented them from making a noise to sell a product and specified the time and place for doing business. These ordinances in effect impeded workers from earning money in order to protect the welfare of the community as a whole. Boston declared that "No person hawking, peddling or selling . . . shall cry his wares to the disturbance of the peace and comfort of the inhabitants of the city." Little Rock outlawed noises "which are so harsh or discordant, or so prolonged or unnatural, or unusual in their use, time or place as to occasion physical discomfort to persons in the vicinity." San Francisco banned racket with "a tendency to frighten horses." Cincinnati disallowed the use of any noise-producing device for the purpose of advertising.

Peddlers maintained that these ordinances infringed on their right to make money and pursue their jobs. But courts upheld the laws. The city of Chicago enacted an ordinance that limited peddlers to certain areas of the city and barred them from advertising their wares on the streets by shouting. Fruit and vegetable venders sought to restrain the enforcement of the law arguing that it discriminated against them and deprived them of their right to engage in the business of peddling on the streets. However, the court sustained the constitutionality of the ordinance. Peddlers had no vested property right to make noise by advertising their goods by shouting, any more than they had a right to advertise by any other means, the result of which disturbed the peace and quite of a neighborhood.[33]

Peddlers in Chicago remained dissatisfied. In 1911, they staged a city-wide strike and then rioted in protest against the anti-noise ordinance. During the five-day strike, peddlers caused thousands of dollars of damage. They beat venders refusing to join their demonstration and engaged in open confrontation with the police. On the third day, "a flying squadron of wreckers" appeared, and police arrested 75 protesters "in a day of rioting and wild disorder such as has not been seen in Chicago since the garment workers' strike."[34] But the peddlers failed to repeal the law. Their actions turned public opinion against them as only one councilman came to their support. The others, including the mayor, reflected the wishes of the majority of the city's residents and insisted that the law stay in force. A judge's $200 fine on one of the leaders of the riot along with individual financial loss in continuing the strike deflated the protest. Peddlers acquiesced in theory if not completely in practice.[35]

Ordinances relating to noise from motor vehicles and animals imposed restrictions on an individual's personal property for the benefit of the entire community. Those concerning vehicles pertained to mufflers and horns. In drawing up New York City's ordinance requiring mufflers on vehicles with internal combustion engines, the Committee on Laws and Legislation reported that "much annoyance is caused by the owners and drivers . . . who habitually fail to use the muffler or silencer on their cars." The muffler "is almost invariably cut out by the mere thoughtlessness and disregard of the rights of others" resulting in the needless disturbance of the peace and quiet of the community. Officials in other cities concurred, and ordinances requiring mufflers were widespread.

Horns and other signaling devices also came under attack. A pedestrian described the motorist's code: "'I am coming: If you do not hear my Gabriel trombone I am afraid I shall run over you.'"[36] To combat the nuisance, cities such as Providence and Kansas City restricted horn blowing except as a warning signal.

Cities took action to regulate irritating pets, especially dogs and fowl. Detroit prohibited people from keeping dogs that "by loud or frequent or habitual barking, yelping, or howling, shall cause serious annoyance to the neighborhood." Spokane barred fowl within the city limits. Washington disallowed "any fowl, parrot, or bird which, by crowing, cackling, talking, or singing, or in any other manner, shall disturb the comfort and quiet of any neighborhood."

Local restrictions on industry took the form of ordinances curbing steam whistles, among the most grating of city noises because of their stridency. William Dean Howells observed that the shriek of factory whistles was "so offensive that it would seem like a crushing weight upon the head."[37] Hartford banned their "excessive and offensive noise." St. Louis declared it illegal for "any saw-mill or factory of any kind, or . . . any machine shop or foundry, or mill of any kind, to blow, or sound . . . for any purpose whatever any steam whistle within the city." Other cities forbade the use of steam whistles on railroads except for emergency warnings or limited the time for blowing whistles and sirens.

The most innovative legislation established zones of quiet—an entirely new concept in urban planning. The zones appeared first around hospitals and then around schools as reformers sought to protect those to whom noise would be most distressing—the sick needing rest and children trying to learn. Some cities eventually permitted temporary zones of quiet around individual residences where a person was dangerously ill. The zones in effect limited the personal freedom of some in order to preserve the health and well-being of others. They were protective shields within the urban environment. As doctors came to recognize the deleterious effect of noise on their patients, they gave their "cordial endorsement" to the concept so that "the ordinary and extraordinary street noises will be hushed and invalids and nervous individuals be relieved from their constant irritation."[38] Ordinances established zones from 250–500 feet around hospitals and schools and provided for the posting of signs reading, Notice—Zone of Quiet.

New York was the first city to set up zones in 1908. Others like Cleveland, Louisville, and Milwaukee quickly followed suit.

Violators of anti-noise ordinances were guilty of a misdemeanor. They faced fines from a few dollars to several hundred dollars, or imprisonment from a few days to six months, or a combination of the two.

While the profusion of anti-noise legislation came at the municipal level, there was one notable exception. Anti-din reformers achieved their greatest legislative success through the efforts of William Stiles Bennet, a lawyer and a member of the advisory board of the Society for the Suppression of Unnecessary Noise in New York City. As a United States Representative, Bennet pushed through Congress without objection the only piece of national legislation against noise during the period. The Bennet Act of 1907 regulated boat whistling in harbors across the country by "prohibiting useless and unnecessary whistling."[39] It was the first noise-abatement bill that Congress ever authorized. Anti-din advocates favored it in order to decrease the cacophony that rampant whistling had been causing in rivers and harbors. Marine interests approved a uniform policy for ships on waterways for reasons of safety. They supported the bill to eliminate the confusion from discordant signaling.[40]

Anti-noise reformers exemplified the progressive confidence in the exemplary power of law. Having a particular law against a specific nuisance seemed to insure its abatement. In amassing their legislative record, they gave little thought to the later task of enforcement. They were slow to learn that laws did not enforce themselves.

Local priorities restricted the enforcement of anti-noise ordinances. A city often had industries of such economic importance to some of its citizens that it would not pass or could not enforce laws that hampered business. Adding to this problem was the fact that a certain amount of noise was frequently inseparable from the ordinary completion of the work or necessary for changing shifts or insuring safety. The difficulty was in trying to decide when the noise became excessive. The *New York Times* saw this problem in relation to the Bennet Act:

> The difficulty of the case lies in the fact that the evil is not intrinsic, but is simply the abuse of a necessary power by men who must be permitted to exercise a good deal of discretion in using that power. Nobody thinks of asking that all or any steamboats shall give up their whistles altogether. . . . The city is too fond of its commerce to attempt any interference with its rights or even its needs.[41]

In the same regard, no one thought of asking an industry or a small business to give up all of its noise.

On a more personal level, individuals hesitated to invoke anti-din ordinances just as they delayed pursuing nuisance suits. No one doubted that racket annoyed millions of citizens. However, violations of existing regulations seemed "to drive no one to the point of making complaints and appealing to the law."[42] The element of periodicity often determined whether a noise was offensive. If the din was brief or occurred irregularly, the individual tolerated it. Only after it became a continual irritant would he venture a complaint. Because each person had a life within his community, he hesitated to grumble about noise since he wanted no one to complain about him. Finally, nearly everyone shifted sides on the issue. One hated clatter unless he made it himself.

Enforcement of anti-noise ordinances was difficult for a more practical reason. Most police forces were understaffed and overworked. Compared with other crimes, violations of anti-noise laws seemed of minor importance and therefore received little more than token attention. The cop on the beat disliked imposing the laws on people whom he knew. Consequently, anti-din ordinances were neglected unless a policeman or other municipal official took particular interest in the problem or until civic protest forced the police to act.[43] In Baltimore, a surge of complaints against roosters resulted in the death or banishment of one thousand of the birds. In 1914, the city boasted the first special anti-noise policeman on earth. His appointment evolved from the popular demand for the enforcement of noise abatement laws. The officer patrolled quiet zones, issued tickets to noise offenders, and confiscated noisy articles and pets.[44]

Public support was essential for successful enforcement. When reformers campaigned for laws against unnecessary racket, the public responded enthusiastically, and cities passed ordinances. Once the laws were on the books, however, reformers let up and publicity lapsed. The public lost interest. People liked to believe that the ordinances would solve the noise pollution problem without further efforts on their part. Without continued public sentiment demanding enforcement, the laws lost the potency they might have had, and the same practices often reappeared.

The growing complexity of the issue also dampened earlier enthusiasm. The sheer growth of numbers defied any easy, cost-free solution for urban noise pollution. Despite mufflers on cars, thousands more automobiles traveled the roads each year. Though manufacturers observed noise regulations, more factories surrounded expanding cities. And swelling tenements and neighborhoods made silence impossible.

And a subtler force was at work. Despite evidence to the contrary, people associated noise with progress. The chairman of Chicago's anti-noise committee observed, "Many of us are still under the impression that noise and lots of it means progress and 'hustle.'"[45] The *New York Times* pointed out that Americans endured urban din in part due to the "fallacious" belief that "the prosperity of a city is directly proportioned to its noisiness."[46] In trying to secure a quiet environment, Americans were not willing to alter basic attitudes about business and production. They wanted the benefits of technology with quiet surroundings. If they could not have both, they were willing to sacrifice environmental quality. Before a new ecological outlook could develop, a process of reeducation would have to take place. Laws alone were not enough:

> The passage and enforcement of anti-noise ordinances will not alone bring about a quiet city. . . . The public must be taught that quiet surroundings as well as pure food, pure water, clean air, and proper methods of sewage disposal are all hygienic measures essential to the health and comfort of all.[47]

After an exhaustive study, the New York City Noise Abatement Commission of 1929, the first group to attack noise pollution as a social problem, came to the same conclusion: "We do not believe that noise can be legislated out of existence."[48] A broader environmental perspective was a necessary prerequisite for effective law enforcement and for lasting improvement of the urban environment.

Despite their somewhat naive faith in the efficacy of legislation, anti-din advocates made important legal gains. As a result of urbanization, they came to believe that individual liberty had to give way to a greater concern for the welfare of the community as a whole. They began to accept, even demand, that law limit the freedom of the individual to protect the majority and to preserve basic rights, especially the right to privacy. The passage of laws against noise reflected the growing belief that government had the responsibility to protect the environment. Theoretically, if not practically, anti-noise legislation demonstrated that environmental regulation was necessary and possible.

Equity, Eco-racism, and Environmental History

MARTIN V. MELOSI

Influenced by European Romanticism, Americans have thought and written about their relationship to the natural world at least since the beginning of the nineteenth century. The earliest works of environmental history—concerning the United States at least—were written primarily in the 1930s and 1940s and focused on the West.[1] But, American environmental history as a distinct field of study—possessing a wide range of nuance and topic—did not take shape until the late 1960s with the emergence of the modern environmental movement.

Although it drew enthusiastic support from college students and others caught up in the political and social turmoil of the 1960s, the modern environmental movement was rooted more deeply in the American experience. Attracting major support from the middle and upper-middle classes, and bolstered by the maturing of ecological science, it functioned politically as a coalition of groups with a variety of interests, including natural-environment issues such as outdoor recreation, wildlands, and open space, and in concerns over public health and environmental pollution.

Older preservationist organizations, such as the Sierra Club and the National Audubon Society, experienced a revival in the early 1970s. Newer groups reflected a range of political and social objectives from the corporate-backed Resources for the Future, to the more militant Friends of the Earth, and later Greenpeace. Their political views, consequently, were not necessarily compatible, nor were their reform tactics similar. Some accepted governmental intervention as a rational way to allocate resources or to preserve wildlands; others were suspicious of any large institution as the sole protector of the environment. Some worked within the existing political and social structures; others blamed capitalism for promoting uncontrolled economic growth, materialism, and the squandering of resources.[2]

Since the emergence of environmental history was so strongly influenced by political and social goals of environmental activism in the 1960s and 1970s, some members of the academic community were quick to dismiss it as a "fad" or to brand it simply as "advocate history." To be sure, many budding environmental historians did not shy away from advocacy, and much of the scholarship rings with conviction. But by compelling its practitioners to study the past through a combination of science, environmentalism, and history, and by asking grand questions of its data, the new works of environmental history had the potential to address important issues long neglected by other fields.[3]

In tone, substance, and topic, much of the scholarship of the late 1960s and 1970s reflected the spirit—if not the breadth—of the new environmental movement, focusing on the cultural and intellectual roots of environmental thinking or sometimes on the political implications of the older conservation movement.[4] Only rarely in this period did historians venture into the realm of ecological sciences as expressed in Rachel Carson's monumental *Silent Spring* (1962).[5]

The young discipline of environmental history, therefore, took much of its inspiration—if not its execution—from the modern environmental movement. In doing so, historians often shared a common set of values, including a biocentric (or more precisely an 'ecocentric') world view, a belief in the intrinsic value of nature, a faith in ecological balance, and skepticism about—if not contempt for—uncontrolled economic growth.

As we stand in the middle of the last decade of the twentieth century, it is important for us to contemplate the degree to which issues and trends over the last quarter century have modified—or even changed—the contours of environmental thinking in the 1960s and 1970s, and to what degree these changes have influenced—or might influence—the scope, scale, and form of our scholarship.

Some of our colleagues already have begun this process. At our meeting in Olympia, Washington, in 1991, I was intrigued—and I must add, a little unsettled—by Donald Worster's presidential address in which he speculated about the potential impact of chaos theory on ecological science, and the implications for our traditional historical perspective on the natural world.[6] I also have been impressed by the work of Carolyn Merchant and others, who have introduced gender issues into our discourse through research on eco-feminism.[7] And I have been particularly enthusiastic about the work of Joel Tarr, Christine Rosen, Harold Platt, Craig Colten, William Cronon, and several others, who effectively have integrated urban, public health, and industrial themes into an environmental history once dominated by the wilderness.[8]

Of the questions which challenge the contemporary world (as well as the human past), race is an issue also confronting the environmental movement—and destined to help reshape it. Yet, it is a topic largely missing from the literature of environmental history. Consideration of racial issues is implicit in a variety of studies, of course, but explicit only in the extensive work on Native Americans.[9] In the literature on the United States, at least, historical treatment of African Americans, Hispanics, and Asians with respect to the environment is marginal at best. In addition, as a scholarly organization, the American Society for Environmental History has a membership which is overwhelmingly white. Indeed, this conference in Las Vegas is the first of our national meetings to sport more than token sessions on racial issues. Four panels have focused in all or in part on race and ethnicity, compared with none at Pittsburgh (1993), two in Houston (1991), and one in Olympia (1989).

Aside from the intrinsic importance of race as an issue for further inquiry, the current public debate over questions of environmental equity, environmental justice, and eco-racism, are changing the focus of the environmental discourse in the United States and in other parts of the world. And just as the environmental movement of the 1960s and 1970s helped to shape the burgeoning field of environmental history, the current public dialogue over equity and environmental justice ultimately may have a similar impact.

The appearance of the Environmental Justice Movement in the late 1970s and early 1980s offers a medium through which to examine the question of how race has been introduced into the debate over environmental goals and policies in recent years. The movement also suggests potential shifts—or even basic changes—in perspective which may challenge traditional notions of environmentalism.[10] Amidst the diversity of contemporary interests and goals, those individuals in the Environmental Justice Movement seem to be most strident in questioning older environmental thinking of the 1960s and 1970s.

The Environmental Justice Movement found its strength at the grass roots, especially among low-income people of color who faced serious environmental threats from hazardous wastes and other toxic material. Women have been key leaders in the anti-toxics effort, including Virginia civil rights activist Cora Tucker, Lois Marie Gibbs of the Citizen's Clearinghouse for Hazardous Wastes, and Sue Greer, organizer of People Against Hazardous Waste Landfill Sites (PAHLS).[11] According to sociologist Andrew Szasz, "The issue of toxic, hazardous industrial wastes has been arguably the most dynamic environmental issue of the past two decades." By 1980, he said, "the American public feared toxic waste as much as it feared nuclear power after Three Mile Island."[12]

The reaction of local groups to toxics—such as lead poisoning or exposure to pesticides—and to hazardous wastes—through landfills and other disposal sites—may have begun as NIMBYism (Not in My Backyard), but has evolved into something much different. Lois Marie Gibbs stated, "Our movement started as Not In My Backyard (NIMBY) but quickly turned into Not In Anyone's Backyard (NIABY) which includes Mexico and other less developed countries."[13]

A radical environmental populism—ecopopulism—has emerged within the larger tradition of American radicalism, according to Szasz, rather than as an outgrowth of the modern environmental movement. One estimate suggests that almost 4,700 local groups appeared by 1988 to oppose toxics. Before the publicity over Love Canal (1978) contact between the groups was scant, but in the 1980s a more vibrant and better networked social movement appeared to be arising.[14] Some scholars argue that the struggle for environmental justice for people of color predates the 1970s, but these efforts generally were contested under the rubric of "social" as opposed to "environmental" problems.[15]

For those defining the goals of the movement, grassroots resistance to environmental threats is simply the reaction to more fundamental injustices brought on by long-term economic and social impacts. According to Cynthia Hamilton, associate professor of Pan African Studies at California State University, Los Angeles, the consequences of industrialization "have forced an increasing number of African Americans to become environmentalists. This is particularly the case for those who live in central cities where they are overburdened with the residue, debris, and decay of industrial production."[16]

For African Americans and other people of color in the movement, struggles against "environmental injustice" are—as sociologist Robert D. Bullard noted—". . . not unlike the civil rights battles waged to dismantle the legacy of Jim Crow in Selma, Montgomery, Birmingham, and some of the 'Up South' communities in New York, Boston, Philadelphia, Chicago, and Los Angeles."[17]

Within this context, activists in the movement are claiming a full range of rights for any social group, including fair public treatment, legal protection, and compensation.[18] Bunyan Bryant and Paul Mohai of the School of Natural Resources at the University of Michigan have taken the argument a step further, contending that the civil rights movement which faltered in the late 1970s and 1980s may be seeing its resurgence in the area of environmental justice.[19]

The Environmental Justice Movement, therefore, has its historic roots in civil rights activism, and its members and leaders have openly disclaimed connection to the traditional, or mainstream, American environmental movement. A focus on more immediate human-oriented—or anthropocentrist—goals, as opposed to more generalized ecocentrist values, is characteristic of the Environmental Justice Movement. For example, there is substantial mistrust over attention that environmental groups give to global population issues (with their racial implications), and frustration over the little attention given to apparently mundane public health issues. As Robert Gottlieb has argued, some alternative environmental groups "have begun to shift the definition of environmentalism away from the exclusive focus on consumption to the sphere of work and production."[20]

In October, 1991, a multi-racial group of more than 600 met in Washington, D.C., for the first National People of Color Environmental Leadership Summit. In its "Principles of Environmental Justice," conference participants asserted the hope "to begin to build a national and international movement of all peoples of color to fight the destruction and taking of our lands and communities . . ." resting upon the re-establishment of "our spiritual interdependence on the sacredness of our Mother Earth," and, among other goals, "to secure our political, economic and cultural liberation that has been denied for over 500 years of colonization and oppression, resulting in the poisoning of our communities and land and the genocide of our peoples. . . ."[21]

The "Principles of Environmental Justice," interestingly, overlap with many of the values found in the literature of other environmental groups. However, leaders in the Environmental Justice Movement are prone to characterize mainstream environmentalism—especially as represented by the so-called "Group of Ten"[22]—as white, often male, middle- and upper-class, primarily concerned with wilderness preservation and conservation, and insensitive to—or at least ill-equipped to deal with—the interests of minorities. The movement's priority issues most often mentioned are predominately urban-based: siting of toxic facilities in minority neighborhoods and public health problems, such as lead poisoning. A concern over the use of pesticides in the produce industry is a link to farm workers and migrants, who represent the rural equivalent of the urban underclass. Bryant and Mohai concluded:

[Environmentalists] are viewed with suspicion by people of color, particularly as national environmental organizations try to fashion an urban agenda in the 1990s. To champion old growth forests or the protection of the snail darter or the habitat of spotted owls without championing clean safe urban environments or improved habitats of the homeless, does not bode well for future relations between environmentalists and people of color, and with the poor.[23]

Token representation of people of color in mainstream environmental organizations is an additional reminder of the gap between the movements.[24]

Clearly, there is much to justify such criticism. Frederick D. Krupp, executive director of the Environmental Defense Fund, noted, "The truth is that environmental groups have done a miserable job of reaching out to minorities."[25] Nevertheless, politics makes strange bedfellows, and within the Environmental Justice Movement there is division of opinion over whether to join forces with mainstream environmental groups, to cooperate with them in areas of common interest, or simply to follow a separate path.

The rift among those committed to environmental reform can be traced in part to the failure of mainstream groups to reach out, fired by the suspicion of those in the Environmental Justice Movement that "people-centered" environmental issues have low priority among the Group of Ten. However, an additional reason for the rift is the once widely held—but largely unsubstantiated—belief that people of color and low-income groups marginalize environmental issues, especially if economic survival is at stake.[26]

To counteract the assumption that people of color lack an interest in the environment, supporters of the Environmental Justice Movement have addressed that issue frontally. Dana A. Alston, director of the Environment, Community Development and Race Project of the Panos Institute in Washington, D.C., situated environmentalism in a larger social context: "Communities of color have often taken a more holistic approach than the mainstream environmental movement, integrating 'environmental' concerns into a broader agenda that emphasizes social, racial and economic justice."[27] In an effort to dispel the notion of environmental advocacy as "a white thing," several studies have pointed to the strong environmental voting record of the Congressional Black Caucus and the commitment of minorities to key clean-air and clean-water legislation.[28]

In analyzing the evolution of the environmental movement, sociologist Dorceta E. Taylor wrote that existing environmental groups have largely failed to attract minorities due to the particular appeals and incentives they promote. For instance, the argument that minorities are struggling to meet basic needs and thus place environmental issues low on a list of priorities assumes that the priorities are permanently fixed:

> The argument does not allow for the possibility that environmental issues could become high-priority issues for minorities by redefining environmental issues in terms of basic needs, or that individuals might seek to meet high-order needs before all of their basic needs are met. Because many of the environmental problems facing minorities are immediate and life-threatening, it is predicted that they will become involved in environmental organizations and groups, if and when these groups deal with issues of survival and basic needs.[29]

In analyzing several studies conducted in the 1970s and 1980s concerning the different levels of black/white involvement in environmental issues, Taylor concluded that the environmental "concern gap" between blacks and whites can be understood by exploring the disparity between "concern" and "action." First, previous studies may mask levels of black concern because of measurement errors.[30] Second, blacks have a history of higher rates of affiliation with voluntary social, political, or religious associations than whites.[31]

A persuasive argument about the relationship between people of color and environmental concern is the notion that environment is culturally constructed, and participation must be understood from that perspective. Barbara Deutsch Lynch's study of Latino environmental discourses sheds light on contrasting views of the environment between U.S. Latino peoples and Anglo-American environmentalists. The study took into account the role of "the garden and the sea" as traditional sources of livelihood for Spanish-speaking peoples—as well as instruments of bondage to dominant economic systems such as plantation life—and contrasted these perceptions with such images as the frontier, wild rivers, and forests in the Anglo-American community. "The ideal or utopian natural landscapes of Latino writers," Lynch observed, "are peopled and productive." Lynch concluded that by "looking at the impact of environmental ills or mitigation programs on U.S. Latinos solely in terms of end points determined by Anglo environmental agendas [siting of toxic waste facilities, for example] only perpetuates the silence of Latino voices on the environment and postpones fundamental changes in the U.S. environmental discourse."[32]

In light of the context in which environmentalism among people of color has been cast, the Environmental Justice Movement's focus on "eco-racism" is not surprising. Some in the movement connect class and race, but many others view racism as the prime culprit. Rev. Benjamin F. Chavis, Jr., former head of the

NAACP, is credited with coining the term "environmental racism" while executive director of the United Church of Christ's Commission for Racial Justice (CRJ).[33]

Rev. Chavis became interested in the connection between race and pollution in 1982 when residents of Warren County, North Carolina—predominantly African American—asked the CRJ for help in resisting the siting of a PCB (polychlorinated biphenyl) dump in their community. The protest proved unsuccessful, resulting in the arrest of more than 500 people, including Chavis, Dr. Joseph Lowery of the Southern Christian Leadership Conference, and Congressman Walter Fauntroy of Washington, D.C.

The Warren County incident and others—some affecting middle-class blacks as well as the poor—convinced Chavis and his colleagues that a national study correlating race and toxic waste dumping was in order. After five years of work, the CRJ produced *Toxic Wastes and Race in the United States: A National Report on the Racial and Social-Economic Characteristics of Communities with Hazardous Waste Sites* (1987). The report was the first comprehensive national study of the demographic patterns associated with the location of hazardous waste sites. The findings stressed that the racial composition of a community was the single variable best able to predict the siting of commercial hazardous waste facilities. Minorities, especially African Americans and Hispanics, were overrepresented in communities with these facilities. Furthermore, the report concluded, it was "virtually impossible" that these facilities were distributed by chance and thus race must have played a central role in location. Supporters of the report's conclusions argued that other, less comprehensive studies conducted as far back as the 1970s generally corroborated the findings.[34]

The CRJ report—especially its strong inference of deliberate targeting of communities because of race—gave powerful ammunition to those interested in broadening a concern over ill-defined "environmental equity" into the movement for Environmental Justice. Recent statements by Rev. Chavis demonstrate the fleshing out of the call for environmental justice:

> Millions of African Americans, Latinos, Asians, Pacific Islanders, and Native Americans are trapped in polluted environments because of their race and color. Inhabitants of these communities are exposed to greater health and environmental risks than is the general population. Clearly, all Americans do not have the same opportunities to breathe clean air, drink clean water, enjoy clean parks and playgrounds, or work in a clean, safe environment. People of color bear the brunt of the nation's pollution problem.[35]

The question of deliberately targeting communities of racial and ethnic minorities is viewed by some leaders of the movement as indispensable in keeping the focus on the relationship between race and pollution. Also critical are efforts to reject the notion that siting decision are most often based on distinction by class not race. The perceived culprit in deliberate targeting is not simply private companies, but also government. "In many instances," Robert Bullard asserted, "government *is* the problem." He argued that a "dominant environmental protection paradigm" has been in operation which, among other things, institutionalizes unequal enforcement of laws and regulations, favors polluting industries over "victims," and delays cleanups.[36]

Efforts by the federal government to address some of the concerns over environmental racism and inequity have been viewed with skepticism by those within the movement. A June, 1992, report issued by the Environmental Protection Agency (EPA)—*Environmental Equity: Reducing Risk for All Communities*—supported some of the claims of the exposure of racial minorities to high levels of pollution, but it linked race and class together in most cases. A study conducted by the *National Law Journal* in 1992 questioned the EPA's environmental equity record, pointing out that in the administering of the Superfund program disparities exist in dealing with hazardous waste sites in minority communities as compared with white neighborhoods. William Reilly, EPA director under presidents Reagan and Bush, was strongly criticized for not attending the People of Color Environmental Summit. And despite convincing President Clinton to sign an *Executive Order on Federal Actions to Address Environmental Justice in Minority Populations and Low-Income Populations,* there is disappointment because an Environmental Justice Act has yet to pass Congress—and likely will not in the current political climate.[37]

There is little doubt that the Environmental Justice Movement has, in the last decade, broadened the issue of "equity" as it relates to environmentalism. The movement has persuaded—or possibly forced—environmental groups, government, and the private sector to consider race and class as central features of environmental concern for Americans as well as for people of color in the Third World. It has helped to elevate the toxics and hazardous waste issue to a central position among a vast assortment of environmental problems. It has shifted attention to urban blight, public health, and urban living conditions to a greater degree than earlier efforts by predominantly white environmental reformers. And it has questioned the demands for economic growth at the expense of human welfare. Whether or not the Environmental Justice Movement grows beyond its current strength, it has altered—and could possibly transform—the debate over the future goals and objectives of American environmental policy.

The movement, however, is not without its limitations, particularly its stance on the issue of race versus class; its underestimation of its friends and sometimes mischaracterization of its foes; and its own exclusivity. After all, the Environmental Justice Movement, although born at the grass roots, is first and foremost a political movement with an agenda questioning many traditional practices and values, and attempting to define new ones in order to change the law and the regulatory apparatus of the nation.

The core view that race is at the heart of environmental injustice is borne of an intellectual and emotional attachment to the civil rights heritage of the past several decades. Few would deny—including the EPA—that poor people of color are *often* disproportionately impacted by *some* forms of pollution. But the qualifiers are significant. Outside the movement, there has been serious questioning: Is the issue really environmental racism or just poverty? Even within the movement there are those who cannot cleanly separate race and class in all cases. Given the political goals of the movement, the unbending assertion of the centrality of race may prove unworkable if broadening the constituency is to be achieved.

The available evidence for the contention that race was the key variable in siting toxic or hazardous waste facilities is not convincing. Michel Gelobter, an assistant commissioner for Policy and Planning for the New York City Department of Environmental Protection, has studied the connection between environmental regulation and discriminatory outcomes. In his studies of exposure to air pollution (total suspended particulates) in urban areas between 1970 and 1984, he concluded that there were some inequities in average exposure by race and only small differences by income. However, in comparing relative improvements in air quality in those years, the poor experienced a much lower relative decrease in exposure than the rich, while nonwhites experienced slightly greater relative reductions than whites.[38] Such findings point to the complexity of the issue of race, class, and environment, especially when different forms of pollution are measured. Air pollution is particularly difficult to evaluate because of its often ubiquitous nature—especially ambient sources such as auto emissions—which fails to discriminate between rich or poor, black or white. Several forms of water pollution are equally ubiquitous.

Most of the attempts to measure pollution and relate it to distribution have been conducted by economists. Leonard P. Gianessi, Henry M. Peskin, and Edward Wolff produced the first study which examined the distribution of pollution on a nationwide basis. They were concerned with government efforts in the 1970s to apply uniform regulations over the entire nation:

> Although the theory of efficient environmental management suggests that the degree of control should reflect local geography and tastes, pressures for uniformity have been brought to bear by local political leaders who are afraid of losing industries to areas with less stringent controls. Thus, there is a clear bias in existing laws toward technologically defined and fixed standards that limit the allowable emissions of a polluter regardless of where he is located.[39]

They concluded that when the distribution of policy benefits are considered with the distribution of costs of pollution abatement, the results are not uniform. The study indicated, however, that lower-income groups gain the most in this instance, with nonwhites (except those in the highest income group) leading the way.[40] Given the difficulty in utilizing all appropriate indicators, this study must be viewed as producing incomplete and inconsistent results. If the study examined more than air pollution—especially types of land pollution—the conclusions may have been more concrete. Yet, by moving beyond anecdotal evidence and intuitive conclusions, Gianessi, Peskin, and Wolff demonstrated the difficulty in determining a conclusive pattern which links race to incidence of pollution.

Even in the case of the widely disseminated study of the Commission for Racial Justice, questions have arisen about the efficacy of its conclusions concerning race. Gelobter discussed one of the study's key findings that approximately 24 percent of minorities have one hazardous waste facility in their zip code area, although they represent only 12 percent of the population. "But because minorities make up approximately 24 percent of the urban population of the United States," he queried, "it's possible that most hazardous waste sites are simply in urban areas." "This study's measure of environmental discrimination," he concluded, "would have been strengthened if it had controlled for urban- versus rural-located facilities."[41]

Law professor Vicki Been also had reservations about the report. While she stated that there was significant evidence of disproportionate siting of the locally undesirable land uses (LULUs)—"sufficient to require legislatures to address the fairness of the distribution of LULUs"—she maintained that the evidence: (1) "does not establish that the siting process, rather than market forces such as residential mobility, caused the disparity"; (2) "does not establish that siting decisions intentionally discriminated against people of color or the poor"; and (3) "is limited by the imprecision of the study's definition of the neighborhoods compared."[42] The mobility issue is critical, for example, because white neighborhoods with polluting industries which eventually are populated with minorities cannot necessarily be considered as "targeted" with respect to race. In future studies, terms such as "minority community," "African American neighborhood," and so forth, need to be defined more clearly. Distinctions between "inner cities versus suburbs" with respect to environmental risk also need to be recast in order to reflect current definitions of urban growth and development. Metropolitan areas especially are a complex of multiple cores, peripheral development, and edge cities. In cities like Houston, for example, "inner city and suburb" does not sufficiently explain the geographic distribution of the races. African American and Latino enclaves exist throughout the metropolitan area and are not simply confined to a single core at the center of the city. Such a perception misrepresents the nature of modern cities and ascribes to "suburbia" concepts which are long outdated.

The issue of intentional placement of environmental risks in minority neighborhoods or communities—deliberate targeting—also is a knotty problem, and clearly the most controversial charge made by those in the Environmental Justice Movement. While anecdotal evidence exists for such practices, uncovering a consistent pattern of behavior or documenting such a racially motivated policy is very difficult.

Oftentimes cited as an example of deliberate targeting is the so-called Cerrell report. In 1984 the California Waste Management Board commissioned Cerrell Associates to advise the state on how to deal with political impediments to siting mass-burn incinerators. The confidential report concluded that the state was likely to meet less resistance in a low-income, blue-collar community as opposed to a middle- or upper-class area. While the report helped to confirm the existence of targeting practices by class, race was not a category in Cerrell's demographic analysis.[43]

A significant impediment to making the case for deliberate targeting is the lack of judicial remedies. In the broadest sense, civil rights law–based approaches are difficult in proving deliberate targeting. At this time, no court ruling has clearly supported claims of civil rights violations in selecting sites for polluting operations. In *Bean v. Southwestern Waste Management Corp.* (1979)—the first lawsuit of its kind—a class-action suit was filed against the city of Houston, the state of Texas, and Browning Ferris Inc. for the decision to site a municipal landfill in Northwood Manor (a middle-income neighborhood populated by 82 percent African Americans). The plaintiffs claimed that the decision was in part motivated by racial discrimination. The court ruled that the decision was "insensitive and illogical," but that the plaintiffs had not demonstrated the intent to discriminate on the basis of race. Similar suits faced the same result.[44]

Legal experts recognize the current lack of legal remedies and have suggested alternatives. Alice Brown of the NAACP Legal Defense and Educational Fund in New York City recommended utilizing environmental and public health laws as opposed to civil rights law to achieve the same ends. Others suggest avoiding abstract calls for fair siting and to utilize instead specific theories of fairness. And still others recognize the value in seeking a federal equity mandate through legislation—an effort that has been mounted in the last few years but without success.[45]

The equivocal relationship with mainstream environmentalism and the sometimes mischaracterization of the goals of other environmental groups also blurs the objectives of the Environmental Justice Movement and its positioning in terms of several environmental issues. Clearly, there is much to justify the criticism of the mainstream environmental groups for ignoring the inclusion of people of color and low-income groups into their ranks. Frederick D. Krupp, executive director of the Environmental Defense Fund noted, "The truth is that environmental groups have done a miserable job of reaching out to minorities."[46] And there is reason to believe that the priorities of the "Big Ten" often have been focused on wilderness protection, preservation, and outdoor recreation at the expense of some urban-based problems.

In some respects, however, some in the Environmental Justice Movement have often mischaracterized mainstream environmentalism to the same extent that they themselves have either been marginalized or ignored by the major environmental groups. The evolution of modern environmentalist thinking and action suggests that there is much common ground with those involved in grassroots protests, including minorities.

While old-style conservationists focused on efficiency, esthetics, and a small dose of equity, the modern environmental movement has not only been involved in natural environment issues, such as outdoor recreation, wildlands, and open space, but in concerns over environmental pollution and the maturing of ecological science. Environmentalists generally share an appreciation for the fragility of ecological balances, a notion of the intrinsic value of nature, a personal concern for health and fitness, and a commitment to self-reliance. Many, if not all, of these values can be found in the "Principles of Environmental Justice" articulated at the People of Color Environmental Summit. In addition, the politics of so-called mainstream environmentalists vary more sharply than their critics claim. They by no means espouse uniform political views or reform tactics. Some accept governmental intervention as a way to allocate resources or preserve wildlands and natural habitats. Others are suspicious of any large institution as the protector of the environment. Some believe that the existing political and social structure is capable of balancing environmental protection and economic productivity. Still others blame capitalism for promoting uncontrolled economic growth, materialism, the squandering of resources, and even the co-opting of the environmental movement for its own ends. Aggressive, often militant, protest and citizen action has been carried out by groups such as Greenpeace, Friends of the Earth, Zero Population Growth, the National Wildlife Federation, Ecology Action, and a host of grassroots anti-nuclear organizations.[47] Mainstream environmentalists also give attention to urban issues through attention to pollution abatement and public health problems, but not to the extent that the Environmental Justice Movement has made urban environmental problems a centerpiece of their program.[48]

Some members of both movements (or both parts of a larger movement) have already begun to seek common ground to form alliances or sponsor joint ventures. Some major groups, for example, have attempted to establish an "Environmental Consortium for Minority Outreach" in Washington, D.C., especially after Benjamin Chavis and others openly protested the lack of minority representation on their staffs.[49] Yet much remains to be done to bridge the gap in a divided environmental community.

Critics of the environmental movement, from the right as well as the left, always have attempted to present it as monolithic, out of touch with the realities of everyday life. The variety of views and objectives within the broad environmental community cannot be so easily characterized, nor can its values be distilled into a homogenous belief system. Likewise, the depths and nuances of minority environmental views have not and will not be so easily gauged. Real accommodation between the various elements in the environmental movement will need to come down to a more concerted effort to understand and appreciate the variety and depth of a host of values and agendas.

The attempt to connect racial issues with environmental issues—whether utterly defensible or not—is significant nonetheless because it is having a substantial impact on changing the current dialogue surrounding environmentalism and pushing questions of equity to the forefront of debate. In addition, the public battles with mainstream environmentalists are having a desired effect by raising serious questions about the inclusiveness of the environmental movement in general. Unlike the ecocentric/biocentric, nature-oriented, and ecologically grounded environmental movement of the 1960s and 1970s—which was overwhelmingly middle class and politically diverse—the Environmental Justice Movement is more grassroots based, racially mixed, urban-oriented, politically radical, and more immediately human-centered in its orientation. Whether such a divergence from traditional environmentalism can force a more general shift in environmental thinking is problematic. Nonetheless, as historians we must take note of the changes and try to gauge their impact on us and our work.

The emergence of the Environmental Justice Movement suggests several points of inquiry worthy of deeper historical analysis: (1) environmental equity, especially as it relates to race, class, and gender; (2) environment as a cultural construct; (3) the clash between anthropocentrism and ecocentrism; (4) the importance of urban environmental problems, especially as they impact human life; and (5) the nature of the environmental movement itself, including its short-term and long-term goals. Of these five areas, more exploration of the first three may offer the freshest insights, since historians have devoted substantial attention to the last two in recent years.

The Environmental Justice Movement, because of its controversial stances on race, class, and the environment, and its skepticism about the goals and objectives of mainstream environmentalism, is playing a historic role in reintroducing "equity" into the public and academic debate over environmental policy. Equity, however, has been transformed into "environmental justice," with a particular focus on the traditional American underside caught beneath the wheels of an avaricious economy. From the historian's vantage point, this is but one aspect of a larger issue—an issue already addressed broadly by philosophers,[50] as well as by social scientists—especially sociologists and economists—concerned mainly with distributional effects.

In the case of the latter, for example, sociologist Allan Schnaiberg argued that the redistributive element (such as a windfall profit fund to provide cost offsets to the poor) has been largely absent from most of the history of environmental movements through the 1970s, "despite rhetorics that have been vaguely populist." And that environmental movements "are simply not welfare-oriented to the degree that a stable sustained coalition-building effort will be possible."[51] Such a conclusion leaves us to speculate if and how concerns over environmental equity can be uncovered in the historical record, especially if they were not a priority in various environmental movements over time as Schnaiberg argues.

In *Forcing the Spring* (1993), Robert Gottlieb points to a few historical episodes of "environmental discrimination" with respect to workplace hazards, especially the Gauley Bridge episode which led to the death of hundreds of miners—white and black—working for Union Carbide in West Virginia during the Great Depression.[52] Industrial accidents, workplace hazards, and community pollution problems offer potentially good data for examining questions of equity with respect to exposure to health risks.

Clayton Koppes has suggested another approach to addressing the equity issue in his article, "Efficiency, Equity, Esthetics: Shifting Themes in American Conservation" (1988). Koppes argued that three ideas dominated the American conservation movement in the Progressive Era: efficiency (management of natural resources); equity (distribution of the development of resources rather than control by the few); and esthetics (the preservation of nature free from development). Of the three, efficiency held the greatest sway. Supporters of the "gospel of efficiency"—proponents of applied science and environmental management—did not want to undermine development *per se,* but questioned short-term private gain at the expense of long-term public benefit. Although this view was not wildly popular among all capitalists, it certainly was less threatening than strict preservationism.

Koppes argued further that for many conservationists of the Progressive Era, "efficiency was not enough; they were also concerned for greater equity." In this context, "equity" implies that natural resources remain in public control so that their benefits could be distributed fairly. "The equity school," Koppes stated, "saw wise use of the environment as a tool to foster grass-roots democracy." By the 1960s, the efficiency school remained dominant, the esthetic school at least had successfully protected the national park and monument system, but the equity branch wallowed. Without grass roots organizations to press for change—and with resistance to redistributive efforts at every turn—equity moved little beyond the conceptual stage.[53]

Equity, in Koppes's study, has clear definitional limits—more in line with the concerns over distributional issues than class or race questions. It is, nonetheless, a useful starting point for asking some key questions about the intent and direction of national policy expressed in terms of the impartial distribution of resources.

Andrew Hurley's recently published *Environmental Inequalities: Class, Race, and Industrial Pollution in Gary, Indiana, 1945–1980* (1995) is a model monographic study which goes to the heart of environmental justice and environmental racism as suggested by the current debate. Applying the twin perspectives of environmental and social history, he argues that industrial capitalists and wealthy property holders had "a decisive advantage in molding the contours of environmental change. Those groups who failed to set the terms—African Americans and poor whites—found themselves at a severe disadvantage, consistently bearing the brunt of industrial pollution in virtually all of its forms: dirty air, foul water, and toxic solid wastes."[54]

The cultural construct of environmentalism opens up another world of possibilities. Leading the way have been several works on Native Americans and women. Also, of particular significance is the biohistory of Alfred Crosby, who identified European models of environmental practice and how they clashed with indigenous approaches.[55] Despite these fruitful efforts, there is room for more and for a wider range of studies, especially in dealing with the environmental values and goals of a wide range of racial and ethnic communities. Since environmental historians have never feared borrowing methodological approaches from other disciplines, the notion of culturally constructed environmentalism begs for greater range of methods including studies on the use of language.

The question of anthropocentrism versus ecocentrism is not new, but can be brought to bear more directly on issues concerning race and class. Also important are questions about the practical objectives of environmental reform. Human-centered issues offer an immediacy and accessibility to environmentalism that global warming, ozone depletion, overpopulation, and so forth do not possess. Because the Environmental Justice Movement is political at its core, the concreteness and immediacy of its environmental agenda is understandable. But there are some significant longer-range issues underlaying the embrace of this brand of anthropocentrism. In a new book, *Who Pays the Price? The Sociocultural Context of Environmental Crisis* (1994), applied anthropologist Barbara Rose Johnston argues that environmental quality and social justice issues are "inextricably linked." "Efforts to protect a 'healthy environment' may, in some cases, result in human rights abuse," she added, "and depending upon subsequent social response, may ultimately fail to meet original environmental integrity objectives. And conversely, responding to human rights needs while ignoring the environmental context infers temporary intervention rather than substantive solution; it may thus serve to initiate or perpetuate a cycle of human rights abuses." To lessen victimization from environmental threats, Johnston and other authors in the book promote the need for more citizen empowerment.[56] While the emphasis in the book is clearly policy-related, the subject of the human-rights dimension of environmentalism begs for the historian's scrutiny.

Of all the faults we may possess as scholars and teachers, for all our personal biases and secret passions we may indulge while pursuing our research projects, no shortcoming is more deadly than complacence. If the emergence of the Environmental Justice Movement shows us anything, it clearly demonstrates that the foundations of environmentalism laid twenty-five years ago are not unshakable; that the connection between environmental rights and civil rights have to be taken seriously. We have an obligation to ferret out what is happening to the theory and practice of environmentalism over time. We have the training and the interest to make a contribution to the incredibly complex interface of humans with their world. We have a duty to expand the historical horizons of our field whenever possible.

WATER WORKS

The human thirst for water is unquenchable, and we expend enormous amounts of time and energy in pursuing it; water is critical to the survival of human civilization. Its flow is essential to the life of the body politic, too: who controls or regulates access to water, and how these forces are manifest in the social order and economic systems are issues of considerable import. Follow the movement of water in any given society and you will learn a great deal about how it is organized.

Two assessments of antebellum America bear this out. In his careful examination of the confluence of "Rice, Water, and Power" in tidewater Georgia and South Carolina, Mart Stewart reveals how the "huge hydraulic machine" that black slaves and white masters built to develop a rice economy enabled the masters to extend their "dominion and control over both the environment and laborers." Yet the evidence also indicates that along this impressive system of canals flowed another cultural system: during their "free time" the slaves plied these waters and exploited their resources, managing to carve out "small spheres of autonomy." These would widen with manumission, so much so that freed men and women ultimately supplanted the planters' "hydraulic landscape of power and profitability."

This reversal of fortune would not occur in the lengthy controversy surrounding what Brian Donahue calls "The 'Flowage' of the Concord River Meadows." The now-serene river was once a battleground between farmers who operated within a common field system that allowed them to harvest the plentiful meadow grasses for forage, and capitalists who dammed the river, thus inundating the meadows, to fill canals, run mills, and provide water for manufacturing plants and thirsty urbanites. This classic tilt between different economies was not all it seemed: Donahue teases out how many of the protesting farmers, through their participation in the development of commercial agriculture, also contributed to the demise of the very crop they wanted to protect.

Such complex consequences are equally evident in Donald Pisani's "Irrigation, Water Rights, and the Betrayal of Indian Allotment." In it, he tracks the history of the competition between the Indian Bureau and the Reclamation Service for control of scarce water supplies in the American West, surely an uneven competition. It resulted in the triumph of the Reclamation Service, which first secured legal sway over water projects on many reservations, a form of political power it later utilized to strip Native American rights to land and water, and then diverted these to a white-dominated system of irrigated agriculture. These policy decisions not only defied a landmark Supreme Court ruling but meant that the Interior Department reneged on its legal obligation as guardians of Native American patrimony, a conflict, he concludes, of "tragic proportions."

Rice, Water, and Power

Landscapes of Domination and Resistance in the Lowcountry, 1790–1880

MART A. STEWART

The production of one of the American antebellum South's chief export commodities, rice, which was grown on plantations in tidewater South Carolina and Georgia, required the large-scale manipulation of the lowcountry environment. The well-engineered rice plantation, one South Carolina planter explained, was a "huge hydraulic machine." "The whole apparatus of levels, floodgates, trunks, canals, banks, and ditches is of the most extensive kind," he said, "requiring skill and unity of purpose to keep in order."[1] The "machines," in their most developed form, were massive engineering achievements that reshaped portions of the tidewater environment and required enormous investments of well-organized labor to make and maintain.

Water by its very nature imposes specific social, economic, and political necessities upon those who would control it, Karl Wittfogel has argued in *Oriental Despotism*, his provocative analysis of the great hydraulic civilizations in the ancient Middle and Far East. As many scholars who have tested Wittfogel's ideas through specific case studies of both past and present hydraulic cultures have discovered, each culture has distinctive features that are not predicted by his model.[2] Yet this basic insight, that in societies that channel water for productive purposes the flow of power and the control of water become harnessed together, applies to all studies of these cultures.

Water and power converged in the tidewater, also, in a unique set of relationships. When rice planters and their slaves wrenched these "hydraulic machines" out of the river-swamp muck of the tidewater and operated them to produce this commodity, rice, they followed and further bound up and refined their relationships with each other. They imposed an orderly design on specific sections of the coastal plain environment and created a landscape within which crops, labor, and plantation communities were organized. Planters rationalized nature in the plantations to an extent that was unsurpassed on any agricultural enterprise in North America previous to the late nineteenth century.

Space is shaped in characteristic ways by every society; it is a social construct and an economic production, but can also be a means of production as well as a political instrument. Humans shape the land according to social, economic, and political agendas, within the constraints and limits of the particular environment in which they live. Planters coerced slaves into constructing systems to harness the flow of the tides to increase the production of rice for the market, and then used these systems to extend their domination and control over both environment and laborers. The landscapes they created expressed and reinforced at the same time the basic social and economic values of lowcountry society.[3]

The planters' landscape of rice, water, and power was not constructed on unshakable ground. Slaves lived differently than their masters in this landscape and differently than their masters expected them to live. Power in the tidewater moved with the manipulation and control of the flow of water on the land, but the waterways were open to African-Americans who could move about on them. The unique feature of rice production work regimens—the task system—gave slaves in the lowcountry their "own time" within which they could moderate or modify the usual order of the plantation. At the same time, they moved in a different perceptual environment than their masters, both within and beyond the plantation banks.

The landscape they created was not as permanently marked as the one planters made. It moved between, rather than along, the inscriptions of property ownership and control. Planters always sought to order the movements of their slaves, and the African-American landscape was shaped on contested terrain. Nonetheless, it was deeply etched into the region. The rich estuarine environments filled subsistence needs and yielded up small commodities to the African-Americans whose movements around the locale were closer to the ground than their masters. Culture and nature were not so sharply distinguished from each other for slaves; the African-American experiences on the land and water were also intricately woven into the cultural fabric of the communities they created.

Two landscapes evolved in the lowcountry. One was clearly marked, and expressed the attempts of planters to manipulate water and dominate labor in the interest of production and profit. The other moved along quarter-drains and canals on the rice plantations but also on the waterways beyond the banks, and followed the attempts of slaves to mark out small spheres of autonomy. In the long duration, this latter, the landscape of African-Americans, became the prevalent one. After the Civil War and Emancipation, as the highly structured plantation landscape began to erode, freedmen made more visible the landscape they had begun to shape as slaves.

Several scholars have described the development of the lowcountry rice agriculture regimen, and a summary sketch will be adequate here. Rice planters in the area first used an impoundment method to grow rice in freshwater swamps. In the mid-eighteenth century they began experimenting with the use of tide-powered hydraulic systems. The development of these systems involved considerable trial and error on the part of the planters and their slaves. By the nineteenth century, methods for growing rice became increasingly standardized in the lowcountry. Planters had an ideal in mind in the 1820s and 1830s when they wrote articles about "rice culture" for agricultural journals. They acknowledged local variations in the application and practice of the system, but regarded fewer and fewer of these as the consequences of chance. By 1850, planters talked about the system being "perfected."[4]

The efforts to install the necessary system of drainage ditches and banks that would allow the planters and their managers and overseers a rational control of water flow and the energy source that was available to them in the tides required large, well-organized slave forces and were constrained by several environmental imperatives. In the first place, the disease environment of the lowcountry made year-round habitation dangerous to whites; Africans who were partially immune to malaria and yellow fever acquired a demographic edge on these plantations and throughout the lowcountry.[5]

Planters also had to observe certain environmental conditions when they decided on the location of their plantations within the tidewater. The sites had to be near the mouth of a river and at a sufficient "pitch of tide," with enough difference between the water levels of high and low tides to facilitate easy flooding and draining of the fields.[6] If the plantation was too close to the sea, the tides could push salt or brackish water into the fields. A location too far upriver would remove the plantation from tidal influence. Such a location, above the floodplain, would also expose the plantation to the undiminished force of freshets coming down the river.[7] Rivers with long watersheds and deep channels—in Georgia, the Savannah and the Altamaha, and in South Carolina, the Combahee and Santee, for example—made the best locations. Rivers that did not have an adequate freshwater discharge to maintain a freshwater current above the heavier incoming salt current, and estuaries that resembled salt lagoons or that had turbulent currents and considerable mixing of fresh and saltwater did not make good environments for tidal plantations.[8]

Once planters, probably with the assistance of skilled slaves, established a proper site for a wet-culture rice plantation, the procedure they used to take in the area that was to be cultivated was usually the same in most places in the tidewater.[9] Plantation-making required a strong slave force to put up the outer banks, construct the "check" banks, and install trunks and floodgates in the banks. These systems required massive investments of labor. An eighth-square-mile plantation had two-and-a-quarter miles of exterior, interior, and "check" banks, and twelve or thirteen miles of canals, ditches, and quarter-drains. In other words, slaves working with axes and shovels had to move well over 39,000 cubic yards of fine-grained river-swamp muck to construct an eighty-acre plantation, in addition to clearing the land and leveling the ground.[10] The regular maintenance of the system also required large applications of labor, especially when freshets or gales damaged the systems. Because of the labor requirements alone, then, the landscape developed and expanded slowly.

The grid of irrigation, drains, ditches, and canals, and of squares enclosed by check banks, all enclosed within an outer embankment, ordered the environment and provided a framework for crop production. Both crop culture and labor were organized within this grid on the rice plantation. The various steps of rice cultivation were delegated to the different hands by task; the standard size of each task from the beginning was the quarter acre, marked off by quarter-drains.

The nature of the task system and its difference from the gang system used on upland cotton plantations is fairly well-known.[11] What is significant here was that the task system and the physical grid within which it was organized became intertwined. By early in the antebellum period, a "task" had become a unit of measurement, about 105 feet long, or a quarter acre, that planters and slaves applied to other situations. They perceived space and time in different ways through the "task." Most significantly, as slaves began to make distinctions between their own time and the time spent completing their daily tasks for the planter, they also began to distinguish between the space marked out by the task, the planter's space, and the space outside the banks and in the interstices of the grid that was their own.

The development by planters and slaves of a hydraulic grid, a rational system with which tidal energy could be manipulated, and within which the crop and labor systems were organized, was not a deliberate attempt by them to structure resources and labor in an industrial model, to make a factory in the field. Such a model may have been expressed by this arrangement, but the arrangement itself was merely a system to manipulate water and to organize the land and the work upon the land. Once in place, however, the grid became a frame through which both planters and slaves organized their perceptions of the land and labor on the land. As the system became "perfected," planters organized the environment on an industrial model and the slave force in a highly disciplined manner. Slaves, for their part, perceived space and time organized by this framework as not their own, and sought to articulate their own landscape in the gaps between tasks and in the space beyond the banks.

Rice does not require an extensive irrigation system and intensive use of the land to produce a crop. Natives grew it in parts of Indonesia, Malaysia, India, and the highlands of Vietnam on swidden plots in forests. Along the Inland Delta of the Niger River in West Africa, Africans grew rice on floodplains with only an occasional alteration of the environment. More significantly, West Africans grew different kinds of rice in different floodplain conditions and manipulated crops and agricultural practices to suit the environment, rather than vice versa.[12] When high yields are crucial, however, wet rice culture utilizing an extensive irrigation system makes an enormous difference in the amount of rice that can be produced.

Tidewater rice-growing, then, expanded with the development of an economy that made this agricultural commodity profitable and was pushed into its fullest development by planters who were interested in maximizing production and profits. Planters simplified nature on the plantations, and created an artificial ecosystem with a massive application of human energy. The creators and managers of this system continued to redefine nature by conceiving of water in the surrounding environment as a source of energy. The raw flow of the coastal rivers and the Atlantic tides became a resource to command for flooding the fields. For planters who installed water mills on their plantations, water also became so many tons per minute of energy to extract for driving plantation machinery. South Carolina and Georgia rice planters selected out certain tidewater resources and amplified them by channeling them into the plantations' production process. With each successful step they took in the direction of channeling and dominating nature, they increased their belief in their power to control it.

The rice plantation landscape appeared to be a successful adaptation to nature. The efficient operation of this hydraulic machine, however, masked a fundamental reality. In order for the simplified and fragile ecological equilibrium of the rice plantation to be constructed and maintained, the equally fragile social equilibrium of the plantation community had to be stabilized through firm managerial authority and strict discipline. A quick repair of damage to the system was necessary, as when Roswell King, the manager of a large plantation on Butlers Island in the Altamaha River, in March 1807 took forty hands, a rice flat full of straw, and a boatload of timber, and shored up a fifty-foot breach—caused by a freshet—in the outer bank of the plantation in six hours. This kind of authority and disciplined organization was a social development, one that went hand-in-hand with the development of the natural environment. The engineering of nature and the engineering of one group of humans by another developed mutually. The slaves, subject to the discipline and order imposed by their owner, became instruments of environmental manipulation. The environment—especially the flow of water—then became an instrument to control the slaves. Slaves on the plantations became subject not only to the authority of the master and his hired managers, but also to the demands of the artificial ecosystem they had labored to establish. The environment did not determine this structure of authority and discipline, but once planters and their managers and overseers created this system, they used the shaped environment to consolidate it.[13]

Neither nature nor enslaved humans were so tractable, easy to regulate, and amenable to discipline, however, and challenges to the domination of masters were common. Some disordering forces blasted the foundations of social and environmental stability on the plantations. Several hurricanes caused such serious damage that some planters had to rebuild portions of their plantations. When British forces left several Georgia sea island plantations after a brief occupation in early 1815, over a thousand lowcountry slaves followed their promise of freedom and left with them. Their flight, in this rehearsal for emancipation, knocked a large hole in the labor forces of some planters and forced them to reduce their production. These were major challenges that temporarily swamped the exertions of some planters to maintain regular operations, and that required extensive salvage efforts, innovative management, and the coerced labor of a large force of slaves to overcome.[14]

To control and prevent smaller perturbations in the usual stream of activity on a well-regulated rice plantation, planters used various means to manipulate and coerce the slaves: bribes and incentives, whippings, extra work as punishment, close supervision by a hierarchy of drivers, overseers, and managers, and the extensions of various "privileges." Though these methods were largely successful in maintaining order on the plantations, slaves were seldom as disciplined and as dependent on their masters as masters imagined or wanted them to be. Not only did they have time, after their tasks were completed in mid- or late afternoon, to pursue their own interests, but their intimate knowledge of the environment both on and off the plantation allowed them to pass through it with ease and to procure from it plants and animals to supplement their provisions and small commodities for their own profit.

On the plantation, their work accustomed them to a much closer view of the cultivated environment than the managers' or masters'. When a storm came they went out on the banks to repair breaches; sometimes the slave settlements themselves were immersed. They were responsible for cleaning out the small but destructive imperfections of the cultures, for ridding the fields of the pests, the weeds and insects, that periodically have population explosions in single-crop environments. If caterpillars or grasshoppers attacked the cotton (cotton was sometimes grown in rotation with rice), the hands had to make an attempt to clean the fields. "About ten days past those large, black and red winged Grasshoppers was so numerous I ordered them picked from the Cotton," Roswell King reported on one occasion. He could describe them at arm's length because he counted the bushels of grasshoppers—about forty—the hands picked from the plants.[15]

At the same time, they were aware, from row to row, of the progress of the plants during the growing season. They put seeds in the ground and covered them with their feet, stirred and tilled the earth when hoeing, and bent down over rice stalks or moved slowly down rows of cotton during harvest. The hands experienced the crop cultures from the ground up.[16]

Perceptions and behavior went hand-in-hand. Slaves translated this keen awareness and precise knowledge of the environment into a landscape of subsistence and small profit. They commonly planted in the interstices of the plantation grid, in the hours after they completed their tasks. They raised small crops of corn, sweet potatoes, benne seed, sugar cane, red peppers, okra, groundnuts, and rice around their cabins, on the canal banks, and in ponds and "bottom places" in the undeveloped portions of the plantations. Lowcountry slaves also commonly owned a variety of barnyard fowl and ran cattle and hogs in the nearby woods and marshes.[17]

They also used to advantage their understanding of the swamps and woods beyond the banks. There they found firewood, logs to make canoes, and Spanish moss to sell for mattress stuffing. They gathered, hunted, and fished the rich resources of the estuarine environment. White-tailed deer, opossum, rabbits, raccoons, alligators, and waterfowl, as well as a wide variety of shellfish, crustaceans, and fish provided them with supplements to their plantation rations. Tidewater and sea island slaves may have procured nearly half the meat in their diets from wild sources.[18]

Masters encouraged slaves to provide for themselves, to reduce plantation expenses, but they wanted slaves to depend on them for the core of their diets. But slaves who had time of their own, could move about easily, in good weather, on the network of waterways that laced the region, and had the skills to exploit the rich estuary and coastal environments for both sustenance and trade had access to unique opportunities that were often threatening to plantation order.[19]

Technically the slaves on the coast were subject to the state slave codes, which restricted all movements without a permission ticket. But unregulated movement on the waterways beyond the immediate environs of the coastal plantations was a common accomplishment for slaves, though they had to go and return "between two days." Their success in exploiting estuarine resources in the surrounding environment provided incentives for continued exploitation, as did close ties with African-Americans on neighboring plantations. Masters often extended certain "privileges" for excursions beyond the banks, including trips to nearby towns, yet they sought to monitor movement carefully within circumscribed limits. When they were unable to restrict the movements of slaves with regulations, they destroyed or locked up their canoes. Slaves simply made new ones. Canoes, knowledge of the waterways, and the slaves' movements beyond the banks in their "own time" made one of the larger gaps through which plantation order and discipline leaked.[20]

Planters worried that slaves who traveled beyond the banks and beyond plantation regulations were doing much more than gathering moss. In 1829, when Roswell King, Jr., attempted to stop the trade the Butlers Island slaves conducted in Darien, a mile away from the plantation by water, he was worried not about the trade, but the other kinds of activities he associated with it. Along with the trade, he wanted to end, he said, ". . . indiscriminate intercourse with other plantations, having wives abroad or strangers having wives here, introducing disease and contention, leaving homes on Saturday night, & returning on Monday morning, all of which tend to produce drunkenness, idleness, thievery, & indisposition."[21] For King, as with others who sought to grow rice in the tidewater, the movements of slaves beyond the banks brought disorder of every kind. Planters saw order in the environment and the regulations of the slaves as intertwined, and sought to manage both. They bound up plantation order and the grid of discipline with their authority and attempted to maintain control, as best they could, over both the actions and the environments of their slaves.[22]

What was "disorder" to planters—they also called challenges to plantation order "contamination" or "pollution"—was a limited autonomy to slaves.[23] When slaves moved beyond the grid of discipline on the plantation, they were not skulking around the waterways with no purpose other than to make trouble. They were creating a community of their own that had its own coherence and that was closely tied to the environment where they lived and labored. The slaves also generalized their unique perceptions of and interaction with the natural environment of the tidewater and their knowledge of its functions into cultural forms that bound up self, community, and nature in specific locales (such as folk tales that reflected their precise understanding of the natural environment and social relations together, and family networks that were closely tied to the locale). And though their culture was hemmed in by the constraints of slavery, it provided them some insulation against the efforts of their masters to control their lives.[24]

As several scholars have explained, after the Civil War and Emancipation planters retained the dominant hand in the control of land along the coast, but they could no longer coerce labor to keep the rice plantations going. If laborers were to work in the rice fields, they had to be bargained with. Freedmen and women whose understanding of the cycles of rice production had been passed down for at least two generations and who had the skills to exploit the tidewater environments to meet subsistence needs had a bargaining edge when contracting for labor with their old masters. Though South Carolina and Georgia planters were able to coax former slaves to labor for them in the rice fields, they often had to make concessions to them.[25]

Planters also attempted to confront new labor shortages and to maintain the rice plantation regimen by reducing the formidable labor needs of rice culture. They used mules to pull drags to clean the ditches, plows and harrows to prepare the ground, rice drills to plant, horse-hoes to weed the crops, and carts to carry harvested rice from the fields to the rice barns. Previously most of these tasks had been done exclusively by hand. Planters also built elevated platforms on the banks for the bundles of recently harvested rice, to protect them from freshets and to eliminate possible salvage costs. Some also changed the growing regimen to circumvent the necessity of certain kinds of labor.[26]

In the 1880s, the development of a modern, more mechanized rice industry on fresh land in Louisiana, Arkansas, and East Texas added to the troubles of the Georgia and South Carolina tidewater planters and put them in a competitive bind. Though most were able to reduce labor costs, it became still more difficult for many of them to maintain the necessary profit margin to keep the hydraulic infrastructure of the plantations in good order.[27]

Planters especially struggled with the high costs of maintaining the banks, drains, and ditches. Every freshet and gale challenged the functioning of the hydraulic system and the commitment of planters to continue operations. The planters' illusion of domination dissolved completely at this point, and the effectiveness of the "hydraulic machine" was challenged conclusively at the same time. The freedmen refused to get into the ditches and go out on the banks altogether. When planters were no longer able to coerce them, ditch and bank maintenance was the one kind of plantation work they refused to do. Rice planters throughout the lowcountry experimented with imported laborers—Irish, English, Italian and Chinese—for the necessary ditch and embankment maintenance, but failed to find a long-term solution to this perennial problem.[28]

This was the strongest sign of the vulnerability of the rice plantation hydraulic system to forces beyond the banks. The War permanently upset the equilibrium of social relations on tidewater plantations. With a change in social relations came a change in environmental ones. This increasingly exposed the vulnerability of the rice plantation landscape to both human and natural forces. What one rice plantation manager in 1880 called the "two essentials" of a rice harvest, good weather and "plenty of labor offering," were needed for maintaining the means of production as well as for production. When the weather was not good, the second "essential" became even more important.[29]

Rice plantations and the culture for which they provided the material foundation were deeply entrenched in the lowcountry. Planters had structured society and economy in the tidewater with the hydraulic infrastructures of the rice plantations; these required large commitments of capital, labor, and other resources. Once in place, the system carried powerful incentives for further commitment. In spite of the dependence of the system on the market and the vulnerability of it to environmental and social challenges, tidewater planters continued producing rice for another thirty or forty years.[30]

Finally, planters proved unable to persist with the illusion of economic viability. Low profits, strong competition, and labor and environmental problems combined to destroy the economy and the landscape that structured it. A series of devastating storms along the Georgia and South Carolina coasts in the 1890s sealed the fate of this kind of land use in the lowcountry. Planters could no longer bear the high cost of maintaining the rice production infrastructure and of producing rice in a restricted labor market. Though a few rice plantations continued to operate into the twentieth century, this hydraulic landscape, which had dominated the tidewater for over a century and had provided the material foundation for a society that had already faded, eroded beyond repair.[31]

In the meantime, tidewater African-Americans shaped their antebellum landscape more freely and in more visible terms, in a patchwork of landholdings that became hubs of their movements around the locale for occasional wage or sharecropping labor and for fishing, hunting, and gathering. They were unusually successful, compared to freedmen elsewhere, in acquiring small parcels of land of their own. The demise of staple-crop agriculture, declining prices for land, and the availability of wage-paying jobs combined to provide conditions that freedmen exploited extensively to purchase land, to acquire some measure of self-sufficiency and to achieve independence from the supervision and domination of their old masters. The rice plantation landscape was no longer contested terrain, as freedmen either gained some leverage to work as they wished or removed themselves from the plantations altogether.[32]

Though the economy lowcountry African-Americans were able to create and negotiate remained a poor one and occupied a peripheral position to the larger American economy, freedmen in the tidewater were on the whole able to remain more independent of white supervision and domination—to be more "free"— than freedmen and women in other plantation areas in the South after the Civil War and the failure of Reconstruction. They did so by using the land in their own ways and then by working for wages or for a share of the crop on a daily, weekly, or seasonal basis. In effect, they expanded the "own time" they had as slaves to fill most of their days. Coastal freedmen founded an economy and community life on perceptions of and relationships with the natural environment they had cultivated most intensively as slaves, after they had completed their tasks. When some researchers from the U.S. Bureau of Soils observed in 1910 that the only labor available for agriculture in coastal Glynn County was from the black population, but that these hands were "unskilled and scarce for farm work, owing to the fact that a living can be made easily in fishing and oystering or by working intermittently at turpentine gathering or lumbering," they were recognizing work patterns, perceptions of the environment, and a land use behavior that had its origin in the area over a century before. The planters' hydraulic landscape of power and profitability had disintegrated, and this, the landscape of the African-Americans they dominated, became the dominant one in the tidewater.[33]

"Dammed at Both Ends and Cursed in the Middle"

The "Flowage" of the Concord River Meadows, 1798–1862

BRIAN DONAHUE

The Great Meadows bordering the placid Sudbury and Concord Rivers of Eastern Massachusetts are a quiet wilderness today. The narrow channel winds along in a broad floodplain of sedges, buttonbush, black willow, and purple loosestrife, a ribbon of wild marshland through the suburbs twenty miles west of Boston, troubled only by a few canoes. For the past century, the Concord River has been famous mainly for its birds and for the naturalists who have watched them, following one another in the ripples of Henry Thoreau's rowboat: William Brewster, Ludlow Griscom, Edwin Way Teale.[1] Through alert conservation, the river has become a haven from the hubbub of rapid economic growth in the surrounding region.

But the Concord River meadows were not always a quiet backwater. Like the surrounding uplands now recovered by forest, these overgrown meadows were once active farmland. For centuries, farmers in the Concord valley turned out by the hundreds on the meadows every August to mow their plots of native hay. From the settlement of Concord and Sudbury in the 1630s until well into the nineteenth century these "fresh meadows" were at the heart of the local agricultural economy. As a result of their agricultural importance, the river meadows were at the center of a long struggle by farmers to protect themselves from the environmental impact of rapid economic growth during the first half of the nineteenth century.

The Concord River "flowage" controversy began in 1798 with the construction of the Middlesex Canal dam in Billerica, and ended in 1862 when the Legislature voted to repeal its own Act of 1860 ordering the removal of the Billerica Dam. In between came a half-century of court cases, capped by investigations of a Joint Committee of the Massachusetts Legislature and a select Scientific Commission. It was, according to valley farmers who lost every round in this struggle except the next-to-last,

> a story of strange complications, not to say chicaneries, resulting in a practical, persistent, and as we fully believe, *preconcerted* denial of justice to an unoffending and loyal people, flagrantly wronged and outraged through two entire generations.[2]

On the surface this was a confrontation between traditional agrarian uses of the river and rising urban and industrial demands on the same resource. A dam constructed to supply water to a canal and later used to power a textile mill apparently flooded farmland upstream, and farmers strove to remove this nuisance. There is a growing literature in these flowage cases. According to Morton Horwitz, interpretation of the common law by the courts changed during the early nineteenth century to accommodate economic growth and the interests of those promoting it. In earlier agrarian society, property rights were strongly conservative: if a dam flooded land, riparian owners could rip down the encroachment. But during the nineteenth century the courts began to allow "reasonable use" of existing land and water resources by new appropriators such as canals, turnpikes and mills, facilitating private economic development which was now judged to be in the public interest. Horwitz calls this the "instrumental conception" of the law.[3]

The old common law protections were also tied to defense of common resources. Gary Kulik argues that farmers in eighteenth-century Rhode Island were able to protect fish runs from milldams by drawing on an agrarian tradition balancing private economic activity with public good. But with the rise of large cotton mills in the 1790s farmers lost political power, and fish runs were sacrificed to the new economic order.[4] Flowage cases seem to demonstrate the triumph of rising capitalist interests over more traditional farmers.

In its rhetoric and outcome, the Concord River flowage case certainly fits this pattern. However, beneath the surface lies a muddier story. Like the Legislative Committee in 1860, this chapter is concerned with the complex causes of increased flooding in the nineteenth century. As environmental history, it examines both the specific hydrological and the broader economic and social causes of ecological change. In particular, it examines the role of the farmers themselves in the economic growth of the region, and in the environmental impact of that growth. For it should not be assumed that these farmers were simply victims of the economic and ecological transformation overwhelming their traditional way of life.

The Sudbury and Assabet Rivers join to form Concord River at Egg Rock in Concord. The Assabet is a swift little river, but the Sudbury and Concord form a stretch of some 25 miles, from Wayland and Sudbury through Lincoln, Concord, and Bedford and on down to the Billerica millpond, where the river only drops two feet—about an inch to the mile. The river itself is only about 100 feet wide, but it is flanked all along the way by marshes (traditionally called "meadows") that stretch out as wide as a mile in places. These meadows are covered with sedges and brush today, but when English immigrants settled Concord and Sudbury during the 1630s the meadows contained native grasses that supplied desperately needed winter fodder.[5] Seventeenth-century English farmers had little experience growing cultivated hay, and farmers in New England continued to rely primarily on natural salt or fresh meadow hay throughout the colonial period.[6]

The meadows as the Puritan fathers found them were providential, but far from perfect: they were soft, and prone to flooding. From the start, farmers in the valley complained about the "badness and weetnes" of the meadows, and labored to transform them into a more reliable agricultural resource. Petitions were sent to the Legislature in 1636, 1742, 1789, and 1816, requesting the authority to incorporate "Commissions of Sewers" to improve the drainage of the meadows by cutting weeds and dredging bars in the river. Numerous ditches were dug across the meadows, cutting off seepage from the uplands and carrying away the flow of springs and brooks, gradually rendering the meadows drier and more uniformly accessible. In short, the bountiful meadows inherited by nineteenth-century farmers appear to have been as much a creation of their forefathers as they were a blessing of nature.[7]

The meadow grasses were a mixture of native species and domestic varieties introduced by cattle turned onto the meadows to graze the "aftermath" in the autumn. The more common meadow grasses were red top (Agrostis alba), bluejoint (Calamagrostis canadensis), fowl-meadow grass (Poa palustris and Glyceria striata), and reed-canary grass (Phalaris arundinacea). Poorer, wetter parts of the meadows usually contained rushes (Scirpus ssp.) and sedges (Carex stricta and other Carex ssp.), but some stretches at the river's edge were covered with prized stands of "pipes," which appear to have been a species of horsetail, *Equisetum fluviatile*.[8] The farmers remarked that meadow hay

> had properties and peculiarities, which caused it to be sought for, in far and near. . . . The yield was from one to two tons to the acre, all good stock hay; and much of it known in this district by the term "pipes," was so nutritious and attractive, that cattle would suspend their chewing, even of good English hay, on hearing the music of the "pipes," as it rustled from the scaffold, and wait to be served with it.[9]

This great source of fodder provided the mainspring for colonial farming in the valley. The winter floods dropped sediment on the meadows, keeping them productive at no cost. Fresh meadow hay supplied 70 to 85 percent of the fodder fed to cattle in these towns before the Revolution,[10] and cattle were central to both subsistence and market production. Besides meat, milk, and locomotion, cattle provided manure to fertilize the principal subsistence crops of corn and rye grown on the easily tilled but not particularly fertile sandy uplands bordering the meadows. As the farmers put it, the meadows

> required no labor but that of reaping the harvest, no fencing, no fertilizing, but . . . on the contrary, they filled our cow-yards and barn-cellars with the best of fertilizers for our uplands, much of which is rather light. . . .[11]

Thus, the inexhaustible river meadows drove the agroecological system in the valley, nourishing the livestock and through them the tilled fields.

Nearly every farmer in these towns inherited, purchased, or leased a lot in the meadows. Some farmers were lucky enough to own a stretch of meadow adjoining the rest of their farm; the rest carted their hay home from one of the strips into which the larger meadows were divided. The common field systems originally established in Concord and Sudbury had broken down into dispersed, nucleated farms within a few decades of settlement, but a vestige of the commons persisted for another two centuries on the great Meadows, because of the concentrated nature of this vital resource. Well into the nineteenth century, when the time to mow the hay arrived in late July or August, a large part of the farming community turned out on the meadows, just as it had for generations. "It was like a vintage," recalled Wayland's minister, who arrived in the valley in 1815, "an occasion of great excitement, in getting such large crops—teams passing up and down. It was like a vintage in the South of Europe."[12]

Henry Thoreau (who was later hired by the meadow owners to help survey the river in preparing their case against the dam) described the scene this way as he rowed by during the 1850s:

We are now in the midst of the meadow-haying season, and almost every meadow or section of a meadow has its band of half a dozen mowers and rankers, either bending to their manly work with regular and graceful motion or resting in the shade, while the boys are turning the grass to the sun. I passed as many as sixty or a hundred men thus at work today. They stick up a twig with the leaves on, on the river's brink, as a guide for the mowers, that they may not exceed the owner's bounds. I hear their scythes cronching [sic] the coarse weeds at the river's brink as I row near. The horse or oxen stand near at hand in the shade on the firm land, waiting to draw home a load anon.

As I return downstream, I see the hay makers now raking with hand or horse rakes into long rows or loading, one on the load placing it and treading it down, while others fork it up to him; and others are gleaning with rakes after the forkers. All farmers are anxious to get their meadow-hay as soon as possible for fear the river will rise.[13]

The river meadows remained an important part of agriculture in the valley throughout the first half of the nineteenth century, still supplying over half the hay in these towns as late as the 1840s, and at the same time providing a strong link to the traditions of the past.

The changes that affected the flow of the Concord River during the first half of the nineteenth century can be divided into two sorts: new demands on the river from outside agriculture, and growth and change within farming itself. New competing uses for the river's water included transportation in the form of the Middlesex Canal; industry in the form of mills proliferating on the river and its tributaries above and below the meadows; and urban water supply in the form of Boston's appropriation of Lake Cochituate and subsequent construction of reservoirs on the river itself. Agricultural expansion accompanied these other agents of economic growth in the region, and was closely bound up with them both in its causes, and in its ecological impact. For the sake of clarity, I will first follow the growth of urban and industrial demands upon the river and the farmers' efforts to defend their meadows against these "encroachments," then turn to a more critical analysis of the farmers' own role in promoting the ecological disruption from which they suffered.

Trouble for the farmers began in 1798, when the Middlesex Canal corporation built a new dam at an ancient millseat it had acquired in Billerica. The Canal, put into operation in 1803, connected the Merrimack River with Boston. Its summit was at the Billerica millpond on Concord River, which was now tapped as feeder for the Canal.[14] The rickety old milldam on the site apparently never caused much problem with flooding upstream, although farmers in the valley did tear it down in 1722 for blocking fish runs—this dispute was resolved by opening the spillway for two months every spring to let the fish up.[15] The rebuilt Canal dam was stouter and higher, and by 1808 had been raised high enough by means of flash-boards to affect the meadows upstream, according to the farmers. They claimed that because the dam slowed drainage, the river flooded more frequently and stayed flooded longer following summer rains, spoiling the hay crop. They also claimed that the meadows had grown softer and wetter, making access more difficult for teams at haying and for grazing cattle in the fall. They noticed a grad-ual degeneration in the quality of the hay, as coarse sedges and water weeds (for instance pickerel weed, *contederia cordata*) replaced more palatable grasses. More meadows deteriorated as the dam was periodically raised over the following decades. Meadow owners had to strap large racquets to the feet of their teams in wet years, and had to carry haycocks out of the lowest meadows on poles between the shoulders of two men, if they could save the hay at all.[16]

Several generations of these outraged farmers brought a "melancholy succes-sion" of unsuccessful lawsuits against the dam owners. Farmers from Wayland (originally part of Sudbury) led the fight into the 1850s. The first suit was lost in 1811 when they decided the dam was not high enough to "stand" the river 25 miles back upstream to the Wayland-Sudbury meadows.[17] Blocked by the bar, the farm-ers turned once more to their traditional efforts to clean the channel, recalling lat-er how with huge implements, swimming oxen, and men neck deep in mud and water, "we wrought, as best we might, and we could spare time, in the channel of the river. It was all in vain."[18]

The farmers abandoned these efforts in 1828, when the Canal corporation re-built the Billerica dam with stone and raised it higher still. According to the farm-ers, the river was now a long lake. They gave up cutting water weeds and dredg-ing sandbars, and went back to the courts. But they found that legal mud was even thicker and deeper than river mud.

The meadow owners brought a suit at common law in 1832 charging that the corporation had no right to build a higher dam because its charter stipulated that construction was to have been completed by 1808. But the Supreme Judicial Court ruled in 1839 that the Canal proprietors could maintain and improve their canal as they saw fit, thus giving them, as the meadow owners saw it, "a perpetu-al right to raise their dam without notice to anyone."[19] The farmers were forced to apply for simple damages under a special adjudication procedure that had been set up by the Legislature to settle claims against the Canal back when it was built—the sort of "instrumental" provision smoothing the way for developers that Horwitz is talking about. Under this procedure complainants had only one year after the injury in which to apply for damages. Accordingly, the meadow owners were dismissed again in 1842, because they had been injured *every* year since 1828.[20]

Meanwhile, the Middlesex Canal itself had fallen on hard times. The Canal had enjoyed a few years of prosperity in the 1830s carrying cloth and cotton for the new mills in Lowell, until they made the mistake of floating up a locomotive for the fledgling Boston and Lowell railroad, which soon put them out of business.[21] In 1852 Charles Talbot bought the Billerica dam and mill privileges and established a textile mill. The meadow owners promptly sued Talbot, claiming that he had no legal right to the top three feet of the dam added since 1798 under the Canal charter. But the Supreme Judicial Court ruled in 1856 that although the Canal itself was moribund, its *charter* was still kicking, and Talbot owned its entire privilege to use as he pleased.[22] Wrote one meadow owner to the *New England Farmer:*

> The great Dred Scott decision has not surprised the people of the free States more than the . . . recent decision, which came booming up here from the Court House in Boston has surprised us poor farmers and owners of . . . meadow lands, which decision we admit to be a dead shot.[23]

These farmers lived with the river all their lives, fishing in it, diving to set bridge piers in its muddy bottom, dredging it and spreading mud on their sandy upland fields, and above all harvesting the diverse native grasses that grew on its broad meadows, keeping a weather eye on the level of the river and the softness of the ground. Now its flow was under the control of outside interests indifferent to their concerns, and apparently beyond their reach. They must have longed for the days when such nuisances could be shifted with a crowbar—a really satisfying legal procedure. But worse was to come: a "fresh and fatal calamity" awaited them.

While the Billerica milldam inundated the meadows from below, new threats came from above. During the second quarter of the nineteenth century the valley was transformed by an explosion of water-powered industry along the Concord River and its tributaries—particularly the swift flowing Assabet, but the upper Sudbury as well. From a handful of colonial grist and sawmills sited mostly on small brooks, the number of mills on the rivers grew to at least 58 by 1862, and many were large cotton, woolen, carpet, and powder mills that ponded the river itself.[24] The effect of these enterprises on the river was noticed in the meadow towns below. Thoreau caught the transformation:

> So completely emasculated and demoralized is our river that it is even made to observe the Christian Sabbath, and . . . on a Sunday morning (for then the river runs lowest,

owing to the factory and mill gates being shut above) little gravelly islands begin to peep out in the channel below. Not only the operatives make Sunday a day of rest, but the river too. . . . All nature begins to work with new impetuosity on Monday.[25]

These millponds had the effect of a reservoir system, holding water in the basin and keeping up the level of water in the meadows below later into the season.

The most dramatic assault from upstream resulted from an arrangement between the Boston Water Board and millowners along the river. In 1844, the City of Boston acquired Long Pond in Framingham and Wayland as its new water supply reservoir, and rechristened it "Lake Cochituate"—a spurious Indian name meant to give the waters the sparkle of native purity in the eyes of urban consumers.[26] Wayland farmers initially approved because Long Pond supplied one-third of the Sudbury River's summer flow, and they hoped diverting this water to Boston would help dry up their meadows. But as they soon learned, "Vain hope! The relentless adversary was again upon us."[27] Anticipating lawsuits from mills below the meadows in Billerica and Lowell that depended on the river's flow,[28] the Boston Water Board constructed "Compensating Reservoirs" at the river's headwaters in 1851, to supply extra flow during the summer. After the suits were settled, Boston obligingly conveyed these reservoirs to millowners.[29] The result was that when the river fell and the meadows began to dry out every July and August, the millowners below opened their reservoirs above and filled the stream up again, spoiling the hay in between. As Col. David Heard of Wayland put it, the farmers' valley was now "dammed at both ends and cursed in the middle."[30]

In 1859 the meadow towns rallied, and sent petitions to the Legislature, invoking the spirits of their forefathers. By removing the dam, they wrote,

> your Honorable Bodies would heal the wounds of the living and quiet the manes of departed inhabitants of this Valley. Our generous and self-sacrificing sires and grandsires, who fought Concord fight, . . . did never seek as a primary object to obtain [damages]. To reach and exterminate the harassing and humiliating despotism of the Dam, was ever their dearest aim, as it is also ours.[31]

In response, the Legislature set up a joint Special Committee to look into the controversy. During the summer of 1859, the Committee took testimony from dozens of aged farmers concerning the deterioration of the meadows during their lifetimes. The millowners responded that "the whole of [the petitioners'] evidence was one enormous traditionary blunder,"[32] and they recalled the farmers' sires and grandsires to refute them—that is, they exhibited the two-hundred-year string of petitions for "Commissions of Sewers" in which valley farmers complained about flooding. The river flooded because natural bars held the flow in a series of long pools, the millowners argued, not because of the dam.

But the Committee found the farmers' testimony overwhelming, and recommended lowering the dam 33 inches to its original 1711 height, stating that it "regarded the rescue of ten thousand acres of arable land as a public use, of full as much general advantage as the creation of mill powers under the Mill Acts of this Commonwealth."[33] Accordingly, at the spring 1860 session an act was passed ordering that the dam be lowered 33 inches, the Commonwealth to pay the cost. "Jubilee!" cried the *New England Farmer*, "The Year of Redemption is at Hand!"[34] Not quite. A few days before the dam was to have been decapitated in September, the millowners obtained a stay of execution on the grounds that the Legislature had no right to appropriate their property for the private benefit of a few farmers. But the Massachusetts Supreme Court ruled against the millowners in November, applying the "instrumental" conception of the law in *favor* of agriculture, writing:

> If it is lawful and constitutional to advance the manufacturing . . . interests of a section of the state by allowing individuals acting primarily for their own profit to take private property, there would seem to be little if any room for doubt as to the authority of the Legislature . . . to make similar appropriation . . . in order to promote the agricultural interests of a large territory.[35]

Agricultural leaders hailed the decision as a long-overdue stroke of justice, and the way again appeared clear for the dam to come down.

But in the meantime a new Legislature had been elected, and manufacturing interests moved quickly to have it stay the Act of 1860 and to appoint a Scientific Commission to re-investigate the whole flowage controversy. The man named to head the Commission was one of the era's foremost engineers: Charles S. Storrow, who had designed the waterworks, mills, and indeed the entire mill city of Lawrence, Massachusetts. The Commission resurveyed the river, and in the summer of 1861 conducted an experiment, lowering the dam 33 inches at the spillway and monitoring water levels in the meadows above for the next few weeks. The survey showed that the top of the dam was higher than any point on the river bottom for 25 miles upstream, which certainly appeared to support the farmers' position. But the experiment was more equivocal, producing only a negligible drop in the water level at the Sudbury meadows after the dam was lowered. The Commission concluded that the primary cause of flooding was natural bars and weeds, just as the millowners had argued. The Commission did admit that it wouldn't do the farmers any good to clear these while the dam was still in place.[36] Of course the farmers hadn't bothered to try since 1828 for precisely this reason.

But the Legislature seized on the Commission's Report and in 1862 repealed the Act of 1860, which it declared if carried out would "disturb and unsettle the existing manufacturing interests of the Commonwealth."[37] The struggle was over. The Billerica dam still stands today, and the meadows are still flooded, for whatever reason. In the following decades the meadows were gradually abandoned as an agricultural resource, and today they are a National Wildlife Refuge.

In retrospect, it seems clear that the Scientific Commission was wrong, and that the dam was the primary cause of increased flooding, though surely not the only cause. By topping all the bars for twenty-five miles upstream the dam held back an immense amount of water, not only in the channel itself but in the vast basin underlying the adjoining meadows. The impact of this raised water table would have been exactly as the farmers testified: softer, wetter meadows, and more rapid and frequent flooding following heavy rains—the effect being the same as running water into a tub that is already half full. With the constant influx of runoff from higher in the watershed, more than a few weeks were required to allow the river to reach a new equilibrium. What the millowners won was a political victory more than a scientific vindication: the farmers' angry retort that the Legislature had effectively denied that water will run downhill appears to have been justified. Economic power triumphed over common sense. But then again, the farmers were hardly innocent of participating in the economic transformation of the valley, nor of disturbing the flow of the river themselves.

Several of the elderly farmers testifying before the Legislative Committee in 1860 recalled making the four-day trip to Boston on the Middlesex Canal with loads of firewood and other valley produce. Others participated in various efforts to cut channels through bars, not so much to improve drainage as to allow passage for boats on the river in low water.[38] For all their later antagonism toward the Canal, valley farmers appear to have initially welcomed and made good use of it. In 1807, a few years after the Canal began operations, four Concord men wrote to the Canal corporation looking for assistance opening the river for canal boats, predicting a large increase in shipments of barreled beef and pork, firewood and ship timber from the valley.[39] Until the railroad took away its freight in the 1840s, the Canal gave farmers more than high water on the meadows—it also gave them a high road to market. Valley farmers played a central role in the commercial expansion of the region during the first half of the nineteenth century. In the process, they dramatically altered the agricultural landscape, and had a significant environmental impact on the watershed.

The river meadows were the foundation of colonial farming in the valley, as described earlier. However, there was a natural limit to this farming system, defined by the maximum amount of hay that the natural meadows could supply. This in turn limited the number of cattle that could be kept, and the supply of manure for tilled crops. There is strong evidence that by the end of the colonial period farmers in the valley were pressing painfully against these limits, and suffering from declining crop yields.[40] In response, a set of closely linked agroecological and economic adjustments were made during the last decades of the eighteenth and early decades of the nineteenth centuries, transforming farming into a much more specialized commercial activity and driving renewed expansion. This began before the industrial expansion of the nineteenth century—it was a response to an internal dilemma in agrarian society, which helped stimulate growth in other sectors.[41] Farmers turned away from subsistence production of grain and began purchasing wheat flour from Pennsylvania, New York, and the West.[42] This allowed the available manure to be concentrated on the remaining tilled fields, improving the yields of corn and oats now being grown primarily to feed cows and horses.[43] At the same time, farmers attacked the manure deficit from the supply side by steadily planting more cultivated "English" hay on their uplands, reducing their dependence on the natural meadows. By specializing in some products and importing others, farmers were able to efficiently exploit more land in the valley, supplying growing urban and industrial markets.

The expanded English hay crop[44] was the agroecological and economic key to the emerging commercial farming system. Cultivated hay joined meadow hay in supporting livestock, but an increasingly large portion of it was sold to taverns, stables, and stage and wagon companies in the booming valley milltowns.[45] An enlarged hay supply was also the foundation of growing dairy production, which fast became the leading agricultural sector in the middle decades of the nineteenth century, especially in Concord. Cultivated hay acreage in the valley surpassed fresh meadows by about 1850. In short, a largely subsistence agroecological system revolving around the river meadows was transformed into a thoroughly commercial system driven primarily by expanding cultivated hay production.

Commercial farming brought a new agricultural landscape as well, with the spread of new farming and land improvement methods. After 1825, an important part of the increase in English hay came from drained wetlands. Many brook and swamp meadows were brought into cultivation by dropping the water table the required two feet with ditches and buried stone and tile drains. In 1841 Henry Colman noted that "the most remarkable improvements in the county have consisted in the redemption of peat bogs and their conversion from sunken quagmires into most productive arable and grass lands."[46] By the 1850s leading Concord farmers were praising drainage before all else: Dr. Joseph Reynolds called it the key to the rise in stock and milk production underway in the region.[47] As a surveyor, Henry Thoreau helped lay out a drainage project, and he gives us a grudgingly admiring picture of one of these improving farmers. While surveying one swamp he

> saw my employer actually up to his neck and swimming for his life in his property, though it was still winter. He had another similar swamp which I could not survey at all, because it was completely underwater, and nevertheless, with regard to a third swamp, which I did survey from a distance, he remarked to me . . . that he would not part with it for any consideration, on account of the mud which it contained. And that man intends to put a girdling ditch round the whole in the course of forty months, and so redeem it by the magic of his spade. I refer to him only as the type of a class.[48]

By 1860 it was estimated that hundreds of acres in Concord alone had been improved by draining,[49] and Ralph Waldo Emerson joined in the applause:

> Concord is one of the oldest towns in the country,—far on now in its third century. The selectmen have once in five years perambulated its bounds, and yet in this year a very large quantity of land has been discovered and added to the agricultural land, and without a murmur of complaint from any neighbor. By drainage, we have gone to the subsoil, and we have a Concord under Concord, a Middlesex under Middlesex, and a basement story of Massachusetts more valuable than all the super-structure. Tiles are political economists. They are so many Young Americans announcing a better era, and a day of fat things.[50]

As the drainage enthusiasm spread in the 1850s, improving farmers cast an instrumental eye upon their under-developed river meadows. As long as the Billerica dam blocked the outflow, it was obviously impossible to drain the meadows sufficiently to make them cultivable. But as Dr. Reynolds wrote:

> If instead of raising a dam in Billerica three feet high, the channel had been lowered three feet, ten thousand acres of rich alluvial soil . . . which now yields only coarse meadow grass . . . would be covered with rich fields of good hay.[51]

In the late 1850s, several Concord residents who were also state agricultural leaders emerged to lead the fight before the Legislature, including Simon Brown, editor of the *New England Farmer,* and Judge Henry French, a crusader for farm drainage and the agricultural cause. They declared that the meadow owners were prepared to invest $20,000 to excavate the channel and drain the meadows, rendering them as richly productive as the Lincolnshire Fens.[52] Judge French argued that industrial water power was archaic and urged millowners to install steam, thus reappropriating valuable resources for farming. In their petitions and testimony to the Legislature, Brown, French, and other Concord farmers gave short shrift to preserving the meadows for their traditional coarse forage, but instead stressed their potential for high-value commercial production once properly drained. What they envisioned was an ambitious project of agricultural development through large-scale environmental manipulation.

Concord River valley farmers were not simply defending a "traditional" way of life against the invasion of a foreign economic system, although strong ties to the traditional role of the meadows in these communities certainly remained. Even up on the Sudbury meadows where these ties appear to have been strongest, the traditional "system of husbandry" centering on meadow hay had evolved into a market-oriented system producing beef, butter, English hay, and even cranberries, which emerged as an increasingly valuable meadow crop after 1840.[53] Many other farmers, led by men in Concord who knew how to influence the Legislature, had little interest in the traditional place of the meadows except perhaps as a sort of moral, legal, and physical muster-field for a new campaign to transform an under-utilized resource to a high state of commercial production. These men held that agricultural improvements of this kind were *more* progressive than encouraging water-powered manufacturing, and ought to be legally favored on that basis. They were not defenders of tradition but champions of a "better era, and a day of fat things."

In practice, however, commercial agriculture in the Concord valley was extractive, unsustainable, and environmentally degrading. Increased production during the first half of the nineteenth century depended upon continuous clearing and reduction of forest. Strong fuelwood demand provided an immediate return to clearing and also led farmers to cut woodlots faster than they could regenerate, but the major cause of progressive deforestation was conversion to hayfields and pastures. Behind this expansion of cleared land lay not only increased demand for hay and dairy products, but also the fact that these upland mowings and pastures had a strong tendency to wear out, so fresh land was always needed. The commercial farming system, with its heavy reliance on cultivated hay and extensive pastures, steadily extracted nutrients from the grasslands. This was especially true when hay was sold off the farm, but even when the hay was kept home and fed to cattle the resulting manure was generally consigned primarily to the tilled crops. Manuring of hayfields, let alone pastures, was universally admitted to be grossly inadequate, and the use of lime, plaster, potash and bones haphazard at best. It was the agroecological role of the grasslands to supply nutrients to the rest of the farm, and since the upland grasses did not receive the annual blessing of the spring floods that fertilized the river meadows, they gradually ran down. Hayfields commonly declined in yield until relegated to ragged pasture; depleted pastures were invaded by brush and finally pines. Throughout the agricultural boom of the first half of the nineteenth century valley farmers followed an internal frontier migration through their own towns, in search of more land to clear. As long as such land could still be found, the system made good economic sense, at least in the short term. But it was not sustainable, and it had environmental repercussions.[54]

The result shows clearly in the tax valuations for Concord valley towns (see note 44): a steady decline in woodland, and a steady rise in "Unimproved" land, most of which was brushy abandoned pasture rather than simply recently cut-over woodland.[55] Of course, much of this scrubland was on its way back to becoming forest, but clearing far outpaced reversion during the first half of the nineteenth century, so that by 1850 only about 10 percent forest cover remained. The valley was in the most deforested condition of its history. One-quarter to one-third of the land in these towns was covered with huckleberry bushes and young pines, another quarter to third was in exhausted, scrubby pasture, while only about a third was in a fair state of cultivation. This was the abused landscape where Henry Thoreau went a-huckleberrying, railing at the farmers. Huckleberrying and philosophizing were about all much of the land was good for. The best part of the farm was gone long before the poet arrived to enjoy what remained.

Rampant deforestation led to an increase in the flow of the river, especially during the summer. During the growing season, forests pump an enormous amount of moisture back into the air, but croplands, hayfields, pastures, and cut-over woodlots transpire substantially less.[56] At the height of the flowage controversy during the middle decades of the nineteenth century, forest had all but disappeared from the valley, so more water passed downslope and into the river—flooding the hay meadows. After mid-century, there was a net reversion of pasture to forest, as farmers relied increasingly on Western grain rather than native grass to feed their cows, and the pace of clearing slackened.[57] A further phase of reforestation accompanied the twentieth-century decline of agriculture, bringing forest cover in the valley up to its present level of about 65 percent by 1950.[58] As it happens, the flow of the Sudbury River has been monitored since the 1860s because of its role in supplying Boston's water, and changes in the relationship between precipitation, forest cover, and flow have been analyzed by forest hydrologist James Patric, using these data. Patric's study shows a steady rate of flow from 1865 to about 1905, and then a marked *decline* in flow over the first half of the twentieth century as the watershed was thoroughly reforested.[59] There are unfortunately no data on river flow before 1860, so there is no direct evidence on the period of the flowage controversy, but it seems highly probable that the flow then was even higher than during the later nineteenth century when forest had partially recovered. Extensive swamp drainage probably also increased runoff. In short, it is quite likely that rapid expansion of farming during the first half of the nineteenth century contributed significantly to river meadow flooding. The river was dammed at both ends, stripped on all sides, and cursed in the middle.

It seems fair to conclude that the Billerica dam was the primary factor in keeping the meadows soft and setting the stage for more frequent flooding by raising the underlying level of water in the basin, while deforestation and upstream reservoirs made for increased summer flow and still more high water on the meadows. But it is no coincidence that all of these factors produced the worst flooding between 1830 and 1860. At bottom they were the same cause: rapid economic growth outstripped the ecological capacity of the valley to provide all the demanded resources, and both land and river were degraded as a result. Agricultural expansion, industrial growth, and urban thirst worked together to disturb the flow of the river. It was perhaps bad luck for the farmers that these disturbances were all in the direction of *increased* summer flowage, it might have been otherwise had Providence still been with them.[60] But the crucial point is that unbridled economic expansion provoked unforeseen, far-reaching, detrimental environmental consequences, and that farmers were not innocent of causing these problems, even if they were its principal victims.

For two centuries from the time the valley was settled, the river meadows lay at the heart of farming along the Concord River. But during the first half of the nineteenth century changes came to the valley and the meadows deteriorated, so that they were largely abandoned during the latter half of the century and left to the wilderness of sedges and buttonbush beloved by naturalists. The farmers blamed their loss on the invasion of the valley by new industrial and urban interests, which came to dominate the flow of the river. The farmers fought back doggedly, demonstrating both their continued economic interest in the meadows and their strong identification with a tradition that had defined their way of life for generations. However, the farmers' position was full of complexity and contradiction. They helped generate the new commercial order along the river, they adjusted their farming patterns and their use of the meadows to take part in that order, they utilized the river and the Canal to reach their markets, and many among them looked forward to transforming the meadows into an entirely new kind of agricultural resource. They themselves over-exploited their farmland in ways that stripped the watershed of forest and worsened the flooding of the meadows. Thus, the "flowage" controversy was less a conflict between agrarian tradition and industrial expansion than between the resource demands of several interdependent interests in an expanding economic system which over-worked the land and waters of the valley.

Of course, it is still significant that of these competing interests, it was the farmers who lost. Farmers may have been active agents in American economic growth from a very early date, but it is much less clear what good this did them. How often have American farmers been able to take part in the market economy in ways that long upheld a desired way of life, or that created a sustainable relationship with the land they farmed? In the Concord River valley, the ecological base of agriculture was undercut as market production increased, and farmers soon discovered that they were second-class citizens in the new economic order they helped create. Like other American farmers at other times, these farmers were both enthusiastic agents and perplexed victims of economic imperatives in the exploitation of land, and apparently never quite understood the devil's bargain they had made. Perhaps only Henry Thoreau, passing quietly up the river in his rowboat, had a real grasp of that.

Irrigation, Water Rights, and the Betrayal of Indian Allotment

DONALD J. PISANI

Alvin Josephy, Jr., published a searing article in June 1970 for the new conservation section of *American Heritage* magazine. The essay, "Here in Nevada a Terrible Crime," was a tract for the times. Pyramid Lake, the second largest desert body of water in the United States and one of the nation's natural wonders, was drying up. Non-Indian farmers had systematically diverted more than half of the Truckee River, the lake's major source of water, into the Truckee-Carson Irrigation Project sixty miles east of Reno. Unlike nearby Lake Tahoe, already a sacrifice to the gods of gambling and tourism, Pyramid Lake was seldom visited. Nestled among stark, austere, forbidding mountains, it offered solitude and contemplation.

The story would have been sad enough if nature had been the only victim. The declining water level threatened to create a land bridge to Anaho Island, a 750-acre National Wildlife Refuge, thus exposing the largest colony of white pelicans in North America to a host of hungry predators. But the lake was also the heart of a Paiute Indian reservation, and in 1967 almost 70 percent of the reservation's residents were unemployed and more than half of its families earned less than $2,000 a year.

The poverty of the Indians, Josephy emphasized, did not result from laziness or inadequate resources, nor were the Paiutes simply human relics or prisoners of an outmoded culture. They had been betrayed by the Department of Interior, the

243

trustee and guardian of their patrimony. For decades the department had consistently supported the interests of the project farmers who cultivated land once part of the first irrigation project constructed under the Reclamation Act of 1902. Meanwhile, the lake had declined eighty feet since the beginning of the century. As a consequence, the cutthroat trout—that once flourished in the lake, spawned in the Truckee River, and served as a major source of tribal income—had disappeared. When representatives of California and Nevada sat down to divide up the Truckee River's "surplus" water in 1955—with the explicit consent of Congress and the acquiescence of the Department of Interior—the Paiutes were not offered a seat on the interstate commission. Indian rights and needs played little part in the negotiations.[1]

The Paiute tragedy was a classic example of the misuse of Indian natural resources. Public disclosure of that poignant story also coincided with the "rediscovery" of special water rights originally promised to American Indians by the U.S. Supreme Court in the 1908 Winters case.[2] That decision was largely ignored or forgotten until 1963 when the high Court suggested that reservation Indians held open-ended or elastic rights to sufficient water to develop their reservations—a right not restricted to existing uses or even to the development of maximum agricultural potential.[3]

In the 1960s and 1970s, water rights received increasing attention from politicians, lawyers, writers, and even a few academics.[4] But, with the exception of land issues, few historians looked at the use of Indian natural resources.[5] This essay explores how irrigation changed from being the handmaiden of the campaign to "civilize" the Indians into one of its greatest enemies in the period from 1891 to the 1920s. It also explains how Indian water rights became captive to experiments in federal reclamation, thereby contributing to the failure of allotment.[6]

In the American Southwest, Indians had practiced irrigation long before the advent of reservations or the establishment of a national arid-lands reclamation program in 1902. Eight-hundred years ago Hohokam Indians watered about 100,000 acres in the area of present-day Phoenix using an elaborate 135-mile network of canals, many of them carved through tough volcanic outcroppings and lined with clay to prevent seepage. Other irrigators, including the Subaipuris and Pimas, succeeded the Hohokam civilization, which vanished around 1400.[7]

Nevertheless, irrigation did not become an important part of national Indian policy until after the adoption of the Dawes General Allotment Act in 1887. Thereafter, one measure of the absorption of the dominant culture was how intensively the members of particular tribes used the land. Irrigation promised the same benefits to whites and Indians—smaller and more compact farms, immunity from drought, higher value crops, and larger yields. It also permitted the cultivation of less fertile soil.[8]

Congress made the first federal appropriation ($76,000) for Indian irrigation in 1867 to help concentrate natives from the lower Colorado River on a reservation in Arizona. That water system was abandoned in 1876,[9] and regular appropriations for irrigation did not begin until the early 1890s.[10] In July 1890, the Indian office recommended an irrigation survey of Navajo land to prevent herdsmen from leaving the reservation; about 9,000 Indians had left their sanctuary in search of water and pasture. Civil war threatened as whites and Indians competed for a scarce resource. The War Department completed its hydrologic investigation in 1892, and Congress voted $40,000 to begin work on the project.[11] At about the same time, Congress set aside $200,000 from money due the Crow Indians—in payment for the sale of part of their reservation—to construct "irrigating appliances for the lands retained by them."

By 1896, $257,599 had been spent on the Crow water system, most of it in wage payments to Indian laborers. The Indian Bureau also began new works on the Fort Peck, Fort Hall, and Blackfeet reservations. It spent additional money to repair primitive canals on the Pima, Yakima, Pyramid Lake, and Walker River reservations, and to sink test artesian wells on the Rosebud, Standing Rock, and Pine Ridge reservations in South Dakota.[12] Agents and superintendents frequently praised the industry, skill, and dedication of native workers, but they soon discovered that for a variety of reasons most Indians did not appreciate the value of irrigation. Its potential benefits remained obscure to people accustomed to hunting or stock-raising, and with good reason. The agent in charge of irrigation on the Navajo Reservation described its residents as "quick, shrewd, and intelligent," but noted that they opposed cultivating their parched and barren land for fear that the improvements would attract covetous whites and rival tribes. Success had its price.[13]

Irrigation offered substantial incidental benefits. Because most reservations were remote and isolated, Indians had little opportunity to work at manual labor—even though agents and superintendents encouraged them to look for jobs on non-Indian farms, irrigation projects, and railroad construction gangs. The Indian service was convinced that its charges would learn thrift only through work and that annuities and payments for surplus land would be squandered unless native people understood the value of money. By setting aside part of the proceeds from reservation land sales to pay for irrigation, and then returning most of the money to the Indians in the form of wages, the Indian office could under-mine what one commissioner called "the racial prejudice against common man-ual labor." Working for wages would promote habits of regularity, industry, and thrift, the bureau believed; it would also encourage independence and individu-alism by teaching a wide variety of vocational skills, including carpentry and ma-sonry.

Many officials went even further; they believed that successful family farming was impossible without training in manual labor. The actual value of irrigation as a way to raise more valuable crops became secondary. For example, in 1900 Con-gress appropriated $30,000 to construct new canals on the Pima Reservation, even though the absence of a storage dam on the Gila River meant there was no water to fill the ditches. "While the ditches may not be of use," the commissioner commented, "it is certainly wise to require Indians to perform labor in return for the appropriation, as otherwise they might be led to abandon their former habits of industry and become pauperized."[14] Floods and thunderstorms frequently washed out the flimsy canals and headgates constructed by the Indian office's un-skilled workforce, but the commissioner continued to justify those public works as essential to character building.[15]

Officials in the Indian service understood the value of larger, more efficient and dependable works. But they must have worried about losing authority over Indian irrigation to the War Department or the U.S. Geological Survey, both well-staffed agencies with hydraulic engineers. During the gloomy drought years of the late 1880s and early 1890s, and the long economic depression that followed, many westerners demanded that the nation launch a massive arid-land reclama-tion program. That move raised the possibility that rival agencies might absorb Indian irrigation as the foundation for their own work. The commissioner re-peatedly appealed to Congress for money to hire skilled engineers. However, the first "superintendent of irrigation," appointed in 1898, was not much of a hy-draulic engineer—although he had been a loyal employee of the Indian service and had supervised construction of the Crow Reservation works beginning in 1891.

Regular appropriations for Indian irrigation and funds charged against annu-ities dramatically increased in the twentieth century—$30,000 in 1892, $100,000 in 1902, $200,000 in 1909, and $325,000 in 1913. By 1914 the reservations had been divided into five districts, each under a supervising engineer directed from Washington. However, the Indian Bureau never enjoyed an adequate engineering staff, and its canals and ditches rarely matched the high quality of the Reclamation Service structures built after 1902. Because heavy construction on Indian projects was contracted to many different private engineering firms, the works lacked coordination and standardization of design.[16]

The Indian projects served a substantial land area. In 1910 ditches covered 376,576 acres, about 48 percent of the land that could be watered within Reclamation Service projects. A decade later, Indians had 568,620 acres under irrigation. At the end of World War I, a period that witnessed a boom in land and crop prices in the United States, the Indian commissioner proclaimed that Indians had responded "nobly to the call for greater production and materially increased the acreage cultivated and the yield per acre." The amount of irrigated land more than doubled between 1914 and 1919, and in the latter year the total value of crops produced on Idaho's Fort Hall Reservation exceeded $1 million, a sum that surpassed the entire cost of the irrigation system.[17]

Unfortunately, those rosy statistics masked a grim struggle between the Indian office and the Reclamation Service for control over Indian land and water. That conflict began in 1902 and was fed by the reluctance of the supremely confident officers of the Service to share their mission with any other agency. The shortages of irrigable public land and water and inadequate revenue to construct the twenty-four Reclamation Service projects authorized by the end of 1906 compounded the difficulties. In the years after 1904 the new agency invaded many reservations. The Reclamation Service drew the boundaries of its Lower Colorado Project in California and Nevada to include the Yuma Reservation, an area embracing the project's most fertile, accessible land. Several hundred Yuma Indians on the reservation still clung to a precarious existence wrested from the alluvial soil adjoining the Colorado. A 1904 federal statute gave the Reclamation Service the right to "reclaim, utilize, and dispose of any lands" within the reservation as long as each Indian received an irrigated farm. The government sold nearly half the reservation land (6,500 acres) to non-Indian settlers. Proceeds from the sale were credited to the Indians, less the cost of their irrigation works. White farms within the reservation averaged about forty acres each, four to eight times larger than the plots assigned to the Indians.[18]

Subsequently, that policy was extended to allotments outside reservations. For example, in the Carson Sink Valley, heart of the proposed Truckee-Carson project, 196 Nevada Paiutes had been granted about 30,000 acres. In the summer of 1906, however, the secretary of the Interior canceled those allotments in exchange for ten-acre irrigated plots. In all, the Indians gave up 26,720 acres. The commissioner of Indian affairs commented bitterly:

The newspapers of Nevada are even urging white settlers to go upon the lands [before the Indians have left], take their choice, build homes, and make improvements, assuring them that the Reclamation Service will supply them with water and the Indian Bureau must give way.[19]

In any contest between whites and Indians, political expediency, if nothing else, dictated that the Reclamation Service support white farmers. The 1902 act, after all, had been written for those homesteaders.

The Reclamation Service proposed many laws to extend its jurisdiction to specific reservations. In 1904 it extended its influence to the Yuma Reservation, the Pyramid Lake Reservation in Nevada, and the Crow Reservation in Montana.[20] Two years later the Reclamation Act was extended to allotted land on the Yakima Reservation.[21] The Reclamation Service proposed taking over the largest Indian Bureau water projects in 1906, and an interagency agreement was adopted in the following year. All money earmarked for construction work by the Indian office was transferred to Reclamation Service accounts, and the latter agency assumed direct responsibility for projects on the Blackfeet, Fort Peck, Flathead, and Pima reservations. Because water was scarce in the arid West, federal reclamation policies inevitably had a profound effect on many other western reservations.[22]

The Indian office gave way to strong political pressure for several reasons. First, Commissioner Francis E. Leupp and his successor, Robert G. Valentine, supported the conservation policies, including federal reclamation, of the Theodore Roosevelt administration. A coordinated irrigation policy would prevent duplication of effort and promote the most efficient use of natural resources. Moreover, the Reclamation Service, experts claimed, could do a much better job building and maintaining comprehensive hydraulic works than the Indian Bureau. Leupp and Valentine also knew that reclaiming the arid West took precedence over Indian affairs—with both the secretary of the Interior and President Roosevelt. Had the Indian office tried to block the institutional imperialism of the Reclamation Service, it would have placed itself in the position of resisting one of the administration's most popular programs. The Service was expected to reclaim as many as 50 million acres—an area half the size of California. As the national irrigation program developed, therefore, the Indian office would not be able to escape the burgeoning power and influence of the Service; better to form a pragmatic alliance at the outset.

Finally, Leupp considered competition between whites and Indians part of a "conditioning process." In his 1905 report he observed:

Perhaps in the course of merging this hardly used race into our body politic many individuals, unable to keep up the pace, may fall by the wayside and be trodden underfoot. Deeply as we deplore this possibility, we must not let it blind us to our duty to the race as a whole. It is one of the cruel incidents of all civilization in large masses that some, perhaps a multitude, of its subjects will be lost in the process.[23]

A good many Americans had called for a final answer to "the Indian question." Leupp shared their view that only the fit should survive.

Nevertheless, persistent water shortages in the West insured that the two agencies would come into conflict sooner or later. The stage had been set even before the Reclamation Service was established. The droughts of the 1880s and early 1890s, coupled with increasing upstream diversions by non-Indian farmers, had dramatically reduced Indian water supplies. White appropriators on the Blackfoot River, which ran through Idaho's Fort Hall Reservation, forced the local Indian agent to recommend hiring a private company (the Idaho Canal Company) to provide supplemental water from the Snake River. Although the latter stream was overappropriated, the water company held an old claim.[24]

In the Southwest, diversions from the Gila River had dried up the water source for the Pima Reservation. During the late 1890s and early twentieth century, authorities considered several plans to increase the Pima supply. Because they had long been an agricultural people and held some of the earliest rights to the river, the Indian office considered the possibility of a suit. However, more than 900 whites, some as far as 200 miles from the reservation, used water from the stream above the Pima diversion. The agency superintendent predicted that it would take a long time to secure a comprehensive decree allocating the stream's water, that it would cost $20,000 to $30,000, and that it probably would be unenforceable. Despite many different hydrographic studies and congressional approval of a major storage reservoir, the problem persisted at least as late as 1920.[25] Similar shortages due to the violation of earlier Indian water rights occurred on the Yakima and Fort Belknap reserves.

The Reclamation Service insisted that the two basic measures of water rights were priority—"First in time, first in right"—and beneficial use, the requirement that claims could be perfected only by applying water to the soil. The first director of the Service, Frederick H. Newell, warned in 1902 that the Indian office could not reserve water in anticipation of future needs; what was not put to use within a reasonable period, he said, would be lost. Although the agency often claimed that it did not have to observe state water laws, it religiously filed claims and willingly subjected Indian rights to state codes.[26]

During the Roosevelt and Taft administrations the Indian office continued to defer to the Reclamation Service, hoping thereby to increase the Indian water supply through the storage of surplus flood water rather than by laying claim to an already fully appropriated streamflow.[27] Nevertheless, by the end of the Roosevelt administration—as the government opened its first projects outside the reservations to settlers—Francis Leupp began to recognize the irreconcilable conflict between federal reclamation and Indian rights.[28]

Initially, the Department of Justice and the courts supported Leupp. Norris Hundley has written two meticulous, carefully reasoned, and well-documented essays discussing the famous case of *Winters v. United States* and its aftermath.[29] The Fort Belknap Reservation in Montana had been set apart, in the words of the 1888 treaty that created it, "for the permanent home and abiding place of the Gros Ventre and Assiniboine bands or tribes." On no part of the reservation could crops be raised without irrigation, and in 1898 the Indians watered 30,000 acres. Beginning in 1900, white farmers upstream from the reservation diverted increasing amounts of water from the Milk River, and in the spring of 1905 the stream completely dried up before it reached Indian pasture and farmland.[30]

The reservation agent at first urged the Indian office in Washington to press suit to uphold appropriative rights, on the grounds that a former agent had filed for 10,000 miner's inches of water in 1898, a claim senior to most others on that stretch of the river. The Justice Department asked U.S. Attorney for the District of Montana Carl Rasch to intervene on behalf of the Indians. Rasch decided that relying solely on prior appropriation was a risky strategy because the agent who filed on water did so in his own name, rather than on behalf of the United States or individual water users within the reservation. Moreover, there was always danger that older non-Indian claims might surface. Therefore, Rasch broadened the government's case to argue that the entire flow of the Milk River had been reserved for the Indians. When the case finally reached the U.S. Supreme Court in 1908, the tribunal upheld the petitioners, arguing that the Indians and the federal government had set aside the water for both existing and future uses.

The purpose of reservations, the Court reasoned, had been to make the Indians a pastoral people and to destroy a nomadic lifestyle. Because lands were worthless without irrigation, the Indians would not have signed the treaty creating the reservation unless they had assumed that the water remained with the land. The absence of any reference to water rights in the treaty spoke volumes to white lawyers, but the Supreme Court ruled that it was bound to interpret the document the way the Indians had understood it. (In any case, treaties explained to Indians through interpreters could not be construed with the same degree of precision as contracts between whites.) The Court concluded that the water had been reserved by both the Indians and the federal government. "The evidence suggests," Hundley observes, "that the Winters court intended for the Fort Belknap Indians to have all the water from the Milk River that they could put to reasonable use." Many questions were left in the wake of the Winters decision: Was the implied reservation of water limited to the needs of the Indians at the time the reservation was created? Or, was an "indefinitely expandable mortgage" held by the United States on behalf of the Indians? And could Indian rights be leased or sold to whites?[31]

Hundley does not discuss the reaction of the Reclamation Service to the case or the intense infighting that took place within the Interior Department. As the Justice Department's argument began to take shape in the last half of 1905, officials of the Reclamation Service carried their case directly to Theodore Roosevelt. The opening shot in the battle between the Reclamation Service and the Justice Department was a memo drafted by Reclamation's chief legal officer and dated December 8, 1905. According to the document, the Justice Department's attempt "to obtain for the Indians rights greater than could be claimed by any white man" threatened the "broad general principles of policy necessary for the development of the entire arid region." By claiming that the Indians held a species of riparian rights, the memo contended, Justice had undermined one of the major arguments in the pending case of *Kansas v. Colorado*.[32]

In that case, the state of Kansas was trying to block upstream irrigation diversions on the Arkansas River in the state of Colorado. Kansas claimed traditional riparian rights to the entire uninterrupted flow of the stream. Government engineers believed that formal recognition of the primacy of riparian rights in downstream states would prevent the nation from impounding water from interstate streams. And, because riparian rights were indefinite by nature—and impossible to quantify—their recognition would also prevent the federal government from determining the surplus water in a stream or from seeking an adjudication or water rights to quiet titles prior to the construction of new projects.[33]

Theodore Roosevelt asked the attorney general to respond to the Reclamation Service memo. The nation's chief legal officer replied:

> I am of the opinion that it was necessary and proper for the Government, in defending Indian rights within the comparatively narrow field of reservation lands, to invoke the doctrine of riparian rights, although elsewhere the Government is recognizing recent ramifications of that doctrine.

The Indian case was an exceptional one, "and their rights antedate modern evolution in the law of waters." The Bureau of Indian Affairs, the attorney general explained, had a solemn responsibility to protect Indian resources. He denied that there was any inconsistency between the Winters case and *Kansas v. Colorado*. Variations in state water laws were bound to dictate different strategies and arguments in different federal courts. Nevertheless, he concluded his letter of December 18, 1905, on a prudent note:

> [If] everything else must give way to the great irrigation work in which the Government is now engaged, further contention on the basis of riparian rights may well cease; and yet I believe that wherever that doctrine can be successfully invoked in behalf of the Indians, especially under their treaties, this should be done.[34]

Secretary of Interior Ethan Allen Hitchcock also issued a report on the Winters case. Because he supervised both the Indian office and Reclamation Service, one could expect his memorandum to be ambiguous. On one hand, he followed the assumption, widely shared by water lawyers, that the riparian doctrine was gradually giving way in the West to the doctrine of prior appropriation. The government should do everything possible, he maintained, to extinguish riparian rights:

> Should the Supreme Court approve of the doctrine which now seems to prevail in Montana and Washington, and the same be strictly enforced throughout the arid region, the policy of the Government as indicated in the Reclamation Act would be defeated and the development of the entire arid West be materially retarded, if not entirely destroyed.

Nevertheless, Hitchcock also argued that the treaty creating the Fort Belknap Reservation had reserved sufficient water "to insure to the Indians [the] means to irrigate their farms." That interpretation, contended the attorney general, was "in accord with the rules which the supreme court has repeatedly laid down in arriving at the true sense of treaties with the Indians." He confirmed that when the Indians accepted the reservation in 1888, they had "reserved the right to the use of the waters of Milk River, at least to an extent reasonably necessary to irrigate their lands." Although Hitchcock upheld the Indian right, he did not define carefully its origin or extent.[35]

Although Commissioner of Indian Affairs F. E. Leupp pushed Hitchcock for an adjudication of Indian water rights "on every reservation where a conflict has arisen between white settlers and the Indians," in the half-dozen years following the Winters decision, the Indian office failed to settle on a consistent course to protect Indian rights.[36] In most cases, it filed on rights just as the Reclamation

Service did for white farmers, and in 1911 Commissioner Valentine gratefully reported that state authorities "frequently cooperate."[37]

Occasionally, native voices could be heard above the legal debate but the Justice Department remained mute and the Supreme Court made no further comment on Indian water rights until 1939.[38] Congress might still have come to the aid of the Indians, but it, too, fell to the power of the Reclamation Service. In 1912, for instance, the chairman of the House Committee on Indian Affairs introduced a resolution to adjudicate the water rights of the Pima and Yakima Indians, but the secretary of Interior blocked the appeal and the House acquiesced.[39]

Nevertheless, a significant congressional protest against the narrow definition of Indian water rights erupted in 1913–1914. It was prompted in large part by the change in administrations that brought Franklin K. Lane to Washington as secretary of interior and Cato Sells as commissioner of Indian affairs. Soon after they took office, Congress—at the urging of the Indian office—formed a joint committee to study the Yakima River conflict. And when Wyoming's state engineer refused to grant water to Shoshone Indians on the Wind River Reservation without evidence of use, Lane asked Congress for legislation to "provide for confirmation and protection of the prior reserved rights [Winters doctrine rights] for such [reservation] lands."[40]

The Senate Indian Committee followed the initiatives of Lane and Sells. When the annual Indian appropriations bill came up for routine discussion before the full Senate in June, several members of the committee—Senators Joseph Robinson (Arkansas), Carroll Page (Vermont), Joseph Lane (Oregon), and Henry Ashurst (Arizona)—were ready. The usual *pro forma* discussion on the floor turned into a weeklong debate as the senators tried to win approval for a series of amendments drafted in the Indian office.[41] Senator Lane offered an amendment requiring the federal government to reserve "so much water as may be necessary" for domestic use, stock-watering, and irrigation on the Fort Hall Reservation. Failure by the Indians to put water to use within the period stipulated under state law—five years in many western states—would not affect the amount of water set apart for use on the reservation, according to the amendment. Lane's amendment, and one pertaining to the Fort Belknap Reservation, lost on points of order. Senator Robinson sought to limit federal irrigation expenditures on Montana reservations to one-third of the appropriation unless the state adopted legislation to guarantee that Indian water rights would be held in trust as long as necessary and date from the time of filing rather than date of actual use. His amendment was rejected on a voice vote.[42]

Members of the committee repeatedly warned that no further appropriations should be made for Indian irrigation until the law insured that Indians would enjoy some of the benefits. On the Flathead Reservation in Montana surplus land that was sold to whites for $4 to $7 an acre was worth $100 to $500 an acre when irrigated; that, the Senate committee argued, constituted a major federal subsidy to white farmers.[43] None of the "unearned increment" benefited native people, although some of the land they retained appreciated in value. (The Indians did not profit from land speculation because most of their allotments, while theoretically irrigable, were beyond the reach of government canals.) They also encountered discrimination in the delivery of water. Often their ditches were dry until July, when swollen mountain streams carried the greatest volume. As the number of non-Indian farmers on the reservation increased, the amount of water provided to Indians decreased.

Worst of all, the Indians subsidized federal reclamation through both sales of "surplus" land and the substantial mortgage placed on the land they retained. They had no say in how the proceeds from their land sales would be spent, even though the irrigation works they paid for served mainly white farmers. White farmers who took up land within the $6 million Flathead project did pay a *pro rata* share of the cost of building the irrigation works. But they had twenty years to pay, without interest. That constituted yet another subsidy to white farmers.[44] If the project was successful, the Indians might get back the money they had loaned to Reclamation's "revolving fund." But because they had no voice in how their money was spent, that revenue could be lavished on project additions and the Indians would receive no return at all. And if the project failed, they had no assurance that the government would make good on the debt of white farmers. In any case, the Indians received nothing for providing the capital to get many government irrigation projects started.

To complicate matters on the Flathead and other Indian projects it supervised, the Reclamation Service made the reimbursement of operation and maintenance costs a cumulative lien on the irrigable Indian land. If an aspiring Indian farmer waited five or ten years to irrigate his land, he had to pay that substantial debt before receiving any water—that is if he still had the legal right to water. Meanwhile, there was no money to fence or level land, buy plows, seed, and cattle, raise a barn, or build a home. Members of the Committee on Indian Affairs urged that individual farmers should pay for the reclamation of Indian land as on most federal reclamation projects; the payment should not be taken from tribal funds. In this way, the committee believed, the Indians could use the money from land sales as a nest egg. Moreover, if federal Indian policies were designed to promote personal initiative and responsibility instead of tribal collectivism, should not each Indian farmer learn to assume the responsibility for individual debts and contracts? "The system is wrong," Senator Robinson concluded. "You would not get a white man standing in the place of these Indians to go into a scheme of this sort as long as time runs." Oregon's Senator Lane was even more critical, charging that the Indians had been

> pressed into it [the Flathead project] by their guardians, and their land is being held for all of it, while they receive but a fraction of the benefit. It is a great wrong that is being perpetrated on them. . . . It is wicked.[45]

The "wicked" nature of reservation reclamation policies was all the more obvious because the motives of the Reclamation Service were so transparent. Senator Myers of Montana candidly explained why Indian money was needed to build the Flathead project:

> It is because the demand on the reclamation fund for [white] projects now underway is so very great and so pressing that the Reclamation Service claims that it has need for all the money in the reclamation fund and more for projects already started and now under headway. . . . It would be hard to get the Government to put this under the expenditures of the Reclamation Service, when the Reclamation Service is already hard pressed for funds.[46]

The Service had overestimated the money it would receive from the sale of non-Indian lands in the West and the amount that project farmers would return to the revolving fund. Here was a dilemma for the senators. A thorough survey of Indian water rights might take years to complete. Meanwhile, all reservation rights would be under a cloud. White settlers would not buy native land and residents on reservation projects, Indian as well as white, would be unable to make their payments or borrow money to improve their farms. Myers concluded:

> If you halt this work you do the greatest possible injustice to the Indian, because the Indian has got to get his money back out of the sale of land to the white man and if the white man does not get water on his land, he can not make his payments. Therefore if you halt this work, the Indian will be the chief sufferer.[47]

The Senate debate was significant both for its theoretical discussion of the Winters doctrine and its exposure of conditions on Indian irrigation projects. Senator Robinson's careful, perceptive analysis, and the comments of other members of the Indian committee demonstrate that the defenders of Indian rights understood the doctrine almost as well as the lawyers who dusted it off in the 1960s and 1970s—after it had been neglected for so many decades.[48]

Critics of the Winters decision also fully grasped what was at stake. Western senators and congressmen, recognizing the threat to the Reclamation Service and non-Indian farmers, insisted that native people could acquire water only by prior appropriation. Though they dismissed the Winters case as bad law, they considered it significant enough to offer ingenious—and sometimes plausible if not facile—arguments to explain that the Supreme Court had not meant what it seemed to mean. Senator Thomas suggested that the case was an aberration because the courts had upheld religiously the principle that water rights depended on beneficial use. Senator Works of California, a lawyer who specialized in water rights, expressed sympathy for the Indians but reasoned that the case had limited application; the courts, he said, had considered only the rights of Fort Belknap Indians and Montana was still a territory when the reservation was established in 1888. Although the Constitution gave the federal government authority over the resources of the territories, Works argued that when Montana achieved statehood in 1889, sovereignty over all surplus property not specifically retained by the nation was vested with the state. Whether the state of Montana had any obligation to honor a contract made with the Indians before statehood was a live question, but the California senator was confident that federal treaties and executive orders could not confer special rights on Indians *after* a state had entered the union.[49]

Senator Myers added that the Winters doctrine applied only to reservation Indians. Once the reservation had been allotted, all special rights lapsed. Shafroth of Colorado maintained that the Winters court had meant only to reserve water in use at the time the treaty was ratified. In effect, the decision simply attached specific grants to parcels of irrigated land; it did not create a new category of rights. That was something of a guarantee, because Indians could not lose water through disuse. Nevertheless, they could not claim more than they used at the time they accepted reservation life.[50]

A special commission of reservation superintendents appointed to survey conditions on the reservations reaffirmed the recommendations of the Indian committee.[51] That group called for direct federal appropriations for Indian water projects rather than advances from land sales. It also recommended the attachment of irrigation charges to specific parcels of land and urged the rapid completion of reservation projects so that payments by white farmers would be returned to the Indians as rapidly as possible. Finally, in recognition of the lack of cooperation between the Reclamation Service and Indian office, the commission suggested that no further work be permitted on the reservations without the approval of the commissioner of Indian affairs. Congress responded by shifting reclamation charges away from tribal lands to individual irrigable allotments.[52] But the Reclamation Service remained firmly in command. The Indian commissioner observed in 1915 that "the conditions under which the cooperative irrigation works on these reservations [Flathead, Fort Peck, and Blackfeet] has been done in the past is not for their [the Indians] best interest, and . . . its continuance would be a great injustice."[53]

The agricultural boom during World War I obscured and exacerbated the native water rights issue.[54] During the war Indian people were all but forgotten in the quest to open new acreage to the plow and to provide food for the allies. On the Blackfeet Reservation the number of irrigated farms increased from eighteen in 1915 to 329 in 1919. Although Indians worked nearly 90 percent of the farms in 1915, by the end of the war white tenants worked more than 70 percent.[55]

Crop prices and real estate values plunged during the 1920s. As more white farmers fell behind in their irrigation payments, they savagely criticized the Reclamation Bureau. Discredited and unpopular in most parts of the West, the once-pampered child of the Roosevelt administration relinquished control over most of its projects to the farmers and returned the reservation waterworks to the Indian office in 1924.

During the 1920s Congress appropriated very little for Indian irrigation, and by the end of the decade conditions on the reservations appeared more bleak and hopeless than ever. The allotment policy had clearly failed. As part of a survey of the reservations, the Senate Committee on Indian Affairs published a remarkable report written by two engineers in the Agriculture Department, Porter Preston and Charles Engle. Their lengthy study represented the most thorough investigation of Indian irrigation agriculture compiled to that date. It clearly showed the failure of the irrigation program.[56]

According to the engineers there were 150 irrigation projects on the reservations in 1927, ranging in size from a few acres to the Yakima Reservation's 89,000 acres under ditch. Nearly 700,000 acres within allotted and unallotted reservations had been provided with irrigation at a construction cost of $27 million and an additional expense of $9 million for operation and maintenance. About 30 percent of this land belonged to white farmers. However, while Indians watered only about 16 percent of their irrigable land, whites cultivated two-thirds. The most unfavorable conditions prevailed on projects constructed by the Reclamation Service. On the Wapato project in the Yakima Reservation, only 6 percent of Indian allotments were actually irrigated by their owners, on the Blackfeet and Flathead reservations the figure was roughly 1 percent. In fact, 40 percent of all the land irrigated by Indians in the West was located within the Uintah Reservation in Utah and the Gila River Pima Reservation in Arizona.

As dreary as those statistics seem, actual conditions were worse. Most Indians who used irrigation, the Agriculture Department engineers said, simply flooded pasture land to stimulate the growth of native grasses for livestock; Indian crops, therefore, returned only about half the per-acre income of white farmers. The report concluded:

> [M]any of the so-called Indian irrigation projects are in reality white projects. . . . The continual decrease in the acreage farmed by Indians is the natural and logical result of the leasing system, and the leasing system in turn is the inevitable result of the allotment system.[57]

The practice of leasing allotments began in 1891, and grew dramatically in the twentieth century as irrigation made Indian lands far more attractive. Often native people—young and old—were not capable of cultivating their land; on many reservations irrigated allotments were too large for a "head of household" to manage. If Indians could not use the land, it might as well be used by whites. Indian officials were reluctant to discourage leasing, because they believed that native people had to learn to make choices and take responsibility for their acts. Leasing served several useful purposes, the bureau believed. Whites could provide Indians with an object lesson in how to farm. By encouraging non-Indians to take up reservation land, Indians could model themselves after successful farmers imbued with the ethic of industry and thrift. Equally important, leases provided cash for farm equipment, livestock, and other agricultural needs. Finally, leasing irrigated allotments insured that there would be water to farm the land if and when the allottee decided to take over cultivation himself.[58]

The Indian office occasionally grumbled about renting native land to whites, as in 1914 when the commissioner called the policy "a poor one at best." But in the end the bureau did more to encourage than discourage the practice.[59] Although Preston and Engle recognized the risk, they recommended that Indians lease as much land as possible "in order to protect the water rights, and to derive some income from which to pay irrigation charges."[60]

Federal reclamation was launched in 1902 as a bold experiment in social engineering as well as a practical program to reclaim desert land.[61] Much was expected and promised. Had the Newlands Act never been passed, irrigation still would have increased dramatically during the first two decades of the twentieth century, and it would have threatened the water supply available to Indians. As it turned out, federal reclamation projects irrigated only a small fraction of western land by 1920; state-sponsored irrigation districts served a much greater area. Nevertheless, the enormous influence of the Reclamation Service within the executive branch, particularly Interior, forced the Indian office and Justice Department to relax their trusteeship over Indian natural resources.

Beginning with Franklin K. Lane, most secretaries of Interior were westerners highly responsive to political pressure, and state control over water was a sacred cow in the arid states. Norris Hundley observes that the secretaries "usually decided not to alienate the non-Indian voters and congressmen." Those decisions, he argues, "have often been facilitated by the availability of greater technical expertise in the other agencies—especially the Bureau of Reclamation—than in the Bureau of Indian Affairs."[62] In any case, the executive officer most responsible for protecting Indian rights became caught in a conflict of interest of tragic proportions.

Many historians have argued that the desire to force Indians to become farmers was short-sighted, impractical, and self-defeating. Perhaps so, although one recent study by an economist suggests that Indians were making progress as farmers before the allotment policy took hold.[63] What is beyond debate is that irrigation offered new opportunities to exploit Indian natural resources, conditions that were absent in the nineteenth century. By the 1920s Indian irrigation programs were at a stalemate. The success or failure of reclamation in the West, at least from the point of view of Congress, hinged on the ability of farmers to repay their debt to the nation. When many of those farmers defaulted, appropriations for irrigation directed by the Indian office also dried up. The Bureau of Indian Affairs had suffered from the success of the Reclamation Service; now it suffered from its failure. The Reclamation Bureau cast a long shadow.

GLOBAL VILLAGE

Bumper stickers demand that we think globally, but to do so requires a scientific language, ecological insight, and political will that did not begin to emerge until early in the twentieth century. Only in the century's last years has such a perspective become influential enough to blossom as a staple of automotive sloganeering.

Conceiving of the world as a series of discrete yet overlapping environments of air, water, land, flora, and fauna, is one thing, but it is another to construct a global and interdisciplinary approach to the study of these complex ecosystems and their historical contexts. Alfred Crosby, among others, has taken up the challenge, and has done so by following in the wake of the European imperial voyagers. "Biotic Change in Nineteenth-Century New Zealand" links the demographic migrations to the antipodes with the "portmanteau biota" these migrants carried with them, including animals, "weed species," and deadly pathogens. So rapidly did these newcomers alter the islands that travelers to them a century after Captain Cook's initial landfall in 1770 were "disappointed at the un-foreignness" of its natural surroundings, an emerging biotic homogeneity Crosby finds wherever neo-Europeans have landed.

In Australia, organic and human migrants uprooted indigenous peoples and plants to such a degree, Thomas Dunlap has found, that some nineteenth-century settlers felt that they could remake the land in a dewy English image, an illusory greensward that evaporated in the Australian heat and drought. If later generations rejected this colonial landscape in favor of one they considered to be more authentically Australian, their crude handling of the land and the language by which they understood their relationship to it was yet another import. Since the conquest, then, the island continent has long been tugged between the twin poles of "Australian Nature" and "European Culture."

When the neo-Europeans arrived in south Asia they encountered yet another rich and strange biota, yet another indigenous people enmeshed in a landscape built to sustain them. Disruption of these ancient patterns of life came with the arrival of industrial colonialism and the demands it placed on the subcontinent's natural resources. Central to this, Stephen Pyne demonstrates in "*Nataraja:* India's Cycle of Fire," was the relationship between the railroad and forests. To establish a rational system to feed the iron horse's vast appetite, the British reconstructed India's wooded lands, regulated timber harvests, regenerated once-cut-over lands, and vigorously suppressed fire. This latter ambition, so in contrast to precolonial Indian conceptions of the function of fire, confused the ecological and social order, forever complicating a reconciliation of "European principle with Indian reality."

Biotic Change in Nineteenth-Century New Zealand

ALFRED W. CROSBY

The descendants of Europeans live in large numbers outside of Eurasia, which requires explanation. In the Americas north of the Rio Grande and south of Paraguay, and in Australia and New Zealand, the Europeans or, if you prefer, neo-Europeans constitute from 80 percent to close to 100 percent of the population. These neo-European majorities are not simply the product of military conquest, a phenomenon common throughout history, but usually without replacement of the conquered people. India is Indian, despite Kipling. Poland remains Polish, despite all sorts of tramping back and forth by neighbors. Zimbabwe is about as Black as it was 100 or 500 years ago. But Nebraska (incidentally, an Osage word), the pampa (a Quechua word), Toowoomba and Waikato (respectively, Australian Aborigine and Maori words) are places where Europeans and their descendants live in majorities as high or almost as high as in the United Kingdom and Spain. What has enabled European emigrants to "Europeanize" Canada, the United States, Argentina, Uruguay, Australia, and New Zealand?

(Let us call these the neo-Europes. Other regions qualify—Siberia, southern Brazil, Costa Rica—but our plate is full enough as it is.)

Let us consider the neo-Europes vis-a-vis location. Of course, they all lie where there is sufficient water for large-scale agriculture, but that information triggers no deeper insight than the thought that the founders of the colonies that grew into these nations were not feebleminded. The fact that the densely populated portions of the neo-Europes are all in the temperate zones also fails to stimulate much in the way of profundity. As one would expect, Europeans—like their plants and animals, techniques of sanitation and preventive and curative medicine, their styles of raincoats, the kind of grease they use on typewriters, and so on—have proved to be most successful in lands like their home countries in climate.

It is provocative that the neo-Europes, although quite like Europe in many ways, are remote from Europe. Two of them are as far from Europe as one can get and remain on the surface of our planet. The neo-Europes all are located on the other side of one or more oceans from Europe, indeed from the entire Old World. Australasia separated from Pangaea, the Mesozoic world-continent, during the halcyon days of the dinosaurs. At about the same time America also split away, but it has been periodically reconnected to Eurasia via the far north whenever ice ages lowered the ocean levels sufficiently.

This remoteness of the neo-Europes has been a matter of time, as well as space, and therefore of biological development. Isolation from the Old World has meant isolation from important streams of evolution. The biotas of the several shards of Pangaea developed in significantly independent ways. The contrast between the biotas of Eurasia and Australasia and southern South America is much greater than between those of Eurasia and North America, but even in the latter case the difference is impressive and important. Said Peter Kalm, the Finnish naturalist and associate of Linnaeus, fresh off the boat in Philadelphia in 1748:

> I found that I was now come into a new world. Whenever I looked to the ground I found everywhere such plants as I had never seen before. When I saw a tree, I was forced to stop and ask those who accompanied me, how it was called.[1]

Geographical separation meant divergent evolution, which accelerated 10,000 years ago when humanity invented civilization (a vague but handy term), creating new environments, massively altering ecosystems, developing new subspecies and, in the case of very short-lived creatures such as many kinds of microlife, entirely new species. During these millennia, humans of the stratum of the Old World that runs east and west from the Middle Eastern heartland of civilization increased and rose into the millions earlier than other peoples; created centers of dense population, cities, sooner than other peoples; intentionally domesticated more species of plants than any other people, except possibly the Amerindians, and many more kinds of animals than other peoples; hemmed them into dense populations; and unintentionally created—or, to be more precise, created the conditions that led to the selection, evolution, and spread of— such weed species as rats, mice, lice, starlings, disease pathogens, and those plants we specifically call weeds. The biota of civilization are different from the biota of the wilderness. For instance, humans as roaming hunters and gatherers are only mildly troubled with large varmints such as rats, because nomads migrate faster and more continually than rats; but the sedentary peoples of Old World civilizations have had a commensal relationship with rats for thousands of years. The biota of humanity's several civilizations are distinctive or, at least, were so until the past few centuries. For instance, the biota of the Spain of Cortes included horses and smallpox viruses. The biota of the Mexico of Montezuma did not. Such differences can be more significant than the presence or absence of the wheel, gunpowder, or the printing press.

The demographic takeover of the neo-Europes by large numbers of European humans is only one facet of a general invasion of these lands by a *portmanteau biota* that European explorers, conquerors, and emigrants brought with them, and of which they were a part. To illustrate, let me turn to New Zealand, where the Old World takeover has not necessarily been more complete than in large regions of the continental neo-Europes, but more recent and, therefore, more completely documented.

When human beings, specifically the Polynesians, first arrived in New Zealand about 1,000 years ago, its biota were stunningly different from any other in the world. There was only one mammal, the bat, and there were flightless birds as tall as three meters and more. When Captains Abel Tasman and James Cook arrived seven and eight centuries later, respectively, the bigger birds were gone and New Zealand's mammalia included humans, dogs, and rats. Otherwise, the dissimilarities between the biota and those of the Old World or even Australia were still vast. Joseph Banks, the naturalist who sailed with Cook on his first Pacific voyage, recognized no more than 14 of the first 400 plants he examined in New Zealand.[2]

Captain Cook, ordinarily a shrewd man, took the measure of this strange land, with its alien flora and fauna and its cannibal population of about 200,000 fierce Maori, and judged that New Zealand, a half a world and at least half a year by sail from Britain, would make a grand location for a European colony. "Was this Country settled by Industrious people, they would very soon be supplied not only with the necessaries but with many of the luxuries of life." Such arrogance can only be justified if it turns out to be validated by experience, as his did. A hundred years later, New Zealand had a population of a quarter-million Whites (or *pakeha* as all, New Zealanders, Maori and neo-Europeans alike, call them), 80,000 horses, 400,000 cattle, and 9 million sheep; the Maori were down to about 50,000 and declining precipitously. The *pakeha* and a number of the more sophisticated Maori agreed that New Zealand's Polynesians, like many other of its species, were well down the road to extinction: "As the white man's rat has driven away the native rat, so the European fly drives away our own, and the clover kills our fern, so will the Maoris disappear before the white man himself."[3]

New Zealand had changed enormously in the two centuries since Tasman's visit, and as a matter of practical fact the time had been a great deal shorter than that. Tasman had barely touched shore in 1642, and there were no other visits (according to the written record) until Cook's in 1769. And from that year to 1810 or 1820, the only *pakeha* in New Zealand were a tiny number of transients; as late as 1839, there were fewer than 2,000 in both islands, as compared with 100,000 Maori then living. Thirty years later, there were over a dozen times as many *pakeha* and less than half as many Maori.[4]

The Europeanization of New Zealand was proceeding too rapidly for the usual explanations, with their tired references to European military superiority, organizational skill, intelligence, energy, and so on. In his *Origin of Species* (1859), Charles Darwin hypothesized that European species were somehow *higher* types than New Zealand's. The Social Darwinists, whose allegiance he did not always acknowledge, went a good deal further. W. T. L. Travers, a British naturalist and settler in New Zealand, announced *ex cathedra* exactly 100 years after Cook's first visit to New Zealand:

> It is indeed a fact, which does not admit of doubt, which is even presented to us a law of nature—as a necessity—that wherever a white race comes into contact with an indigenous dark race, on ground suitable for the former, the latter must disappear in a few generations.[5]

Neither Darwin nor Travers—not even Travers—was a fool; but for a while, and for New Zealand in particular, their scientific skepticism and normally modest biotic chauvinism were overwhelmed by an onslaught of evidence that was mindboggling and, at the same time, unquestionably valid. They were obliged to deal with proofs—proofs by the bale—of an ongoing biological invasion and, to all appearances, conquest of New Zealand by European organisms.

Let us consider, as they did, the quicksilver advance of their home continent's weed species in the two islands of New Zealand. By "weed species" I mean not only dandelions, clover, and such, but also the aggressive, opportunistic animals and microlife that came along with the White folks from the antipodes. This history begins not with Tasman, who glanced off New Zealand like a musket ball off a granite wall, but in all probability with Cook, circa 1770. On the English captain's second visit in 1773, George Forster, one of the scientists on board, found the first European weed naturalized in New Zealand, canary grass, which Cook had probably left behind three years before. It continued to spread, and in the early decades of the nineteenth century was common in the crop fields of the Maori and, later, of the *pakeha*. More and more weeds arrived every decade. Darwin saw a number of them in the Bay of Islands area in 1835, including common dock, the seeds of which a transient English villain had brought ashore and sold as tobacco. As the *pakeha* disembarked by the thousands after British annexation in 1840, the naturalization of European plants—from willows to watercress—accelerated. *Pakeha* newly come from Europe and braced for alien sights had to look hard to find them. Alexander Bathgate wrote from the South Island in 1874: "A stranger on his first arrival from Britain is usually disappointed at the un-foreign appearance of his surroundings as regards natural objects. The weeds he finds growing in the streets are identical with those he had left in his native town; the grass in the fields is English." In 1882, there were 387 naturalized plants in Auckland Province alone, and of these, 280 were natives of Europe.[6]

Animal immigrants, as well as plant, were swamping New Zealand. The process may have begun as early as 1769, when ship rats may have deserted Cook's *Endeavor* for a luxurious life ashore, where they had no rivals. Cook and other *pakeha* purposely introduced a number of animals in the 1770s, the most immediately successful of which were pigs: the wild ones are still called Captain Cookers. The Maori quickly adopted these four-footed invaders, and many went wild as well. They spread rapidly, domesticated and feral, and existed in populations of many hundreds of thousands by the middle of the nineteenth century. They were a major source of meat for Maori and *pakeha*, and were one of the land's major pests. As of the 1860s, there were large tracts where

> The soil looks as if ploughed by their burrowing. Some station holders of 100,000 acres have had to make contracts for killing them at 6d per tail, and as many as 22,000 on a single run have been killed by adventurous parties without any diminution being discernible.[7]

Cattle, introduced successfully in 1814, thrived under both *pakeha* and Maori masters, and many went wild as well. Cattle wandering in the forests were one of the chief sources of protein for *pakeha* settlers in the nineteenth century. (Cattle don't have to have meadows, and can prosper as forest dwellers.) Other exotic mammals that have gone feral in New Zealand and that still live in its wilderness are horses, goats, sheep, cats, dogs (which became the land's most dangerous carnivores), stoats, ferrets, hedgehogs, rabbits, rats, and mice. Some—opossums and wallabies, to cite a couple—are from Australia, and a very few—wapiti and moose, for instance—are from the Americas; but the great majority both in number of species and number of individuals are Old World, specifically European, in origin. None of New Zealand's mammals, except the bat, was living there when Charlemagne ruled in Europe.

The most spectacularly successful of the immigrant life forms that came to New Zealand by *pakeha* ship were the pathogens of Old World diseases. It is likely that the first *pakeha* explorers left disease organisms among the Maori, but the first epidemics did not appear, according to the indigenes' oral tradition, until the years around 1800. The most famous—or infamous—of these was the *rewharewhu*, sometimes identified as dysentery and sometimes as influenza. Perhaps the name, which means foreign disease, refers to several different epidemics or perhaps to one epidemic of what we bewildered laymen still call intestinal flu. The tradition is that it killed so many that the living were not able to bury the dead. The Maori called another early population explosion among imported pathogens by the name of *papareti* which means a sort of toboggan. They compared the dying of so many with the swift slide of a toboggan. Another they called *toko-tokorabgi*, the spear of heaven.[8]

As the nineteenth century proceeded, our knowledge of the advance of Old World pathogens among the Maori improved. Measles struck in the 1830s and 1850s, whooping cough in 1829 and 1846, influenza again and again from 1826 to 1842. These all decimated the Maori, sometimes in both islands, sometimes in only one part of one island, and are worthy of careful attention, but we don't have space here. We must turn immediately to the demographically most important infections: tuberculosis and venereal disease. By no later than the 1820s, tuberculosis was widely prevalent among the Maori of at least the Bay of Islands region of the North Island, and within a generation spread throughout the Maori of both islands. Many died of pulmonary tuberculosis, a lot of it of the galloping variety: "It is Atua, the Great Spirit, coming into them, and eating up their inside; for the patient can feel those parts gradually go away, and then they become weaker and weaker till no more is left; after which the spirit sends them to the happy island." More suffered the disease as scrofula, an infection of the lymph nodes, especially those of the neck. Said physician Arthur S. Thompson in the 1850s:

Scrofula is the curse of the New Zealand race. In some districts twenty percent, and in others ten percent of the population bear on their bodies the mark of the King's Evil, although all scrofulous persons have not this outward sign of scrofula. Scrofula is the predisposing and remote cause of much of the sickness among the New Zealanders [Maori]; in childhood it causes marasmuses, fevers and bowel complaints; in manhood, consumption, spinal disease, ulcers and various other maladies.[9]

As late as 1939, tuberculosis still accounted for 22 percent of all Maori deaths. Venereal disease, or at least the distinctively Old World kinds, probably arrived in New Zealand exactly as early as European sailors had access to Maori women. The Maori were preadapted to acquire and spread such infections swiftly—they practiced polygamy, accepted premarital intercourse as normal, and often utilized sexual favors as an expression of hospitality. These traditions neatly meshed with the needs of the *pakeha* sailors, especially of the visiting whalers, condemned to celibacy for months, even years, by the nature of their lonely business; venereal infections spread throughout Maori society, killing many and rendering sterile many more. By the 1850s, the Maori birth rate had dropped far below replacement rate. F. D. Fenton, author of the quantitative study *Observations on the State of the Aboriginal Inhabitants of New Zealand* (1859), found the Maori to be remarkably infertile and in a "state of decrepitude."[10]

The spread of imported pathogens among the Maori, in addition to the wars that the importation of thousands and thousands of muskets set off among the native New Zealanders, drastically reduced their population. There were probably at least 200,000 of them in 1769—certainly no fewer than 100,000 because about that many were alive when the first informed estimates were made of their number in the 1820s and 1830s, by which time, it was widely agreed, the population was already decreasing. The Maori reached their nadir at the end of the century. According to the census of 1896, there were only 42,113 of them left in a land that by then contained over 700,000 *pakeha*. In that decade, according to the raw public health statistics, *pakeha* New Zealanders may well have enjoyed the lowest general death rate and lowest infant mortality rate in the world.[11]

No wonder the *pakeha* Social Darwinists thought that laws of nature were in operation to make them the dominant people in New Zealand, as they already were in the neo-Europes of North and South America and Australia. Travers observed that the takeover of New Zealand by invaders appeared to be not only a matter of European humans' superiority but that of European species in general. New Zealand's ecosystem, he said, "had reached a point at which, like a house built of incoherent materials, a blow struck anywhere shakes and damages the whole fabric." He felt that even if every last *pakeha* left New Zealand, the Old World organisms left behind would go right ahead and depose the indigenous flora and fauna.[12]

We might argue with this conclusion—the experience of the hundred years since shows it to be a wild exaggeration—but we cannot deny the reality of the changes he was quite literally witnessing, as were similarly Darwinistic observers in the other neo-Europes. The impact of the portmanteau biota—from the spread of the weed *Plantagomajor* in North America, where the seventeenth-century Algonkin called it "Whiteman's foot," to the sudden appearance of millions of Old World mice in the small town of Lascelles, Australia, where 544 tons of them were swept up in June 1917—was titanic in scope and permanent in effect. Biological convergence was replacing divergence, which had been the rule on this planet for scores of millions of years.[13]

Why were the portmanteau biota so triumphant in the neo-Europes? Several general answers come to mind. First, the portmanteau biota succeeded because they were composed, for the most part, of weed species, by which I mean organisms, plant or animal, macro or micro, that thrive in *ephemeralized* environments: for instance, European grasses and forbs spreading swiftly across the pampa after the Amerindians and Gauchos burned off the dry native needle grasses in the spring; or European grasses, forbs, animals, germs, and people moving swiftly into the biotic chaos and vacant eco-niches created in all the neo-Europes by their very own advance.

The portmanteau biota, moreover, included species from large ecosystems characterized by a high degree of competition, which produced fecund, tough organisms. For instance, feral Old World cattle, whose ancestors had survived hundreds of millennia being trailed by wolves, found Australia's most important carnivore, a medium-sized dog called the dingo, little danger to their survival as a species.

The portmanteau biota also included many small and invasive organisms—most important, micro-organisms—that had taken on many of their most significant characteristics in the unique crucible of Old World civilization, in which large concentrations of humans and animals—pigs, sheep, chickens, rats, mosquitoes, and so on—lived in intimate contact and in dense concentrations. I speak, for instance, of the pathogens of smallpox, malaria, influenza, measles, rinderpest, and distemper—micro-organisms first developed their present characters in the special conditions of Old World civilization, with its intermingling crowds of humans and other creatures. Such conditions were rarely common outside the Old World, and so the neo-Europes lacked many Old World diseases or their equivalents. The New World, for instance, had nothing like the Old World's smallpox, so closely related to and probably of the same origin as cowpox; and the Aztecs fell victims to the disease in windrows, while Cortes lived to a ripe old age.

Finally, the portmanteau biota were made up of many mutually supportive organisms. The Europeans brought cattle and horses to the pampa in the sixteenth century; the tens of millions of feral cattle and horses that occupied the pampa from the sixteenth to the nineteenth centuries provided the resident neo-Europeans with mounts, food, leather, rope, and so on, and all for free. The Old World quadrupeds stripped the pampa of the native palatable plants, and deposited the seeds of Old World grasses and forbs, along with fertilizer, and then thrived upon the resulting forage; and the Argentineans and Uruguayans thrived on them.

The coming of white clover to New Zealand provides a fine example of the invaders' mutual support system. In the early nineteenth century, the clover would grow there luxuriantly, but would not seed and spread. It would last one season and then disappear because this country lacked an efficient pollinator with a taste for clover nectar. In 1839, a certain Miss Bumby imported hives of European honey bees, and the bees prospered. Since that day they have been the chief pollinators of white clover in New Zealand. In turn, white clover is the bees' chief source of nectar. Domesticated and wild the bees have spread widely, outcompeting the nectar-dependent native birds in some regions and driving them back into the surviving native forests. White clover, dependent on immigrant honey bees, is one of the chief forage plants for New Zealand's scores of millions of sheep, which provide the *pakeha* with their chief exports, and for whom and by whom enormous areas of New Zealand have been transformed into European meadowlands. Honey, along with wool and mutton, is one of the nation's chief exports. A full account of the interdependency of white clover, honey bees, livestock, and *pakeha* and of their interweaving roles in changing New Zealand's biota and language would require a long book with appendices of three-dimensional kinetic graphs.

The neo-Europes were conquered by teams of organisms associated with and in many cases essential to Old World civilization—organisms that often, as species, possessed special talents for survival, especially in the kind of environment that Europeans and their animals and microlife created. The immigrant organisms seldom possessed a conqueror's might in and of themselves, but as a team possessed enough power to permanently change the biota and, thereby, the human populations of great expanses of the world.

In many important ways the fragments of Pangaea are drawing together again, and the supercontinent is biologically if not geographically being restored to its ancient unity. The chief beneficiaries thus far have been the members of Europe's portmanteau biota, and the chief victims have been many of the native organisms of the Americas and Australasia, including Amerindians, Australian Aborigines, and New Zealand Maori.

Australian Nature, European Culture

Anglo Settlers in Australia

THOMAS R. DUNLAP

The story of whites and the natural environment of Australia is the tale of a European culture in a non-European land. Recent studies—Geoffrey Bolton's *Spoils and Spoilers: Australians Make Their Environment, 1788–1980* (1981), Derek White-lock's *Conquest to Conservation: History of Human Impact on the South Australian Environment* (1985), and William J. Lines's *Taming the Great South Land: A History of the Conquest of Nature in Australia* (1991)—concentrate, as their titles and sub-titles indicate, on one overwhelming fact of that history: the changes Europeans made in Australia.[1] For Bolton the settlers made all: the built environment by construction, the natural one by destruction. Whitelock concentrates on human destruction and the more recent rescue of South Australia's natural environment. Lines is the bluntest. "The British Empire and Australia's tethers to the post-En-lightened industrial world—human constructions all—not nature, created mod-ern Australia."[2] This chapter takes another perspective. "Nature does not dictate, but physical nature does, at any given time, set limits on what is humanly possi-ble. Humans may *think* what they want; they cannot always *do* what they want, and not all they do turns out as planned."[3] In the long run Anglo Australians could not even think what they wanted. They went out with axes, plows, fence wire, mining drills, and bulldozers, guided by ideas of conquest and domination, the efficacy of technology and the beauty of the English countryside. The land

273

battered the ideas as much as it blunted and broke the tools. The only difference was that the tools went quickly and could be replaced; ideas changed more slowly and new ones had, sometimes painfully, to be made, taking the land into account. Anglo Australian environmental history is a story of continuing reciprocal influences between human societies and the land. The disproportion between humans' quick and evident impact and natural systems' slower and often hidden response and effect conceals the mutuality, but it is there and, in the long run, dominant.

Australia's tale is its own, but also one variant of the history of the Anglo settler lands, the countries (Australia, Canada, New Zealand, and the United States) where the invaders killed or displaced the earlier inhabitants and set about reproducing their old society.[4] Within this group Australia is the extreme physical case. It is the least like the European homeland and the most resistant to attempts to remake it to the familiar pattern. It is "the lowest, flattest, and, apart from Antarctica, the driest of continents. Unlike Europe and North America, where much of the landscape dates back to 20,000 years ago when great ice sheets retreated, the age of landforms in Australia is generally measured in many millions of years."[5] Sixty percent receives less than 400 mm (16 inches) of rain a year, 80 percent less than 600 mm (24 inches), and rainfall is in most places so erratic as to be unpredictable.[6] Lakes are few and small; streams and rivers often run only seasonally and to sinks and claypan rather than the sea. Sun, wind, and rain, not countered by volcanic action or crustal collisions throwing up new minerals or mountains, have leached the soils and flattened the land. The coast was not rich, and Australian explorers found the land more barren and desolate as they sought the center.[7] Mackenzie and Lewis and Clark crossed North America and returned to tell of mountains and at least the riches of furs. Burke and Wills crossed Australia and starved to death on the return. Two hundred years after the first white settlement (and with the aid of modern technology) Australians farm only the southeast, part of Tasmania, and pockets in Western Australia and Queensland. A "frontier line," drawn on the old standard of two people per square mile, would hardly be 150 miles inland at any point, and even much of the coast would lie outside the populated enclaves. Canberra, the only inland city, is an artificial creation, a purpose-built national capital.[8]

Not only was the land harsh, the plants and animals were so strange that Australia entered the European consciousness as the home of freakish animals and strange plants. Here, as a Mr. J. Martin complained in the 1830s:

trees retained their leaves and shed their bark instead, the swans were black, the eagles white, the bees were stingless, some mammals had pockets, other laid eggs, it was warmest on the hills and coolest in the valleys, [and] even the blackberries were red.[9]

Isolated for millions of years, the flora and fauna had evolved and radiated into ecological niches in ways unlike those on any other large land mass. The monotremes (egg-laying mammals) were unique. Marsupials, absent or rare in other lands, were common and assumed forms—the kangaroos—unlike anything the Europeans had seen. Large placental mammals were conspicuously absent; there were no deer, bear, wolves, or even rabbits and foxes. The plants were equally odd. Eucalypts, rare elsewhere, dominated, their lacy, irregular silhouettes giving, early visitors complained, no shade and a strange look to the landscape.[10]

The first colonists brought little but their cultural baggage, but they did bring that. They found it a poor fit. Their first struggle was to see the new for what it was rather than the familiar distorted. Early drawings show kangaroos with mouse-like muzzles, horses' ears, and forepaws like strange hands. Echidnas look like modified hedgehogs. Gum trees and billabongs appear off-color, off-center, and not quite right.[11] Experience bred familiarity, and practice made, if not perfect, at least pictures like Australia and not Britain through the looking glass. Finding the picturesque or the sublime in the new landscape was more difficult, but the settlers tried. Even the convicts tried. Thomas Watling, transported for forgery, wrote home to Scotland in the 1790s of the "Arcadian shades" and "romantic banks" around Sydney Harbor. He had to admit, though, that these were but "the specious external."[12] An aesthetic that found beauty in flowing water, verdant vegetation, high mountains, bosky glens, and "noble" stags withered in a land of little relief, less water, and strange animals.[13] There were no charming vistas, no peasantry steeped in the lore of the land. Even when drovers and shepherds came to occupy the "outback" they did not fit the formulas of Gilbert White's *Natural History of Selbourne,* which set the early-nineteenth-century tone for nature appreciation.[14]

Their culture provided another approach to nature, natural history. Donald Fleming, comparing science in the United States, Canada, and Australia, called it "a fundamental part of the quest for a national identity in societies where the cultural differentiation from Britain was insecure and the sense of the land correspondingly important for self-awareness." It provided individuals a niche in science and a "psychological identification with pioneers" that rooted them in their new culture.[15] The settlers lost no time entering the field, forming their first scientific society, the Philosophical Society of Australia, in Sydney in 1821. It had the imprimatur of the elite; its first president was Governor Sir Thomas Brisbane, himself a fellow of the Royal Society. Each succeeding colony founded its own group and in the second half of the century each boasted a natural history museum and a royal scientific society. The lack of population and resources, though, told. The early societies were all short-lived, the later ones small and poor, and the museums peaked under the first generation of founder-directors.[16] The thrill of discovery, if not the sense of identification with the pioneers, had to be tempered by the knowledge that Australians were second-class citizens in the world of science.[17] Experts in Europe described and named the specimens, for only they had access to the necessary collections, and Australian faunas and floras were all printed in London. Europeans even confirmed field observations about Australian fauna. In 1884 an English scientist created a sensation by reporting that the echidna did indeed lay eggs. An Australian's report, the same week, was ignored.[18]

The settlers though, were less interested in understanding the land than remaking it. During most of the nineteenth century they sought to make Australia a "new England" in the South Seas.[19] Finding the native birds voiceless or unmusical they imported, as one put it, "those delightful reminders of our English home," expecting them to spread through the plains, bush, and forest, which would "have their present savage silence, or worse, enlivened by those varied touching joyous strains of Heaven taught melody."[20] The native animals were not appropriate for the fashionable recreation of sport hunting, and the settlers brought in deer (some seventeen species or sub-species), foxes, rabbits, grouse, and pheasants.[21] Introductions began with the First Fleet, but the flood was concentrated between 1860 and 1880, driven by the European fad for importation (usually termed "acclimatization"), Australians' desires for familiar and useful creatures, their new wealth from gold and wool, and advances in ships and shipping. Sentiment and sport were the dominant motives, and economic rationales conspicuous by their absence. Enthusiasm outran judgment, and the desire to make a new England coexisted with more general desires for things that were useful, beautiful, or novel. The colonists tried, among other creatures, alpacas, llamas, ostriches, camels, peafowl, guineafowl, mynahs, and red-whiskered bulbuls.[22]

Statutes reflected the scale of social values. The first game laws, passed in the 1860s, occasionally noted the expense incurred by public-spirited gentlemen in bringing in game for the improvement of the colony. They always gave landowners property in these species and sometimes part of the fine imposed on violators. Legislators sought to save native birds from "wanton destruction," but with notably less vigor. Fines were lower, seasons long, bag limits and wardens nonexistent. That enforcement depended upon someone's bringing a complaint further biased the system toward the imported species.[23] Animals unsuitable for sport the law divided into pests, which the landowner had to control, and everything else, which it ignored.[24] Settlers shot platypuses for their value as zoological rarities, then for their fur. They killed koalas for fur, birds for their plumes, and kangaroos for meat or hides, or just to save the grass for sheep.[25]

The drive to make the bush the English countryside began waning in the 1880s. For one thing it was not working. Roses withered in the dry heat, the skylark found only a foothold, and the nightingale not even that. Recurrent, devastating droughts slowly killed the dream of small farms for all (though it persisted into the twentieth century). Some of acclimatization's successes were even worse than the failures. In 1860 a shipment of European rabbits brought in by Thomas Austin, "an English tenant-farmer who had made a fortune in Australia and wished to play the part of a sporty squire," survived and multiplied.[26] In twenty years rabbits went from being a valued and interesting import to a major economic disaster. As the "grey blanket" overran the continent increasingly frantic legislatures tried poison, bounties, and fences (one over a thousand miles long). New South Wales offered a £25,000 reward for an effective remedy. Nothing stopped the hordes, and the case quickly became notorious. By 1898 a USDA biologist, discussing "The Danger of Introducing Noxious Animals and Birds," could say that "the results of the experiment are so well known that anything more than a brief reference to them is unnecessary."[27]

The dream of transforming the land also died because Australian society was changing. By 1880 the native-born outnumbered the immigrants, and for them Shelley's nightingale and English hedges were as alien as the woolen school uniforms they wore in Sydney's heat. Their childhood memories were the bellbird's chime and the kookaburra's laugh, the ragged silhouette of eucalyptus trees, the crunch and smell of their leaves and bark underfoot. Nationalism reinforced this current. As the colonies gained responsible government and moved toward independence and Confederation (1901), they sought a national myth to unify and justify their new nation. They found its materials in the bush. Life in the outback, it was said, had made the new man, the Australian—strong, straightforward, loyal to his mates, ready to defend the Empire and Australia.[28] Australian nature became a matter for national pride, and Australians began to celebrate their landscape. The "Heidelberg School," a group painting in the open air around Melbourne, produced in the 1880s images that, one art critic said, have "mediated the relation of the bush of most people growing up in Australia. . . . Perhaps no other local imagery is so much a part of an Australian consciousness and ideological make-up."[29] "And out of that work," said another, "the Australian eucalypt arose as a dominant symbol of the Australian bush."[30]

Between 1880 and the 1920s Australians also assimilated new currents of opinion from the Anglo world: nature education, outdoor recreation, wildlife preservation and parks, and conservation of natural resources. Each movement, though, assumed a distinctive form in Australia, shaped by nationalism, the continuing cultural ties to Great Britain, a growing attraction to the United States, and their own life and land. Imitation gave way to a more selective adaptation.

Nationalism and nature consciousness combined in the nature essays and stories that began appearing in the 1880s. Donald Macdonald's *Gum Boughs and Wattle Bloom* (1888), a collection of his newspaper essays on Australian scenery, marked the start of a self-conscious attempt at an emotional relation with the Australian landscape as a home. Macdonald and his followers, who formed a group spanning two generations and dominating Victorian nature study, used both British and American models, Richard Jeffries, W. H. Hudson, and Henry David Thoreau.[31] The group was also interested in children's nature education. Macdonald started a newspaper column for boys, which Alec Chisholm took over when Macdonald died, and Chisholm wrote a book for school children urging nature and particularly bird study, *Mateship with Birds*.[32] The writers also were enthusiastic proponents of outdoor recreation as a training ground for young soldiers.[33] There was a boon in children's nature stories that followed the English model, in which animals spoke and acted as humans, were moral exemplars and bad examples, but not creatures in themselves.[34] Didactic, humane, and national, they brought the creatures of the bush to a largely urban population in a mixture of fantasy, nature instruction, and nationalism. Ethel M. Pedley's *Dot and the Kangaroo* (1899), one of the first of the genre, recounts the tale of little Dot. Lost in the bush, she is saved and instructed by a mother kangaroo, who finally risks her life to restore the girl to her family. The dedication is to "The children of Australia, in the hope of enlisting their sympathies for the many beautiful, amiable, and frolicsome creatures of their fair land; whose extinction, through ruthless destruction, is being surely accomplished."[35] The frontispiece of May Gibbs's popular fairy tale of the bush, *Snugglepot and Cuddlepie* (1918), shows one of her gumnut babies writing on a leaf: "Humans, Please be kind to all Bush Creatures and don't pull flowers up by the roots." Her illustrations, though containing fairies and other fanciful creatures, are exact enough that botanists can identify genus and in some cases the species.[36] Dorothy Wall's *Blinky Bill: The Quaint Little Australian* (like *Dot and the Kangaroo* and May Gibbs's work, still in print), instructs little children that since koalas can only eat a certain kind of gum leaf, it is not a good idea to keep one as a pet.[37] One woman, recalling her childhood in a small Australian town in the 1950s, said that while she and her playmates led lives much like those of city children these books, "for the most part written in the early decades of Confederation," made them think of themselves "as bush children, tough little individualists expert in the lore and perils of the outback."[38]

Early in the century the states added nature study to the elementary school curriculum, providing a formal introduction to Australian nature and, through school bird clubs, organizations that supported nature preservation. Nature study was a common Anglo enthusiasm, driven by a child psychology that emphasized a child's immediate experience as material for learning and a concern that the urban children learn about something besides city sidewalks.[39] The British model was available, but Australians drew on the United States. One early text paid homage to the mother house of nature study, Cornell; South Australia's proclamation of school bird day was "backed by a recommendation from the United States" and the Gould League of Bird Lovers, the school bird clubs, was formed as an "equivalent to the Junior Audubon Society of the United States" (though named after the British naturalist-entrepreneur who produced the major nineteenth-century works on Australian biota).[40] The continuing domination of British ideas and materials in the schools meant that until after World War II nature was the only subject in which Australian students studied their own country—an unexpected linking of nature and nationalism.

Some Australians were interested in nature preservation, but the impulse was not urgent, and it did not run deep. The states set aside what they confusingly called "national parks" (a title they still retain) for outdoor recreation for city dwellers. The idea of a park for monumental scenery, the norm in the other Anglo settler nations, did not develop. Australia had its own beauty, but it was not of the snowcapped peak or bottomless canyon that was the ideal. The result was that while Australians wanted parks, they wanted them in no specific place. Their criteria were accessibility to city populations and the principle that, as one modern commentator put it, "if the scenery is good and nobody else wants it, then it could be a park."[41] Only the island state of Tasmania, with rugged mountains and well-watered valleys, set up a Scenery Preservation Board (1915) and actively promoted scenic parks.

Australians preserved wildlife, but their concern was, compared with that in the United States or Canada, muted. The perception of continued abundance diffused worries, as did the lack of horrific events. There was nothing like the slaughter of the North American bison that stirred the American and Canadian publics and governments.[42] In addition, Australians believed well into the twentieth century that their native animals, however wonderful, were doomed. They could not compete with "more advanced" forms, and their demise was as inevitable "as that the aboriginal should give way to the white man."[43] As did other Anglo jurisdictions, the states restricted methods of hunting to those deemed sportsmanlike, provided legal protection to nongame animals, and protected a few, like the platypus, that were distinctive and becoming rare, but their action was erratic. Market hunting continued (it was banned in the United States about this time) and enforcement was minimal.[44] Even extinction did not cause much concern, at least if the visible animal was a menace to sheep. Tasmania continued to offer bounties on the thylacine, the "Tasmanian tiger" or "marsupial wolf," until 1912, three years after the last one was presented for payment. It first offered protection in 1929, when the Tasmanian Animal and Bird Protection Board set a one-month closed season. The board outlawed all killing in 1936, but by then it was too late.[45] When the thylacine vanished there was little public concern and few calls for research and protection; not until the 1960s would the public see its disappearance as tragedy.[46]

Conservation in Australia was little and late. The states all formed forestry commissions, usually hiring a forester from another part of the Empire, but they studied far more than they managed. Despite the relative scarcity of water only Victoria made a state-wide effort to develop and manage its system, hiring Elwood Mead from the United States to build an agency.[47] Conservation lacked an institutional base, and there was little demand that one be created. There were few experts, even fewer of them permanent residents, schools or professional organizations. Their low and scattered population encouraged Australians to think of the continent as still largely uninhabited and unexploited, muting the fears of timber famine which drove the development of American forestry. Even if they had been concerned, the lack of a national public domain or Commonwealth authority over natural resources precluded the kind of movement that made conservation a crusade in the United States.[48]

One Anglo current that did take hold was outdoor recreation. In the 1880s

Australians began hiking and camping and by the 1920s had fused these activities into a distinctly Australian activity—bushwalking. Its prophet was Myles Dunphy, who grew up in New South Wales and began hiking in the Blue Mountains in his teens. In 1914 he helped found the Mountain Trails Club, and in 1927 the Sydney Bushwalkers, the club that, he said, coined the word bushwalking. He wrote trail guides, developed equipment, and proposed a set of parks whose acquisition and development kept New South Wales park advocates busy into the postwar era. His ideas and program were the common currency of the Anglo world. His call to "overworked clerks with lackluster eyes and twitching nerves" might have come from Muir, Roosevelt, or Baden-Powell. "You are short-winded, effete, stale, and partly dead," he said. Go camping. "To be out with a swag is good medicine. . . . Without realizing it you will learn initiative, honesty, industry, cleanliness, trustworthiness, order, cheerfulness and how to remain healthy. These make character." Bushwalking was "the Cult of the Outdoors—craft, plus Appreciation plus Homage to the Cause of Life and Nature: a sort of sensible religion, in the practice of which the bushman acknowledges his Maker and justifies his existence" by using his talents for self, his comrades, the cult, and his country.[49]

Dunphy's ideas, though, had a distinctly national cast. The true bushwalker, he proclaimed, disdained packs and pack frames. He carried the gear of the pioneer bushwalker, a swag (a canvas roll with gear in it), a tucker (food) bag, and a billy can (to boil water). The swag was the symbol of Australian brotherhood, and "the swagman is welcome wherever human hearts are right and simple."[50] In 1931 he described his club as the association of those "who instinctively reject roads and beaten tourist routes in favour of the canyons, ranges, and wildest parts of this country. It is for those who love the forests and the broad, open life of the Bush, and who prefer to make their own trails, who have a definite regard for the welfare of wild life and the preservation of the natural beauties of the country." The group aimed "to combat the destruction of all things naturally Australian which casually goes on everywhere and for which posterity will hold this generation to blame."[51]

Until after World War II there seemed room enough for mining, lumbering, bushwalking, parks, and nature. Only when the postwar economic boom put more, and more obvious, pressure on wildlands and recreational areas were Australians forced to choose. Two forces have dominated the ensuing, and continuing, debate: ecology and a second wave of Australian nationalism. Discarding ideas of efficient use of individual natural resources and the separation of areas to be preserved from ones to be developed, wilderness and nature advocates argued that everything had to be preserved if anything was to be saved and that human survival was at stake. World War II reinforced Australian nationalism, and the postwar collapse of the British Empire forced Australians to reconstruct their national identity. This combination of increased development (which made conflict apparent), ecology (which provided a framework to critique the effect of industrial society on the land), and nationalism (which encouraged identification with the country), drove and helped define the popular environmental movement that has begun to reshape Australian ideas and policy.

The first reaction to postwar development was concern but not alarm. In 1949 state wildlife officials, meeting in Hobart, worried that development was threatening scientific study of natural populations and even some wild populations.[52] They agreed that science should guide policy, agreed as well that it did not. Their remedies, though, were new state regulations and greater cooperation between states. The Commonwealth showed little inclination to become involved. Just after the war Victoria and South Australia sought federal support for their koala restoration programs; federal officials rebuffed them on the grounds that wildlife was constitutionally a state responsibility.[53] There was a difference, though. In the interwar period people had lamented the destruction of the native wildlife, now they seemed to think it should and could be saved.

The Commonwealth took its first steps into wildlife work in 1949, when it established a Wildlife Survey Section in the Commonwealth Scientific and Industrial Research Organization (CSIRO), under the direction of an English biologist, Francis Ratcliffe, who had come to Australia in the 1930s to work for CSIRO studying flying foxes (fruit-eating bats), in Queensland and New South Wales orchards, then soil erosion in South Australia. He had fallen in love with the country and stayed.[54] As director of federal wildlife work he shaped field biological studies. He was as well mentor and guide to the first generation of wildlife researchers trained in Australia, the government's representative to the states on wildlife issues, a founder of the Australian Conservation Foundation (1956), and until his death in 1970 the national authority on Australian nature.[55]

The agency's original mission, rabbit control, seemed overwhelming, but in December, 1950, the rabbit disease myxomatosis began spreading from CSIRO's test areas in the Darling River Valley. In the next three years it produced a continent-wild epizootic as remarkable as the rabbit's spread, with death rates in some areas running over 90 percent.[56] This gave the section a great reputation and freed time and resources for other work. There was much to do. The Australian states had been concerned with economically important animals, chiefly the imported pests. Little was known of most native species beyond basic taxonomy. H. H. Finlayson, exploring the "Red Centre" in 1934, worked areas no other scientist had even seen, and the next biologist into the area, Alan Newsome, came twenty years later. John Calaby, who did biological surveys for CSIRO in the 1950s around what is now Kakadu National Park, recalled forty years later the experience of being in land no scientist had ever closely examined.[57] Under Ratcliffe research moved "away from economic pests to the larger biological context," a trend his successor continued.[58] The Commonwealth was accumulating a base of knowledge on which management decisions could be made.

While this was going on scientists and naturalists were taking on another task: warning Australians that they were destroying their environment (a word hardly used before 1940). Works like *The Great Extermination: A Guide to Anglo-Australian Cupidity, Wickedness, and Waste* (1966), lamented the destruction of Australian biota, called for an appreciation of the native beauty of the country, and sought even the preservation of "our heritage of native fishes. . . ."[59] Public personages or experts in resource management had been behind earlier arguments (books like *Save Australia: A Plea for the Right Use of Our Flora and Fauna* [1925]), and they had appealed to economic efficiency. The authors of *The Great Extermination* were interested in nature, not resources, all wildlife, not just birds and mammals, and they concentrated on the damage humans were causing to the continent. They did not think that native wildlife was doomed to extinction. They believed it could be saved and should be, as a symbol of the country and a national heritage.

These attacks meshed with and reinforced public worries. Just after the war, particularly in New South Wales and Victoria, they had centered on the parks, but by the early 1960s included Australia's unique ecosystems, pollution, and environmental degradation.[60] When an infestation of Crown-of-Thorns starfish appeared to threaten the Great Barrier Reef, a 2000-mile-long chain of islands, atolls, reefs, and other coral formations along Australia's northeast coast, there was a public outcry. A national committee was formed to direct research and urge preservation of the reef. CSIRO began supporting reef research, and in 1969 the new Australian Conservation Foundation (ACF) held a seminar on reef problems and research needs.[61] Consciousness spread, and local groups sprang in many areas to protest various mining ventures, factory wastes, beach pollution, sand-mining, and the destruction of native forests for woodchips.[62]

The rapid evolution of the ACF, one of the first groups formed to take action on the environment, shows how, and how quickly, the movement changed. It began in the mid-1960s as a "top-down" operation (an Australian mode that coexists with the culture's strong egalitarian streak). In 1964 Prince Philip expressed hope that Australia would be a member of the World Wildlife Fund when he arrived for a royal visit the following year. That required an organization to petition for membership, and Prime Minister Menzies promptly assembled a group of scientists and businessmen, who formed the Australian Conservation Foundation. The Prime Minister provided £1000 pounds of seed money. The ACF was not open to public membership, and, like the early scientific societies, tied to the elite. Its president was Sir Garfield Barwick, Chief Justice of Australia's High Court, and Francis Ratcliffe, the group's senior scientist, saw it as a group of wise men, supporting important research and advising ministers from behind the scenes.[63] It proved unable to attract support on that basis. In the late 1960s it opened its ranks to the public and moved from research into public education (which caused Ratcliffe to resign).[64] In October, 1973, a group of activists used parliamentary procedures to stall a meeting until their opponents left, packed the Council, and elected a new director. Seven Council members who were scientists resigned in protest and sent a report to the membership, "How the ACF Was Taken Over."[65] Popular environmentalism displaced backroom advice from experts.

The rise in environmental organizations matched the ACF's transformation. Its *Conservation Directory,* issued in 1973, listed 206 groups in the most populous state, New South Wales. (The Australian constitution, by leaving most powers with the states, encouraged action at this level.) Of those that gave the date they were founded, seven were from the 1940s, twenty from the 1950s, thirteen in the first half of the 1960s, thirty-three from the second half of the decade, and twenty-two were formed between 1971 and 1973.[66] The list was necessarily biased toward groups that survived and included many (the Catholic Bushwalking Club, for example) that had other than environmental aims, but it is a rough measure of interest. It showed, not surprisingly, the greatest concern in the cities and the southeast. Though there were active groups in Queensland and Western Australia, the two fully self-governing states outside that area, their greater dependence on grazing and mining blunted environmental sentiment.[67]

The controversies that most shaped the environmental movement occurred in Tasmania, for it was a national recreation ground that could draw on a national constituency to defend its rugged scenery and wild rivers. The first phase began in the early 1960s, when the Tasmanian Hydro-Electric Commission unveiled plans for a dam project that would drown Lake Pedder. This small scenic gem of a mountain lake had been made a national park in 1955, and its defense was a test of park integrity as well as wilderness sentiment. Pictures showed what was at stake, and the HEC's heavy-handed tactics aided the environmentalists, but in the short run that hardly mattered. Opponents delayed the dam into the early 1970s, but plans were approved, it was built, and the lake submerged.[68] The reservoir destroyed Lake Pedder, but not environmental sentiment or organizations. In 1972, in the last phase of the battle, activists formed the United Tasmania Group, the world's first Green political party. Four years later they organized the Tasmanian Wilderness Society.[69]

In the late 1970s the Hydro-Electric Commission's plans to dam the Franklin River set off a second set of protests. These quickly grew into a battle over preservation in the rugged southwest that strengthened the environmental movement and made wilderness a national issue. The Tasmanian Wilderness Society organized rallies and protests (including some guerrilla theater). The most visible was a campaign of nonviolent resistance to block the construction site. It drew on the American civil rights movement and Mahatma Gandhi, attracted national attention to the cause, and filled Tasmania's jails. The society also moved on other fronts that were, ultimately, more effective.[70] It backed candidates for office, appealed to the Commonwealth government to stop the state, and sought World Heritage status from UNESCO for the southwest. In 1983 national electoral pressure, the World Heritage issue, and a crucial decision by the High Court halted the dam. Within a few years there were parks encompassing much of the southwest, now a World Heritage Area.[71] Wilderness and environmentalism were national issues, and the Tasmanian Wilderness Society, now with many mainland branches, changed its name to the Wilderness Society.

Wildlife policy also shifted. Between 1967 and 1977 six of the states established consolidated agencies to manage parks, wildlife, and plants, while the seventh, Victoria, remodeled its existing bureaus.[72] This was in part a move to rationalize administration, but where earlier laws had dealt with scenery or species, these spoke of a "national heritage" and the need to preserve habitats. The Commonwealth took on new responsibilities, pushed by activists and dragged in by issues that the states could not or would not resolve. The issue that drew the government in was the plight of the larger kangaroo species. In the late 1960s drought and shooting (in part for pet food, to replace rabbits lost to myxomatosis) created what wildlife activists believed was a threat to entire populations. When the states proved reluctant to curb the killing, a coalition of environmentalists, humanitarians, and ordinary citizens turned to the national government. Between March 1968 and May 1970 Parliament received eighty-four petitions protesting the commercial harvesting of kangaroos.[73] In response it appointed a seven person Select Committee on Wildlife Conservation (known after its chair, E. M. C. Fox, as the Fox committee), and gave it a broad mandate. It was to examine the need for an immediate survey of wildlife and its ecology as a guide to conservation measures, the adequacy of existing reserves to protect wildlife, and the problems posed by pollution, pesticides, and feral animals. It was as well to advise Parliament on the Commonwealth's proper role in wildlife policy.[74]

The Fox report, delivered in 1972, charted a new course. Its boldest recommendation was continuing federal involvement in wildlife affairs through two new Commonwealth agencies. One, a national biological survey, would unify the now-scattered federal specimen collections, undertake "on a continuing basis surveys of birds, mammals, and reptiles and their ecology and . . . establish a national collection of wildlife species."[75] The other, a parks and wildlife office, would administer preservation programs in Commonwealth Territories and take responsibility for international treaties on migratory and endangered species. It would, "in cooperation with the States," survey park needs and wildlife populations, set national guidelines for policy, monitor endangered species, and establish field study centers and a ranger training school. The committee recommended protecting all habitat areas by including them in national (that is, state) parks, with Commonwealth grants to help the states acquire areas "which are of national significance." It proposed studies on everything from park expansion to the environmental impact of industry and agriculture. With regard to kangaroos, it was more cautious, suggesting that the Commonwealth "approach" the states "with a view to obtaining greater uniformity of laws relating to the taking of kangaroos." It justified all this by appealing to national prestige. Australia, it said, has a unique fauna, which had been little studied and was being affected by economic development, and it was the only technically advanced nation in the world without a national institute for faunal investigations.

Within three years the Commonwealth established a National Parks and Wildlife Service (ANPWS), a Biological Resources Survey, and a Council of Nature Conservation Ministers.[76] Though ANPWS administers directly only parks in the Northern Territory and on a few offshore islands, it provides, under a cooperative agreement, funds and overall direction for the Great Barrier Reef Marine Park (Queensland is responsible for day-to-day management) and money to keep the Tasmanian parks up to the standard required by their World Heritage status. Australia has joined the Convention on International Trade in Endangered Species (CITES) and is protecting migratory waterfowl under treaties with China and Japan.[77] Through export controls, the Commonwealth regulates market hunting of kangaroos. Wildlife remains a state responsibility, but the federal government now has a large role.

Australians have also begun to grapple with ideas of ecology, wilderness, and a sustainable economy, and while they have often appealed to American examples—invoking Thoreau, Muir, and Leopold, American parks, and environmental policies (not always favorably)—the movement there has its own dynamic, shaped by the society and the land. With few earlier organizations dedicated to conservation or preservation, the movement has relied, far more than the American, on new groups rather than old ones transformed. It is far more ideological; no leading environmental group in the United States would have called, as did the Tasmanian Wilderness Society in the early 1970s, for the formation of an Australian environmental "left" outside the conventional Marxist-capitalist dichotomies or for discarding hierarchical organization in favor of communal decision making. The Australian constitution and the lack of a federal public domain have made the states, rather the federal government, the locus of action, and Parliamentary government and an electoral system that makes third parties viable has given advocates a very different relation to politics and policy makers than in the United States.[78] Obstacles to environmental action have their own twist. Australians accept economic development as an unalloyed good with a fervor that has vanished in the United States. No American official would, as the head of Tasmania's mining interests did in 1990, call for mining in all the national parks. He might believe it and wish fervently for it, but knowledge of public reaction, if nothing else, would keep him quiet.[79] Australians are also more embarrassed than Americans to openly defend nature. Dick Johnson, calling for a Victorian Alpine park, wrote in frustration that:

> The crucial spiritual components of wilderness have been ignored in that horror of emotional articulation which is the national gaucherie. In attempts to justify our stance we grope about for utilitarian explanations which simply don't exist. And for want of someone to say that Australians love their wild places, we stand about in embarrassed silence while our masters rip the living guts from the wilderness that is left.[80]

The elements of the land that killed Romantic nature continue to influence Australians' ideas about and defense of nature. In the other Anglo settler countries nature preservation began with monumental scenery, added wildlife, then redefined scenery to include wildlife habitat and unusual and neglected areas like swamps or deserts.[81] In Australia outdoor recreation led and scenery, even when it was an element, entered on different terms. Australians, lacking rallying points like Yosemite Valley, have either embraced distinctive regions—the Blue Mountains, the Tasmanian southwest, or the northern rainforests—or a local patch of bush. They have paid attention to a few animal species—kangaroos, platypuses, and koalas—but mammals and birds have less emotional resonance than in the United States or Canada. In contrast, plants have been a much larger part of Australians' picture of nature (a characteristic shared with New Zealanders). They have a fierce attachment to the distinctive landscape formed by the dominant eucalypt species. In recent years botanical nationalism has crept even into the suburbs.[82] The early Canberra developments, built in the 1950s, have exotic trees lining their streets, the new ones Australian species, and some Victorian municipalities now make it policy to plant native species.[83] The flood of exotics has made Australians, also like the New Zealanders but unlike Canadians or Americans, very aware of what is native and what imported. There have been a few calls, in fact, for deer to be eliminated because they are "not really Australian." A counterblast by an exasperated sportsman caught the note of ecological nationalism in this argument. "It is," he said, "sheer humbug today for a white exotic human, yearning for a Dreamtime environment, to point the bone at a particular wildlife species he or she has been told is in the bush, but which they are never likely to see."[84] The sportsman, it might be noted, condemned the backward-looking environmentalists with terms ("Dreamtime," "pointing to the bone") from Aboriginal Australia.

The interaction of land and culture is most prominent in Australians' new relation to the center of the continent. As explorers' reports and early experience killed hopes of a paradisal, then even pastoral, interior, Australians retreated. The national myths of the "bush" and the "outback" were in country well short of the central desert. The sheep on whose backs the country rode to prosperity grazed on ranges that circled it, and the Heidelberg school's images came from the rural areas around Melbourne.[85] Into the 1920s boosters made rhetorical raids on what J. W. Gregory had in 1906 termed the "Dead Heart" of the continent, speaking of inexhaustible "underground rivers" and modern technology, but their enthusiasm only underlined Australians' failure to come to terms with the land.[86] In 1935 H. H. Finlayson argued in *The Red Centre: Man and Beast in the Heart of Australia* that the land had its own beauty, but his was a lonely voice.[87] In the 1940s some artists turned from the coast to paint "the fantastic and primeval landscape of the dry interior and soil-eroded margins of white settlement," but only in the last generation has the center become part of the popular imagination.[88] There are now bus tours; Uluru has become almost a place of pilgrimage; and there is a tourist guide entitled *The Living Centre of Australia*.[89]

To date the story of Anglo Australians' adaptation to their land has been of using, then understanding, industrial technologies. The settlers expanded into and across the continent with dreams of domination, of changing the land to their needs. Successive waves of industrial technology encouraged them in the belief that they need not think of limits or consider nature's reaction to their "conquest." In the last generation that has changed. Development came to affect amenities just as Australians acquired time and money to enjoy them.[90] Their new ecological knowledge sharpened and focused their experience of pollution and environmental degradation, and provided the intellectual framework for a critique of development as usual. This story is, in outline, that of the settler countries, the industrial societies of Europe, and much of the rest of the world. The parallels and differences between Australia and these wider societies make the island continent a mirror for observers.

Nataraja
India's Cycle of Fire

STEPHEN J. PYNE

In the center dances Shiva, a drum in one hand and a torch in the other, while all around flames inscribe an endless cycle of fire.

This—the *nataraja,* the Lord of the Dance—is more than one of Hinduism's favored icons. It is a near-perfect symbol of Indian fire history. The drum represents the rhythm of life; the torch, death; the wheel of flame, the mandala of birth, death, and rebirth that fire epitomizes and makes possible. In this confrontation of opposites the dance replaces the dialectic; Shiva holds, not reconciles, both drum and torch. Considered ecologically the *nataraja* thus expresses in graphic language the great polarity of India, the annual alternation of wet and dry seasons by which the monsoon, with faint transition, imposes its opposing principles on the subcontinent. India's biota, like Shiva, dances to their peculiar rhythm while fire turns the timeless wheel of the world.

Perhaps nowhere else have the natural and the cultural parameters of fire converged so closely and so clearly. Human society and Indian biota resemble one another with uncanny fidelity. They share common origins, display a similar syncretism, organize themselves along related principles. Such has been their interaction over millennia that the geography of one reveals the geography of the other. The mosaic of peoples is interdependent with the mosaic of landscapes, not only as a reflection of those lands but as an active shaper of them. Indian geography is thus an expression of Indian history, but that history has a distinctive character, of which the *nataraja* is synecdoche, a timeless cycle that begins and ends with fire.

The cycle originated with the passage of India as a fragment of Gondwana into a violent merger with Eurasia. The journey northward, through the fiery tropics; the violence of the great Deccan basalt flows and of the immense collision with Asia; the installment of seasonality in the form of the monsoon—all this purged the subcontinent of much of its Gondwana biota, and tempered the rest to drought and fire. The populating of India came instead by influx from outside lands, followed by varying degrees of assimilation. Here, in the choreography of the *nataraja*, east met west, Eurasia confronted Gondwana, wet paired with dry, life danced with death.[1]

What endemics remained were, like India's tribal peoples, scattered or crowded into hilly enclaves. Only 6.5 percent of India's flowering plants are endemic, compared with 85 percent in Madagascar and 60 percent in Australia. The residual biota thrived most fully to the south; Peninsular India holds a third of the subcontinent's endemic flora. Some species, Asian in character, entered from the northeast. A diffuse array emigrated from the eastern Mediterranean, the steppes, and even Siberia, the Himalayas serving less as a barrier than a corridor. More recently weeds, largely European, have established themselves. The composition of its biota thus recapitulates the composition of its human population—the tribal peoples, their origins obscured; the Dravidians who persevered on the Deccan plateau and to the south; the Southeast Asians, migrating through Assam and Bengal; the Aryans, Huns, Turks, Persians, Pathans, Mongols, and others, entering from the northwest; and Arabs and Europeans, mostly Portuguese and British, arriving by sea.[2]

The geographic ensemble that emerged from this vast convergence was both familiar and unique. Of course there were broad divisions, Asians here, Dravidians here. Of course there were mosaics of field, grassland, and forest, in part because of human influence. But even beyond such matters, this syncretic biota assumed the character of something like a caste society. It is probable that this was no accident. The organization of Indian society impressed itself on the land, with ever greater force and intricacy. Tribal people gathered into disease-ridden hills, better shielded genetically from malaria and other ills. They then reworked those hills in ways that conferred on them a biotic identity. It is no accident that the species most commonly found in habited areas are those most abundantly exploited by the human inhabitants, and are often those best adapted to fire. European weeds, like forts and factories, gathered into specially disturbed sites, then spread along corridors of travel or secondary disturbance. The intricate division of Indian society by caste ensured that different peoples did particular things at particular times, and this was reflected in the landscape of India, not only between regions but within areas that different groups exploited at different times in different ways for different purposes.[3]

The intensity of the monsoon assured—demanded—a place for fire. The sharper the gradient, the more vigorous the potential for burning. Some of the wettest places on Earth, like the Shillong Hills, could paradoxically experience fire and even fire-degraded landscapes. The biota, already adapted to rough handling by India's passage north, responded to fire readily. The flora and fauna that humans introduced, or that migrated into India coincidental with them, also had to be fire-hardened because humans added to and often dominated the spectrum of environmental disturbances and they certainly exploited fire. Explorers and ethnographers reported the practice among southern tribal groups (and in the Andaman Islands) of habitually carrying firesticks, a practice relatively rare outside of Australia and a few other regions. Probably Radcliffe-Brown's peroration on fire and the Andaman Islanders could stand for most tribal peoples on the subcontinent. Fire, he concluded,

> may be said to be the one object on which the society most of all depends for its well-being. It provides warmth on cold nights; it is the means whereby they prepare their food, for they eat nothing raw save a few fruits; it is a possession that has to be constantly guarded, for they have no means of producing it, and must therefore take care to keep it always alight; it is the first thing they think of carrying with them when they go on a journey by land or sea; it is the centre around which the social life moves, the family hearth being the centre of the family life, while the communal cooking place is the centre round which the men often gather after the day's hunting is over. To the mind of the Andaman Islander, therefore, the social life of which his own life is a fragment, the social well-being which is the source of his own happiness, depend upon the possession of fire, without which the society could not exist. In this way it comes about that his dependence on the society appears in his consciousness as a sense of dependence upon fire and a belief that it possesses power to protect him from dangers of all kinds.
>
> The belief in the protective power of fire is very strong. A man would never move even a few yards out of camp at night without a firestick. More than any other object fire is believed to keep away the spirits that cause disease and death.

A veteran Conservator of Forests, G. F. Pearson, noted that even the Ghonds, a long-enduring tribe of Indian central forests, "never go into the jungle now, where tigers are supposed to live, without setting it on fire before them, so as to see their way." Almost certainly India tribal peoples used their firesticks as Australia's Aborigines did. The prevalence of anthropogenic burning in the tropical north of Australia, where the Asian monsoon also dictates wet and dry seasons, is another likely analogue.[4]

But more than aboriginal fire practices from India's "tribal" peoples shaped the land. Agriculture needed fire for clearing, converting, and fertilizing. In India, as throughout monsoonal Asia, slash-and-burn agriculture *(jhum)* became dominant outside of floodplains, ensuring that routine fire would visit even remote sites. Where insufficient forest fallow existed, alternatives were found in *ra'b* cultivation by carrying wood to the site for burning, or mixing it with other refuse and manure prior to conversion into ash. Some peoples fired the hills "with almost religious fervor," observed one disbelieving Briton, in the hope that the ash would wash down to waiting fields. By all these means (and others) a subcontinent of extreme wetness switched, when the polarity reversed, into a land of ubiquitous fire. The *nataraja's* drum became a torch.[5]

The coming of the Vedic Aryans is an event of special interest. Beyond their role in establishing hierarchy as an informing principle of Indian society, beyond their heroic literature, beyond their infusion of Indo-European language and customs into the subcontinent, they introduced two items of special consequence to Indian fire history. They imported livestock, and they installed Agni, the god of fire, as first among the pantheon of Vedic deities. Fire and livestock interacted like a self-reinforcing dynamo. Together flame and hoof reshaped the landscape into grasslands and savannas sufficient to sustain the herds. Where *jhum* was also practiced, its abandoned fallow could be made to evolve into grass and browse through repeated burning. Without fire the process of reducing jungle and reordering landscapes was slow if not prohibitive.

It is no accident that the *Mahabharata,* part of the Hindu canon, describes the burning of the Khundava forest. It has been argued further that the story is an allegory of Vedic colonization. It begins when a Brahman appears to Krishna and Vamuna, then enjoying the forest. They grant his plea for alms, and he immediately shows himself as Agni and requests that he be allowed to feed himself on the forest. They grant this desire too; Agni rewards them with a chariot and weapons; and together they consume the Khundava and its creatures. The city of Delhi rises from the site today. The Brahman, presiding over his fire ceremony, was in fact an important pioneer into new lands, provoking by broadcast and ceremonial fire a new order.[6]

Thus the special status granted to Agni went beyond coincidence. Agni was the originating god, and it is to Agni that the *Rig Veda* opens its invocation; Agni of the two heads, one harmful, one helpful; Agni of the three arms, the manifestation of fire in the heavens as the sun, in the sky as lightning, and on the earth as flame; Agni, the medium between the gods and humanity, the mediator between humans and the earth; Agni, the Indian avatar of the hearth god (Atar) fundamental to other Indo-European peoples, best known through the vestal fire of Rome. Soon, however, Agni was supplemented by Indra, the king of the gods, and eventually absorbed into that bewildering genealogy of deities and heroes, as overgrown as jungle fallow, that is the wonder and curse of Hindu theology.[7]

But the special status that Agni lost within a proliferating Hindu pantheon, he retained through rite. For the Vedic Aryans the fire ceremony remained at the core of ritual existence. It was to Agni that they sacrificed, and through Agni, as burnt offerings, that sacrifices to other deities became possible. Fire accompanied birth, marriage, and death, if possible flame from the same fire serving all through the liturgical life cycle. Agni was thus both means and end, beginning and end, a continuous ring around the affairs of the world.

> Agni, the all-knower, the first one
> Looked out over the beginning of the dawns,
> Out over the days,
> And out in many ways alone, the rays of the sun,
> He spread over sky and earth.

Through the centuries the ceremony mutated, and Agni's unique standing declined before its many challenges. Buddhism confronted it directly, demanding a less violent and extravagant practice, preferring useful gifts (*dana*, or donations) in place of burnt offerings. At Gaya the Buddha, perhaps inspired by the fires that annually burned along the flanks of the Vindhyan Mountains, identified fire as a central metaphor of life. "Everything, brethren, is on fire." Passions and desires afflicted human life as flames did the land. They had to be quenched, the Buddha declared, just as the fire ceremony had to be replaced by a less extravagant rite. Nirvana literally meant extinguishing, the blowing out of fire. Hinduism responded by tempering the fire ceremony, relocating it to indoor temples, and granting it a more symbolic, less consumptive role.[8]

Fire remained fundamental, however, as it does yet today. The *puja*, the central ritual of Hindu life, revolves around a fire that stands for the gods, carries sacrifice to them, and purifies the supplicant. Fire begins the day, as it does the world. It ends life in the form of cremation, as the world will end upon Vishnu's final return. Until then fire powers the cycling of birth and death that is the essence of the *nataraja*.

It is no surprise to learn that, for India, the spiritual interacts with the practical and that what organizes society also organizes nature. The installment of Agni and the Vedic fire ceremony, and the way this acted on Hindu society, had its parallel in the way by which Aryan fire worked on the Indian environment. Fire ordered the landscape as caste did people. The sacrifice to Agni took the form of burning India's forests, or rather of reworking them in somewhat newer ways to support an economy dependent on livestock. The slashed-and-burned Ghats of Karnataka were thus the environmental equivalent to the corpse-burning ghats at Benares. Interestingly the Buddhist revulsion against the fire ceremony had its counterpart in a reaction against the destruction of trees and animals, particularly through fire. The Buddhist king, Ashoka the Great, thus decreed that forest fires should not be lit "unnecessarily" or with the intention of killing or sacrificing living beings.[9]

The new fire practices folded into the old, much as immigrant peoples and ideas enfolded into India's caste-layered society and its mosaic-wrought landscapes. By the time Enlightenment Europeans began studying India, fire was so prevalent that it merged seamlessly with the natural history of the subcontinent. Writing retrospectively in 1928, E. O. Shebbeare recalled that "every forest that would burn was burnt almost every year." Worse, the fires were chronic throughout the dry season, seizing whatever cured fuel presented itself. Joseph Hooker described how, during his descent from the Himalayas in the early 1850s, he saw the plains of Bengal immersed in smoke, the product of fires "raging in the Terai forest" and elsewhere, and observed particles of grass charcoal descending like black snow around him. F. B. Bradley-Birt marveled in 1910 how the "hills round Gobindpur form a wonderful line of light every night during the hot weather," the outcome of native-set fires that smolder for days, and "creep on in zigzag lines from end to end of the hills, invisible by day, but standing out clear and distinct, a brilliant line of light, by night." Benjamin Heyne explained that the "hills here are all on fire, and present a spectacle, the magnificence of which is easier conceived than described." Less enchanted, Inspector-General Ribbentrop fumed in a treatise published in 1900 that the profusion of fire was matched by a "most marvellous, now almost incredible, apathy and disbelief in the destructiveness of forest fires."[10]

A summary of fire causes for the Ghumsur Forest in Orissa tabulated by "Mr.

S. Cox," the District Forest Officer, nicely captures the spectacle, and the disbelieving outrage with which the British witnessed it:

> All the State forests on the borders of the taluk are subject to fires crossing from the numerous surrounding zamindari forests. . . . The latter, if they are in a condition to burn, are always burnt, and the boundary lines are so extensive and run over such difficult country that it is out of the question for us at present to protect them all.
>
> Then in the large hill forests frequented by the Khonds the jungle is fired as a matter of course to facilitate tracking and for other well-known objects.
>
> In the lower hills and more accessible country bamboo cutters and permit-holders generally are responsible for a great deal of the mischief. Wherever a hill is frequented for bamboos there are always constant fires.
>
> Other causes are the practice of smoking out bees for honey—a very common origin of fire—of burning under mango and mohwa trees to clear a floor for the falling fruit and flowers; the roasting of Bauhinia seed; the burning of under-growth round villages and cultivation which might harbour tigers and panthers—this will probably prove one of our most serious obstacles to restocking the sal forests; and the spread of fire from banjar lands under clearance for cultivation. . . .
>
> The long list of causes is almost complete if to the above are added the burning of forest by graziers, and for driving out game or finding a wounded animal.

Not least perplexing (and infuriating) was the fact that out of 53 cases of illegal fire investigated within the protected forests, "no less than 27 were caused by the protective staff itself." The native staff recognized, if their baffled masters did not, that the proper use of fire was the best protection against its misuse.[11]

It was in fact the British who did not understand. It was their belief in fire's necessary destructiveness that was, within the context of India, incredible. The indigenous people knew how fire supported *jhum* cultivation, converted organic residues into fertilizer, kept woodlands and prairies in grass, assisted hunting, cleansed soil of pathogens, and supported foraging for flowers, bees, tubers, and herbs. Fire sustained metallurgy. Fire kept tigers away from villages and opened sites that might otherwise hide cobras. Fire structured the intricate ensemble of biomes that was made by, and that in turn made possible, Indian society. Alone among the elements fire illuminated the complex choreography that bound life with death, the human with the natural. Fire framed the *nataraja*.

The dance missed beats as British rule extended over more and more of Greater India. The British raj imposed not only imperialism but industrialism. Britain linked India with lands beyond the reach of monsoon winds, connected it with economic cycles greater than the rhythms of annual growth and decay, and shrank the encircling fire into the combustion chamber of steam engines. The tempo of the *nataraja* picked up. A ceaseless cycle wobbled, then spun uncertainly into a spiral.

British influence extended piecemeal, as opportunity and necessity presented themselves. Change became serious—and reform deliberate—after the Revolt of 1857 when the Crown replaced the British East India Company as the governing authority. Britain then applied to colonial India the same processes that had restructured Britain over the preceding century. Industrial capitalism and a global market began redesigning the Indian economy. Land reform, or at least the rationalization of land ownership, exploitation, and tax collection, inspired a kind of enclosure movement or revenue "settlement" that gradually spread over the newly acquired lands. "Forest settlement" was a part of this process, and quickly brought European-style forestry into conflict with traditional, communal exploitation of Indian woodlands.

The new ruling caste brought their laws, their language and literature, and their sciences. Agronomists sought to modernize Indian agriculture, as political theorists sought to modernize Indian government. Hydraulic engineers erected dams, dug canals, and designed irrigation works. Mining engineers explored for geologic wealth. Cartographic engineers surveyed the subcontinent, imposing a mathematical order on the land, even measuring the anomalous gravity of the Himalayas. Above all civil engineers laid out the grid that would be the means and symbol of Indian industrialization, the railroad. From 32 km laid down by 1853, the system exploded to 7,670 km by 1870, and then continued to grow. Each reform demanded others, however, if it was to succeed. The railroad, for example, was inextricably dependent on wood—for construction, particularly ties ("sleepers"), for fuel, for cargo. The rationalization of India through the railroad required the rationalization of India's forests.

Indian forestry became one of the great sagas of British rule, however improbable its origins. Britain, after all, had no tradition of forestry and precious little of anything that could be called a forest. But it was clear that the reconstruction of India was doomed without some deliberate intervention. Without forests railroads would run down, agriculture would suffer from drought and flood, soil would degrade, and a timber economy based on the export of teak would collapse. Even by the mid-nineteenth century it was clear that economic and political forces were, like an acid, dissolving the grout that held together the Indian mosaic. If something did not reglue them, nothing would remain but a pile of broken tiles. Besides, the rationalization of the "jungle" (as the uncultivated wildlands were called) was an ideal symbol of liberal reform. If India's jungle could be reordered according to scientific principles, so could the rest of India.[12]

Britain went to the heartland of European forestry for help. In 1856 it appointed Dietrich Brandis as Conservator of Forests for Burma. A botanist subsequently educated in forestry in the grand European manner, Brandis was the archetype of the transnational forester, Humboldtean in ambition, an indefatigable agent of empire, a Clive of natural resource conservation in Greater India. Two years later Brandis became Inspector-General of Forests for all of British India, a dominion that grew dramatically not only as Britain added more provinces to its Indian domain but as the practice of reserving forests proceeded in conjunction with the reorganization of the Indian landscape through revenue settlement.

Brandis pushed for the establishment of the Indian Forest Service, achieved in 1864, one of the compelling institutions of British rule and the centerpiece for forestry throughout the British empire. Cadets received formal instruction in Franco-German forestry at Nancy, France, then served field apprenticeship in India. From there they might proceed to Sierra Leone, Cape Colony, or Tasmania. This was the same regimen the founders of American forestry, men like Gifford Pinchot and Henry Graves, experienced. In 1906 the facility relocated to Cooper's Hill at Oxford, and later a separate school and research institution was established for India at Dehra Dun. The Indian Forest Service, meanwhile, became a part of the civil service and after critical conferences in the early 1870s assumed its modern form. On the recommendations of the conferees the IFS in 1875 launched the *Indian Forester*, for 50 years probably the premier forestry journal in the world.

Enthusiastic foresters—Sir David Hutchins reminded them that they were "soldiers of the State, and something more"—entered into the reconstruction of India, attempting to regulate timber harvesting, to control traditional forest uses by pastoralists and villagers, to regenerate felled or degraded woodlands, and to control fire. They as much as anyone pioneered the shock encounter between Britain and India, between the institutions of the West and the environments of the East. The encounter mixed in equal proportions high drama, absurdity, grit, the irony of noble purpose and practical stupidity. Rudyard Kipling captured something of all this in his story "In the Rukh," a sequel to *The Jungle Books*. "Of all the wheels of public service that turn under the Indian Government," he intoned, "there is none more important than the Department of Woods and Forests." On it depended the reforestation of India. And it is to the Indian Forest Service that Mowgli, now grown but still conversant with his brothers the wolves, goes as a forest guard. Among his duties are "to give sure warning of all the fires in the *rukh*." Those fires needed to be suppressed. The globe-encircling fire engines of the British raj would replace the encircling flames of the *nataraja*.[13]

Here was something new. While over the centuries forests had ebbed and flowed with wars and population pressures, fires had come and gone with the monsoons. Fire practices had changed, but fire had endured. Some years Shiva's drum beat louder than the torch, some years not; the ring of fire expanded and contracted; but always the circle held. It was unimaginable that fire could cease. Without fire the land was inaccessible, India uninhabitable, and life unknowable. Without fire the cosmos faced extinction. Without the encircling fire the *nataraja* would end.

The pioneers of Indian forestry, Shebbeare recalled, saw fire as "their chief, almost their only enemy." The extravagance of fire that seeped, simmered, probed, flared, and raged annually throughout India made a shambles of any presumption to reorder those forests along European models. Fires infested the land like malaria or packs of wild dogs. But the challenge went beyond their damage to pasture and woods, beyond the wanton sacrifice of India's immense wealth of forests. Those fires appeared as an environmental superstition, a taunt that mocked the possibility of remaking India in ways that would serve Britain and serve to legitimate British rule. Britain could justify redirecting India's forests to new purposes only if those purposes had higher standing if they were part and parcel of a more rational order. It had to remake India's "irregular" forests—its tangled "jungles"—into "rational" institutions. It could harvest forests only if it demonstrated how to regenerate and protect them according to some larger principles.[14]

So in addition to the compelling economic reasons that linked forest to rail, and to the political logic that demanded the subordination of rural villages to a central, industrial authority, the British added the symbolism of science to their justification for fire control. The power to control village life resided in the power to control forest and range, and that depended on the power to control fire. Because Britain's claim to impose a modern ecological rule on India relied on its sanction by scientific silviculture, the British had to oppose "primitive" practices with a "rational" agriculture and a scientific forestry. In European agronomy the divide between the primitive and the modern was fire. Fire had to go.[15]

The experiment began in 1863 when Brandis urged Colonel Pearson of the Central Provinces to try to stop the burning. No one believed it was really possible. "Most Foresters and every Civil Officer in the country," Pearson observed, "scouted the idea." Edward Stebbing recalled matter-of-factly that in every province "the officers of the Department had to commence the work of introducing fire conservancy for the protection of the forests in the face of an actively hostile population more or less supported by the district officials, and especially by the Indian officials, who quite frankly regarded the new policy of fire conservancy as an oppression of the people." Even forest officers, Stebbing noted, however much they approved of fire control in principle, "were openly sceptical" of its practical possibility. Had his attempt failed, Pearson affirmed, "any progress in fire protection elsewhere would have been rendered immeasurably more difficult." Pearson shrewdly selected a site protected by natural barriers, a biotic counterpart to the fortresses at Ranthambore and Jaipur. He then laid out fuel-breaks, sent out patrols, exhorted locals to give up burning, and enjoyed a couple of exceptionally wet seasons. To everyone's astonishment, the experiment succeeded. The Bori Forest became a showcase of fire conservancy. At the Forest Conference of 1871–72, based on these experiences, Pearson declared that "there can be no doubt that the prevention of these forest fires is the very essence and root of all measures of forest conservancy." Brandis added his imprimatur. "There is no possible doubt," he wrote, as to its "immense value and importance." Fire conservancy was, not accidentally, the first topic addressed by the first conference on forest administration.[16]

Not completely, not without considerable debate and second-guessing, but thanks to militant enthusiasm and patience and favorable weather, this improbable experiment in fire control evolved into a demonstration program, and then into a prototype suitable for dissemination throughout Greater India. At the great Forest Conference of 1875 Brandis reaffirmed that for the improvement of Indian forests "there is no measure which equals fire conservancy in importance." It is, he continued, "the most important task of the Forest Department in most provinces of the empire, and for that reason was awarded first place in conference discussions." Pearson's successor, Captain J. C. Doveton, detailed the ways and means of fire conservancy and observed sourly that these measures were only necessary because "nearly the whole body of the population in the vicinity of forest tracts have, or imagine they have, a personal interest in the creation of forest fire." Not least of all because of that hostility, three classes of state forests evolve, each committed to a different level of use and protection.[17]

Once confirmed the idea spread, promulgated from the top down. As with the native principalities, so with the native forests; more and more were reduced to British rule by fire protection, for to control fire was to control the native populations. Regardless of the legal status of forests, without fire the local populace had no biological access to the resources of those reserves. By 1880–81 the Indian Forest Service had reduced some 11,000 square miles to formal protection; by 1885–86, some 16,000 square miles; and by 1900–1901, an astonishing 32,000 square miles that spanned the spectrum of Indian fire regimes, from semiarid savanna to monsoonal forest to bamboo groves and montane conifers. Fire control grew as rapidly as the railroads with which it was indissolvably linked. Fire protection targeted particularly the great timber trees of the subcontinent, sal, teak, chir pine, and commercial bamboo. What emerged was a robust exemplar, an adaptation of European techniques to exotic wildlands and colonial politics.

But skeptics were not easily stilled. Pearson spoke dismaying that "it is strange how slow even some, who possess very considerable practical acquaintance with the forests, are to recognize" the intrinsic merit of fire exclusion. In the *Report on the Administration of the Forest Department for 1874* B. H. Baden-Powell echoed and scorned that disbelief:

> Strange to say, that, obvious as the evils of fire are, and beyond all question to any one acquainted with even the elements of vegetable physiology, persons have not been found wanting in India, and some even with a show of scientific argument (!), who have written in favor of fires. It is needless to remark that such papers are mostly founded on the fact that forests *do* exist in spite of the fires, and make up the rest by erroneous statements in regard to facts.

On the matter of fire conservancy science admitted no doubt, and neither did colonial administrators bent on imposing a new order on a very old and complex land.[18]

Like a fire in a punky log, however, the matter would not go out. Soon field men voiced ever greater doubts about the wisdom of "too much fire protection." In wet forests fire protection seemed to retard natural regeneration, and it allowed fuels to accumulate that, once dried, exploded into all-consuming conflagrations. In drier forests, years of seemingly successful protection would be wiped out by massive fires during exceptional years. Exhortations and bribes with goats could not extinguish all the native firebrands who knew from daily experience what burning meant. Villagers refused to resettle or remain in unburned sites for fear that tigers, hiding in the tall grasses, would seize child herders. (Unlike the American or Australian experience, Indian natives would not melt away, vastly outnumbering the ruling caste, and their fires could not be banished into the past or sequestered onto reservations.) Hunting clubs in the Nilgiri Hills noted the deterioration of game where fires had been excluded. In the absence of suitable fire regimes natural regeneration failed in sal, teak, bamboo, pine—and failed consistently, particularly in wetter sites. Field officers began posting querulous memos about increases in diseases, pests, weeds, and other signs of a forest going feral. An agronomic memoir on Indian grasses noted how "an unforseen result of the policy of noninterference with the vegetation" was the accumulation of dead straw that defiantly withstood "rotting" and eventually had to be burned, an act which quickly yielded a variety of useful results. Forest guards surreptitiously burned surrounding lands, including the lower-grade forests, to improve their chance of fire control on class I sites. Upon his retirement in 1952 a native Indian forester commented that in his 41 years of service he had never known a forest to withhold fire for more than three years.[19]

In what might serve as a cameo, an Anglo forester who signed himself "An Aged Junior" described for the *Indian Forester* the puzzling situation in which, through more or less successful fire protection, the forest had acquired a tiger problem. It is apparent that fire had not been random and ravenous, as it appeared to the British, but had been applied to particular sites at particular seasons for particular purposes by particular peoples. Those selective burns had ordered the landscape. Thanks to fire, fresh browse appeared at the proper place at the proper time; deer migrated to those sites; tiger followed the deer; and hunters knew where to find rogue tigers. But eliminating fire, or smearing it, affected that land as the abolition of caste would Indian society. Boundaries blurred. The ecological order became confused. Tigers no longer kept to their place—their place being scrambled and overgrown. They began to menace local communities, follow rangers, and generally make themselves "disagreeable." The forest now had "much fire conservancy and many tigers." Whether successful or not, the *attempt* at fire control was sufficient to unbalance the Indian biota. Changing from small fires set annually to large fires that came every three or four years did not preserve the old order. It was not simply fire that India needed but its syncretic order of fire regimes.[20]

It was not so easy to reconcile European principle with Indian reality. Critics argued for a hybrid program in which controlled burning could supplement fire suppression. In 1897 Inspector-General Ribbentrop, Brandis's successor, had to intercede. To protect regeneration and forest humus (the twin obsessions of European forestry)—to say nothing of saving imperial face—he ruled for the further expansion of systematic fire protection. Edicts, however, did not suppress fires, or doubts. By 1902 the debate rekindled within the pages of the *Indian Forester* and the annual reports of the provincial conservators. In 1905 a compromise was proposed by which controlled burning could be brought into working plans. Meanwhile *sub rosa* burning in Bengal, Burma, and elsewhere scorched the landscape like a people's rebellion.

In 1907 protest boiled over into a Burmese revolution. In the absence of traditional fire—slash-and-burn cultivation, routine underburning—teak simply refused to regenerate. Fire control had drained away the economic lifeblood of the Asian monsoon forest; foresters had prescribed a harsh cure where there had been no disease. Faced with a choice between excluding fire and excluding fire protection, the Inspector-General began withdrawing fire control from prime teak forests. One after another working circles that had subscribed to fire protection now withdrew it—Pyu Chaung and Pyu Kun in 1906, Kan Yutkwin in 1910, Bondaung, Kabaung, and Myaya Binkyaw four years later. By 1914 conservators of sal forests likewise recognized that regeneration "had ceased throughout the fire-protected forests of Assam and Bengal and that no amount of cleanings and weedings would put matters right." They tried to reintroduce fire, but fuels had so changed that it was no longer possible to run benign light fires through the understory; the *taungya* system by which swidden fields were restocked with planted timber trees evolved as a partial compromise. Chir pine, too, was found to be reliant on routine fire, so that nearly everywhere field foresters introduced some form of "early" (that is, spring) burning of grassy understories for fire protection, and integrated regeneration burns into silvicultural cycles. Whatever the causes for the failure of natural regeneration, Shebbeare concluded for an audience of foresters drawn from the British empire, "fire appears to be the only real cure."[21]

By 1926 the cycle of fire practices had come full circle. Imperial resolve retreated before an unscorched earth, the passive disobedience of Indian silviculture. A conservator's conference amended the rules of the *Forest Manual* to make early burning the general practice and to extend complete protection only to special sites on a temporary basis. With nice irony that new regime included the Central Provinces. Some critics wanted even more. Writing from Siran Valley, E. A. Greswell noted that "up to 1922 the [chir] forests had been subjected from time immemorial to periodic summer firing," probably burned once every three to four years. The cessation of those fires damaged regeneration and put the forest at risk from wildfire. The reintroduction of fire was "merely re-establishing a modified form of the environment to which the forests owe their origin." Greswell knotted practice to philosophy when he concluded that "we talk glibly about following nature and forget that the nature we are visualising may be an European nature inherited from our training and not an Indian nature." The fire of Europe was not the fire of India.[22]

But by this time Britain, never fully recovered from the wastage of World War I, was receding in imperial power and enthusiasms, its hold on India becoming steadily more tenuous. Protests increased, often focused on forestry and typically assisted by outbreaks of incendiarism. In 1916 and again, with even greater force, in 1921 political protest in Kumaon inspired a wave of woods arson that brought the regional administration to its knees. Administrators openly admitted their helplessness before the protest of incendiarism, another argument in favor of co-opting burning. But such spectacular outbreaks paled besides the relentless insurgency of small firings. Writing in 1926 M. D. Chaturvedi observed that "prosecutions for forest offences, meant as deterrents, only led to incendiarism, which was followed by more persecutions and the vicious circle was complete." Inevitably, grudgingly concessions followed. Compromises remained compromises, however, the best one could do under troubled circumstances. With few exceptions—but among them some of the best minds in Indian forestry like R. S. Troup—foresters continued to insist that fire was intrinsically bad. They saw it, as they did the native elites, as a necessary evil, not as a powerful ally. Fire remained an impermeable divide in the worldview of European agronomy and silviculture. If a system used fire, it was by definition primitive; if it found surrogates for fire, it could qualify as rational.[23]

In India there were few surrogates possible. Where officialdom approved fire it did so reluctantly, with some embarrassment, and only because fire was seemingly part of an ineffable (and exasperating) East. Fire reduced rational plans to a kind of ecological astrology, and the practices of a scientific forestry to a flame-lit *puja*. Fire persisted as an untouchable caste within the society of silviculture. The Indian Forest Service burned because it was forced to, not because it wanted to. Where fire was used, it was often not sanctioned, and where sanctioned, often not used properly. As British rule met further resistance, that split widened; theory and practice diverged; the landscape was neither old nor new nor some workable compromise between them. The cycle of fire broke.

What had been a circle became a spiral. The process began well before Independence, and it has continued after the British were expelled. What Britain had done with imperial arrogance, independent India claimed it would do with a social conscience; but whatever their sanction the practices continued, and then accelerated; and this acceleration was itself quickly exceeded by a horrific explosion in the subcontinent's population. However incomplete or mismatched, the reforms of the British raj had initiated a population rise that continues its exponential growth to the present day. In 1800 the estimated population of India was 120 million; in 1871, 255 million; in 1950, 350 million, despite the upheaval of partition; in 1990, 890 million. Until the 1970s the numbers of livestock swelled in almost equal proportion. Much of the human increase gathered into cities; a substantial fraction was absorbed by industry; but the rest (over 70 percent) remained on the land, and one way or another, this maelstrom of peoples and beasts sucked down the Indian environment in its vortex.

The upward spiral of human numbers powered a downward spiral in land abuse. Some 16 percent of the world's population crowded into 2 percent of its landmass. India's forests felt the pressures keenly, particularly the *terai* and hill forests that had, because of endemic diseases like malaria, been shielded from use other than by those tribal peoples who had acquired some degree of immunity. Disease control, the construction of dams and roads, intensive logging, clearing for additional farmland, and a redefinition of reserves to serve the tenets of "social forestry" eroded away India's woodlands, and often their soils. The commitment to industrial forestry that British rule had established, the Indian state reaffirmed; previously unexploited indigenous forests were opened by roads, logged, and often replaced by exotics like eucalypts that provided pulp but little of the other products India's woods had supplied Indian society. Although the Indian constitution stipulated that 25 percent of India should remain forested in some form (and the Forest Law of 1952, 33 percent), the reality was closer to 19 percent, and critics thought even that number too high; much of the reserved jungles were too degraded to classify as productive woodland. Once placed under state care, forests had required the coercive power of the state to survive. As further political unrest threatens the nature of the Indian polity as a secular state, that power promises to recede and to leave India's forests exposed to everyone's grasp and no one's care.[24]

The intensity of use has disturbed the character of Indian fire. There remains plenty of burning of course. Agricultural fire is common where cotton, sugar cane, and wheat are grown, and among the crop residues of hill farming. *Jhum* cultivation persists in the northeast, the Ghats, the outer Himalayas, and among tribal peoples in Andhra Pradesh, Orissa, and elsewhere. An estimated 122,000 km² of permanent pasturage is burned annually. Among reserved and protected forests controlled burning assists the regeneration of chir pine, sal, and teak; fuelbreaks are burned early each dry season; particularly where forests plantations are at risk, underburning is practiced to reduce fuels and prevent against wildfire. Altogether this amounts to 5–6 percent of the reserved forest area. Still wildfire, either from "accidental" or incendiary causes, affects an estimated 10,000 km² yearly, as officially reported. Satellite inventories, however, calculate that 80 times this amount burns annually, some 33 percent to 99 percent of the protected forests in different states. These numbers do not account for forests subject to less strict regulation. The forest area affected by fire may reach 37 million ha. Even so the biomass burned as firewood in villages and urban centers exceeds that of all these other sources combined. Increasingly India's woods are being burned in its stoves.[25]

The quest for a suitable regimen of fire continues. It is pointless to argue for a restoration of traditional practices—the circumstances are too much changed to allow them. What had once rested as forest fallow for 30 years is now slashed and burned in five years, and sometimes as little as two. What formerly experienced small fires that percolated through the jungle over the course of five or six months now suffer from no fire or fire that crowds into short, violent events. Even traditional burning no longer recycled nutrients through a subsistence economy but siphoned them off into a global market; where tribals had traditionally burned once under *mowah,* they now burned twice, and the harvested flowers did not go to the village but the metropolis. The complex of fires that once fused the human and the natural together through the layered intricacies of a shared caste is gone. More and more India's fire regimes are defined by a global economy in which the forest exists as cellulose and wood is valued as an export commodity; less and less, by the traditional usage of the forest as a medley of usable plants and animals. The beat of pistons, powered by fossil-fuel combustion, replaces the rhythms of seasonal growing, curing, and burning. Artificial fertilizer replaces *rab;* the tractor and electric pump, the long fallow of *jhum;* autorickshaws, the bullock cart. Sometime around 1980 India crossed an industrial threshold of sorts when deaths from traffic accidents exceeded those from snakebites.

Yet no surrogate complex of fire practices has fully replaced it. To the extent that Indian scientists receive training from Europe or look to European scholarship for guidance, they continue to distrust burning, as though it were still a stigma of primitiveness, a leprosy on the landscape. No one has transformed India's unique experience into a new exemplar for "Third World" firepowers, a model of nonalignment in the dialectic between those who would base fire management on fire control and those who base it on fire use. India's elite still viewed fire as an inevitable if necessary evil, like cobras. If it were possible to escape from the endless cycle of fire, to lay down the burden of burning, they would. That would be release, forestry's nirvana. But the cycle had not vanished: it had become a more vicious spiral.

Instead, with assistance from the U.N. Food and Agriculture Organization (FAO), India launched a "modern forest fire control project" in 1984 that sought to install an integrated fire management system in two demonstration areas, Chandrapur (Maharashtra State) and Haldwani (Uttar Pradesh). The first contains extensive natural and planted teak forests; the second, hills dominated by sal and chir pine. Both projects, that is, intend to apply fire control to support the ambitions of industrial forestry. Incorporated into India's Eighth Five-Year Plan, Phase Two will expand the technologies into ten states and 40,000 km². Whether the project becomes a latter-day Bori Forest, a misinterpreted experiment; or whether it evolves into another showcase of international aid with airtankers and helicopters taking the place of high dams and nuclear reactors; or whether it begins the process of reconciliation between new and old, fusing a uniquely Indian style of fire management, all remains to be seen.[26]

It may be that reconciliation is impossible, that as in the *nataraja* India must simply hold and live with the opposites. This time, however, fire is not part of a cycle of endlessly reincarnating landscapes, but a spiral, propelling the biota to one extreme or another at an ever-quickening tempo. Without those encircling flames, the boundaries are broken, drum and torch no longer link, their rhythms no longer balance, and the dance must end in either frenzy or exhaustion.

Notes

Introduction

1. Donald Worster, interview by Hal K. Rothman, February 26, 1996; Aldo Leopold, *Sand County Almanac* (New York: Oxford University Press, 1949), 129–33.

2. Despite this boom in ecological awareness, scholars nonetheless encountered continued resistance to their work in an as yet unestablished academic subdiscipline. Alfred Crosby discovered this reality when he submitted the book manuscript of *The Columbian Exchange: The Biological and Cultural Consequences of 1492* for publication in the early 1970s. Legend has it that he approached more than twenty publishers before he found one, Greenwood Press, willing to risk its publication. Crosby and Greenwood have had the last laugh: *The Columbian Exchange* remains a standard not only for a full range of historians, but for scholars throughout the humanities and sciences.

3. Worster, interview.

4. John Opie, "The View from Pittsburgh (and Canyonlands)," *Environmental Review* (Fall 1982): 2–4.

5. *Environmental Review* became *Environmental History Review* in 1990 with the publication of vol. 14, no. 1–2, a name change designed to represent better "the contents of the journal to our readers, researches and libraries, as well as potential authors, and a wider audience." The title changed again in 1996, when the ASEH and the Forest History Society combined their publications, *EHR* and *Forest & Conservation History,* into the new *Environmental History,* edited by Hal Rothman. Commentaries on these changes and other relevant issues can be found in J. Donald Hughes, "Editorial," *Environmental Review* (Summer 1983): 133–34; William Robbins, "Editorial" (Spring 1986): 1–2; Hal K. Rothman, "Editorial," *Environmental History* (January 1996): 6.

The Ecology of Order and Chaos, by Donald Worster

1. Paul Sears, *Deserts on the March,* 3rd ed. (Norman: University of Oklahoma Press, 1959), 162.

2. Ibid., 177.

3. Donald Worster, *Nature's Economy: A History of Ecological Ideas* (New York: Cambridge University Press, 1977).

4. This is the theme in particular of Clements's book *Plant Succession* (Washington, D.C.: Carnegie Institution, 1916).

5. Worster, *Nature's Economy,* 210.

6. Clements's major rival for influence in the United States was Henry Chandler Cowles of the University of Chicago, whose first paper on ecological succession appeared in 1899. The best study of Cowles's ideas is J. Ronald Engel, *Sacred Sands: The Struggle for Community in the Indiana Dunes* (Middletown, Conn.: Wesleyan University Press, 1983), 137–59. Engel describes him as having a less de-

terministic, more pluralistic notion of succession, one that "opened the way to a more creative role for human beings in nature's evolutionary adventure" (150). See also Ronald C. Tobey, *Saving the Prairies: The Life Cycle of the Founding School of American Plant Ecology, 1895–1955* (Berkeley: University of California, 1981).

7. Sears, 142.

8. This book was co-authored with his brother Howard T. Odum, and it went through two more editions, the last appearing in 1971.

9. Eugene P. Odum, *Fundamentals of Ecology* (Philadelphia: W.B. Saunders, 1971), 9.

10. Odum, "The Strategy of Ecosystem Development," *Science* 164 (18 April 1969): 266.

11. The terms "K-selection" and "r-selection" came from Robert MacArthur and Edward O. Wilson, *Theory of Island Biogeography* (Princeton: Princeton University Press, 1967). Along with Odum, MacArthur was the leading spokesman during the 1950s and 1960s for the view of nature as a series of thermodynamically balanced ecosystems.

12. Odum, "Strategy of Ecosystem Development," 266. See also Odum, Trends Expected in Stressed Ecosystems, *BioScience* 35 (July/August 1985): 419–22.

13. A book of that title was published by Earl F. Murphy, *Governing Nature* (Chicago: Quadrangle Books, 1967). From time to time, Eugene Odum himself seems to have caught that ambition or lent his support to it, and it was certainly central to the work of his brother, Howard T. Odum. On this theme see Peter J. Taylor, "Technocratic Optimism, H. T. Odum, and the Partial Transformation of Ecological Metaphor after World War II," *Journal of the History of Biology* 21 (Summer 1988): 213–44.

14. A very influential popularization of Odum's view of nature (though he is never actually referred to in it) is Barry Commoner's *The Closing Circle: Nature, Man, and Technology* (New York: Alfred A. Knopf, 1971). See in particular the discussion of the four "laws" of ecology, 33–46.

15. Communication from Malcolm Cherrett, *Ecology* 70 (March 1989): 41–42.

16. See Michael Begon, John L. Harper, and Colin R. Townsend, *Ecology: Individuals, Populations, and Communities* (Sunderland, Mass.: Sinauer, 1986). In another textbook, Odum's views are presented critically as the traditional approach: R. J. Putnam and S. D. Wratten, *Principles of Ecology* (Berkeley: University of California Press, 1984). More loyal to the ecosystem model are Paul Ehrlich and Jonathan Roughgarden, *The Science of Ecology* (New York: Macmillan, 1987); and Robert Leo Smith, *Elements of Ecology*, 2nd ed. (New York: Harper & Row, 1986), though the latter admits that he has shifted from an "ecosystem approach" to more of an "evolutionary approach" (xiii).

17. William H. Drury and Ian C. T. Nisbet, "Succession," *Journal of the Arnold Arboretum* 54 (July 1973): 360.

18. H. A. Gleason, "The Individualistic Concept of the Plant Association," *Bulletin of the Torrey Botanical Club* 53 (1926): 25. A later version of the same article appeared in *American Midland Naturalist* 21 (1939): 92–110.

19. Joseph H. Connell and Ralph O. Slatyer, "Mechanisms of Succession in Natural Communities and Their Role in Community Stability and Organization," *The American Naturalist* 111 (November/December 1977): 1119–44.

20. Margaret Bryan Davis, "Climatic Instability, Time Lags, and Community Disequilibrium," in *Community Ecology*, ed. Jared Diamond and Ted J. Case (New York: Harper & Row, 1986), 269.

21. James R. Karr and Kathryn E. Freemark, "Disturbance and Vertebrates: An Integrative Perspective," *The Ecology of Natural Disturbance and Patch Dynamics*, eds. S. T. A. Pickett and P. S. White (Orlando, Fla.: Academic Press, 1985), 154–55. The Odum school of thought is, however, by no means silent. Another recent compilation has been put together in his honor, and many of its authors express a continuing support for his ideas: L. R. Pomeroy and J. J. Alberts, eds., *Concepts of Ecosystem Ecology: A Comparative View* (New York: Springer-Verlag, 1988).

22. Orie L. Loucks, Mary L. Plumb-Mentjes, and Deborah Rogers, "Gap Processes and Large-Scale Disturbances in Sand Prairies," ibid., 72–85.

23. For the rise of population ecology see Sharon E. Kingsland, *Modeling Nature: Episodes in the History of Population Ecology* (Chicago: University of Chicago Press, 1985).

24. An influential exception to this tendency is F. H. Bormann and G. E. Likens, *Pattern and Process in a Forested Ecosystem* (New York: Springer-Verlag, 1979), which proposes in chap. 6 the mod-

el of a "shifting mosaic steady-state." See also P. Yodzis, "The Stability of Real Ecosystems," *Nature* 289 (19 February 1981): 674–76.

25. Paul Colinvaux, *Why Big Fierce Animals Are Rare: An Ecologist's Perspective* (Princeton: Princeton University Press, 1978), 117, 135.

26. Thomas Söderqvist, *The Ecologists: From Merry Naturalists to Saviours of the Nation: A Sociologically Informed Nnarrative Survey of the Ecologization of Sweden, 1895–1975* (Stockholm: Almqvist & Wiksell International, 1986), 281.

27. This argument is made with great intellectual force by Ilya Prigogine and Isabelle Stengers, *Order Out of Chaos: Man's New Dialogue with Nature* (Boulder: Shambala/New Science Library, 1984). Prigogine won the Nobel Prize in 1977 for his work on the thermodynamics of nonequilibrium systems.

28. An excellent account of the change in thinking is James Gleick, *Chaos: The Making of a New Science* (New York: Viking, 1987). I have drawn on his explanation extensively here. What Gleick does not explore are the striking intellectual parallels between chaotic theory in science and post-modern discourse in literature and philosophy. Post-Modernism is a sensibility that has abandoned the historic search for unity and order in nature, taking an ironic view of existence and debunking all established faiths. According to Todd Gitlin, "Post-Modernism reflects the fact that a new moral structure has not yet been built and our culture has not yet found a language for articulating the new understandings we are trying, haltingly, to live with. It objects to all principles, all commitments, all crusades—in the name of an unconscientious evasion." On the other hand, and more positively, the new sensibility leads to emphasis on democratic coexistence: "a new 'moral ecology'—that in the preservation of the other is a condition for the preservation of the self." Gitlin, "Post-Modernism: The Stenography of Surfaces," *New Perspectives Quarterly* 6 (Spring 1989): 57, 59.

29. The paper was published in *Science* 186 (1974): 645–47. See also Robert M. May, "Simple Mathematical Models with Very Complicated Dynamics," *Nature* 261 (1976): 459–67. Gleick discusses May's work in *Chaos*, 69–80.

30. W. M. Schaeffer, "Chaos in Ecology and Epidemiology," in *Chaos in Biological Systems*, eds. H. Degan, A. V. Holden, and L. F. Olsen (New York: Plenum Press, 1987), 233. See also Schaeffer, "Order and Chaos in Ecological Systems," *Ecology* 66 (February 1985): 93–106.

31. John Muir, *My First Summer in the Sierra* (1911; Boston: Houghton Mifflin, 1944), 157.

32. Prigogine and Stengers, 312–13.

33. Much of the alarm that Sears and Odum, among others, expressed has shifted to a global perspective, and the older equilibrium thinking has been taken up by scientists concerned about the geo- and biochemical condition of the planet as a whole and about human threats, particularly from the burning of fossil fuels, to its stability. One of the most influential texts in this new development is James Lovelock's *Gaia: A New Look at Life on Earth* (Oxford: Oxford University Press, 1979). See also Edward Goldsmith, "Gaia: Some Implications for Theoretical Ecology," *The Ecologist* 18, nos. 2 and 3 (1988): 64–74.

The Theoretical Structure of Ecological Revolutions, by Carolyn Merchant

1. Thomas S. Kuhn, *The Structure of Scientific Revolutions*, 2d ed. (Chicago, 1970). The theory and illustrations presented here are drawn from my *Ecological Revolutions: Nature, Gender, and Science in New England*.

2. Karl Marx, "Preface to *A Contribution to the Critique of Political Economy*," (1859) in Karl Marx and Friedrich Engels, *Selected Works* (New York, 1968), 182–83.

3. Elizabeth Ann R. Bird, "The Social Construction of Nature: Theoretical Approaches to the History of Environmental Problems," *Environmental Review* 11 (Winter 1987); Karin D. Knorr-Cetina and Michael Mulkay, eds., *Science Observed: Perspectives on the Social Study of Science* (Beverly Hills, 1983); and Karin D. Knorr-Cetina, *The Manufacture of Knowledge: An Essay on the Constructivist and Contextual Nature of Science* (New York, 1981).

4. Claude Meillassoux, *Maidens, Meal, and Money: Capitalism and the Domestic Community* (1975; English trans., Cambridge, 1981). Critiques of Meillassoux include Bridget O'Laughlin, "Production

and Reproduction: Meillassoux's *Femmes, Greniers et Capitaux,*" *Critique of Anthropology* 2 (Spring 1977), 3–33; and Maureen Mackintosh, "Reproduction and Patriarchy: A Critique of Claude Meillassoux, *Femmes, Greniers et Capitaux,*" *Capital and Class* 2 (Summer 1977), 114–27.

5. Meillassoux, *Maidens, Meal, and Money,* 36, 39.

6. Abby Peterson, "The Gender-Sex Dimension in Swedish Politics," *Acta Sociologica* 27, no. 1 (1984), 3–17. Peterson's fourfold taxonomy of political interests included (1) Issues related to the interests of intergenerational reproduction; (2) Issues related to the interests of intragenerational reproduction in the family; (3) Issues related to the interests of intragenerational reproduction in the public sector; and (4) Issues related to the interests of reproduction workers (women), i.e. so-called women's liberation issues. Peterson also applied her taxonomy to the politics of reproduction in the Swedish environmental movement. See Abby Peterson and Carolyn Merchant, "'Peace With the Earth': Women and the Environmental Movement in Sweden," *Women's Studies International Forum* 9 (1986), 465–79, esp. 472–74.

7. Renaté Bridenthal, "The Dialectics of Production and Reproduction in History," *Radical America* 10 (March–April 1976), 3–11. For a feminist analysis of reproduction in American culture, see Women's Work Study Group, "Loom, Broom, and Womb: Producers, Maintainers, and Reproducers," *Radical America* 10 (March–April 1976), 29–45; and Veronica Beechley, "On Patriarchy," *Feminist Review* 10 (March–June 1980), 169–88.

8. Charles Taylor, "Neutrality in Political Science," in Alan Ryan, ed., *The Philosophy of Social Explanation* (London, 1973), 139–70, see 144–46, 154–55.

9. On mimetic, participatory consciousness, see Morris Berman, *The Reenchantment of the World* (Ithaca, 1981); Eric Havelock, *Preface to Plato* (Cambridge, Mass., 1963); and Max Horkheimer, *The Eclipse of Reason* (New York, n.d.), 92–127. On the gaze, see *Compact Oxford English Dictionary,* s.v. "gaze": "said of a deer, also of persons, especially in wonder, expectancy, bewilderment." "The hart, stag, buck, or hind when borne in coat-armour, looking affrontée or full faced is said to be at gaze . . . but all other beasts in this attitude are called guardant." William Berry, *Encyclopedia heraldica,* s.v. "gaze." On the Koyukon Indian versus white methods of hunting the deer, see Richard K. Nelson, "The Gifts," in Daniel Halpern, ed., *Antaeus,* no. 57 (Autumn, 1986), 117–31, esp. 122. On imitation of animals by humans in hunting, see Randall L. Eaton, "Hunting and the Great Mystery of Nature," *Utne Reader* (January/February 1987), 42–49.

10. On the dominance of vision in Western consciousness see Hans Jonas, "The Nobility of Sight," *Philosophy and Phenomenological Research* 14 (1954), 507–19; Evelyn Fox Keller and Christine Grontkowski, "The Mind's Eye," in Sandra Harding and Merrill B. Hintikka, eds., *Discovering Reality* (Dordrecht, Holland, 1983), 207–24; James Axtell, "The Power of Print in the Eastern Woodlands," *William and Mary Quarterly* 44 (2) 3rd ser. (April 1987), 300–09.

The Trouble with Wilderness: Or, Getting Back to the Wrong Nature, by William Cronon

1. Henry David Thoreau, "Walking," *The Works of Thoreau,* ed. Henry S. Canby (Boston: Houghton-Mifflin, 1937), 672.

2. Entry on "wilderness," *Oxford English Dictionary;* see also Roderick Nash, *Wilderness and the American Mind,* 3rd ed. (New Haven: Yale University Press, 1967, 1982), 1–22. For other important discussions of the history of wilderness, see Max Oelschlaeger, *The Idea of Wilderness: From Prehistory to the Age of Ecology* (New Haven: Yale University Press, 1991).

3. Exodus, 32:1-35, KJV.

4. Exodus, 14:3, KJV.

5. Mark 1:12-13 KJV; see also Matthew, 4:1-11; and Luke, 4:1-13.

6. John Milton, "Paradise Lost," *John Milton: Complete Poems and Major Prose,* ed. Merrit Y. Hughes (New York: Odyssey Press, 1957), 280–81, lines 131–42.

7. I have discussed this theme at length in William Cronon, "Landscapes of Abundance and Scarcity," in Clyde Milner, et al., eds., *Oxford History of the American West* (New York: Oxford University Press, 1994), 603–37. The classic work on the Puritan "city on a hill" in colonial New

England is Perry Miller, *Errand Into the Wilderness* (Cambridge, Mass.: Harvard University Press, 1956).

8. John Muir, *My First Summer in the Sierra* (1911), reprinted in *John Muir: The Eight Wilderness Discovery Books* (London: Diadem; Seattle: The Mountaineers, 1992), 211.

9. Alfred Runte, *National Parks: The American Experience*, 2nd ed. (Lincoln: University of Nebraska Press, 1987).

10. John Muir, *The Yosemite* (1912), reprinted in *John Muir: Eight Wilderness Discovery Books*, 715.

11. Scholarly work on the sublime is extensive. Among the most important studies are Samuel Monk, *The Sublime: A Study of Critical Theories in XVIII-Century England* (New York, 1935); Basil Willey, *The Eighteenth-Century Background: Studies on the Idea of Nature in the Thought of the Period* (London: Chattus and Windus, 1949); Marjorie Hope Nicolson, *Mountain Gloom and Mountain Glory: The Development of the Aesthetics of the Infinite* (Ithaca: Cornell University Press, 1959); Thomas Weiskel, *The Romantic Sublime: Studies in the Structure and Psychology of Transcendence* (Baltimore: Johns Hopkins University Press, 1976); and Barbara Novak, *Nature and Culture: American Landscape Painting, 1825–1875* (New York: Oxford University Press, 1980).

12. The classic works are Immanuel Kant, *Observations on the Feeling of the Beautiful and Sublime* (1764), trans. John T. Goldthwait (Berkeley: University of California Press, 1960); Edmund Burke, *A Philosophical Enquiry into the Origin of our Ideas of the Sublime and Beautiful*, ed. James T. Boulton (1958; Notre Dame: University of Notre Dame Press, 1968); and William Gilpin, *Three Essays: On Picturesque Beauty; On Picturesque Travel; and on Sketching Landscape* (London, 1803).

13. See Ann Vileisis "From Wastelands to Wetlands," unpublished senior essay, Yale University, 1989; and Alfred Runte, *National Parks*.

14. William Wordsworth, "The Prelude," Book VI, in Thomas Hutchinson, ed. *The Poetical Works of Wordsworth* (London: Oxford University Press, 1936), 536.

15. Henry David Thoreau, *The Maine Woods* (1864), in *Henry David Thoreau* (New York: Library of America, 1985), 640–41.

16. Exodus 16:10.

17. John Muir, *My First Summer in the Sierra* (1911), in *John Muir: The Eight Wilderness Discovery Books* (Seattle: The Mountaineers, 1992), 238. Part of the difference between these descriptions may reflect the landscapes the three authors were describing. In his essay elsewhere in this book, Kenneth Olwig notes that early American travelers experienced Yosemite as much through the aesthetic tropes of the pastoral as through those of the sublime. The ease with which Muir celebrated the gentle divinity of the Sierra Nevada had much to do with the pastoral qualities of the landscape he described.

18. Frederick Jackson Turner, *The Frontier in American History* (New York: Henry Holt, 1920), 37–38.

19. Richard Slotkin has made this observation the linchpin of his comparison between Turner and Theodore Roosevelt. See Slotkin, *Gunfighter Nation: The Myth of the Frontier in Twentieth-Century America* (New York: Atheneum, 1992), 29–62.

20. Owen Wister, *The Virginian: A Horseman of the Plains* (New York: Macmillan, 1902), viii–ix.

21. Theodore Roosevelt, *Ranch Life and the Hunting Trail* (1888; New York: Century, 1899), 100.

22. Wister, *Virginian*, x.

23. On the many problems with this view, see William M. Denevan, "The Pristine Myth: The Landscape of the Americas in 1492," *Annals of the Association of American Geographers* 82 (1992), 369–85.

24. Wilderness also lies at the foundation of the Clementsian ecological concept of the climax.

25. On the many paradoxes of having to manage wilderness into order to maintain the appearance of an unmanaged landscape, see John C. Hendee, et al., *Wilderness Management*, USDA Forest Service Miscellaneous Publication No. 1365 (Washington, D.C.: Government Printing Office, 1978).

26. This argument has been powerfully made by Ramachandra Guha, "Radical American Environmentalism: A Third World Critique," *Environmental Ethics* 11 (1989), 71–83.

27. Bill McKibben, *The End of Nature* (New York: Random House, 1989).

28. McKibben, *End of Nature*, 49.

29. Even comparable extinction rates have occurred before, though we surely would not want to emulate the Jurassic-Cretaceous boundary extinctions as a model for responsible manipulation of the biosphere!

30. Dave Foreman, *Confessions of an Eco-Warrior* (New York: Harmony Books, 1991), 69; italics in original. For a sampling of other writings by followers of deep ecology and/or Earth First!, see Michael Tobias, ed., *Deep Ecology* (San Diego: Avant Books, 1984); Bill Devall and George Sessions, *Deep Ecology* (Salt Lake City: Gibbs M. Smith, 1985); Michael Tobias, *After Eden: History, Ecology, and Conscience* (San Diego: Avant Books, 1985); Dave Foreman and Bill Haywood, eds., *Ecodefense: A Field Guide to Monkey Wrenching*, 2nd ed. (Tucson: Ned Ludd Books, 1987); Bill Devall, *Simple in Means, Rich in Ends: Practicing Deep Ecology* (Salt Lake City: Gibbs Smith, 1988); Steve Chase, ed., *Defending the Earth: A Dialogue Between Murray Bookchin & Dave Foreman* (Boston: South End Press, 1991); John Davis, ed., *The Earth First! Reader: Ten Years of Radical Environmentalism* (Salt Lake City: Gibbs Smith, 1991); Bill Devall, *Living Richly in an Age of Limits: Using Deep Ecology for An Abundant Life* (Salt Lake City: Gibbs Smith, 1993); and Michael E. Zimmerman, et al., eds., *Environmental Philosophy: From Animal Rights to Deep Ecology* (Englewood Cliffs, N.J.: Prentice-Hall, 1993). A useful survey of the different factions of radical environmentalism can be found in Carolyn Merchant, *Radical Ecology: The Search for a Livable World* (New York: Routledge, 1992). For a very interesting critique of this literature (first published in the anarchist newspaper *Fifth Estate*), see George Bradford, *How Deep is Deep Ecology?* (Ojai, Calif.: Times Change Press, 1989).

31. Foreman, *Confessions of an Eco-Warrior*, 34.

32. Ibid., 65. See also Dave Foreman and Howie Wolke, *The Big Outside: A Descriptive Inventory of the Big Wilderness Areas of the U.S.* (Tucson: Ned Ludd Books, 1989).

33. Foreman, *Confessions of an Eco-Warrior*, 63.

34. Foreman, *Confessions of an Eco-Warrior*, 27.

35. It is not much of an exaggeration to say that the wilderness experience is essentially consumerist in its impulses.

36. Muir, "Yosemite," in *John Muir: Eight Wilderness Discovery Books*, 714.

37. Wallace Stegner, ed., *This Is Dinosaur: Echo Park Country and Its Magic Rivers* (New York: Alfred A. Knopf, 1955), 17; emphasis in original.

38. Katherine Hayles helped me see the importance of this argument.

39. Analogous arguments can be found in John Brinckerhoff Jackson, "Beyond "Wilderness," *A Sense of Place, a Sense of Time* (New Haven: Yale University Press, 1994), 71–91; and in the wonderful collection of essays by Michael Pollan, *Second Nature: A Gardener's Education* (New York: Atlantic Monthly Press, 1991).

40. Wendell Berry, *Home Economics* (San Francisco: North Point, 1987), 138, 143.

41. Gary Snyder, quoted in the *New York Times*, "Week in Review," 6.

The Earliest Cultural Landscapes of England, by I. G. Simmons

1. The classic though dated account is W. G. Hoskins, *The Making of the English Landscape* (Harmondsworth, 1970). See also M. Jones, *England Before Domesday* (London, 1981).

2. A new series on landscape history largely by systematic topic is being edited by Michael Read at the Loughborough University of Technology. The first to appear is L. Cantor, *The Changing English Countryside, 1400–1700* (London, 1987). Most standard historical geographies have allusions to landscape. See H. C. Darby, ed., *A New Historical Geography of England* (Cambridge, 1973).

3. A standard prehistory is P. Phillips, *The Prehistory of Europe* (London, 1980); but see also interpretations by R. Bradley, *The Prehistoric Settlement of Britain* (London, 1978); and the environmental context in I. G. Simmons and M. J. Tooley, eds., *The Environment in British Prehistory* (London, 1981).

4. A summary of vegetation changes is in R. G. West, "Pleistocene Forest History in East Anglia," *New Phytologist* 85 (1980), 571–622; detail in R. G. West, "The Quaternary Deposits at Hoxne, Suffolk," *Philosophical Transactions of the Royal Society* B 239 (1965), 265–356; C. Turner, "The Middle Pleistocene Deposits at Mark's Tey, Essex," *Philosophical Transactions of the Royal Society* B 257 (1970), 373–437.

5. C. Turner, "Der Einschluß großer Mammalier auf die interglaziale Vegetation," *Quär-tarpaläontologie* 1 (1975), 13–19.

6. See A. Morrison, *Early Man in Britain and Ireland* (London, 1980); D. A. Roe, *The Lower and Middle Palaeolithic Period in Britain* (London, 1981); and J. J. Wymer, *The Palaeolithic Age* (London, 1982).

7. J. B. Campbell, *The Upper Palaeolithic of Britain: A Study of Man and Nature in the Pleistocene*, 2 vols. (Cambridge, 1977).

8. I. G. Simmons, G. W. Dimbleby, and C. Grigson, "The Mesolithic," in Simmons and Tooley, eds., *The Environment in British Prehistory*, 82–124.

9. The classic site of Star Carr (in the vicinity of which new excavations by T. Schadla-Hall and by P. Mellars will certainly amplify the evidence) is in J. G. D. Clark et al., *Excavations at Star Carr: An Early Mesolithic Site at Seamer, Near Scarborough* (Cambridge, 1954), and *Star Carr: A Case Study in Bioarchaeology* (Reading, Mass., 1972). There have been many subsequent reinterpretations of the site by other authors.

10. A. G. Smith, "The Influence of Mesolithic and Neolithic Man on British Vegetation," in D. Walker and R. G. West, eds., *Studies in the Vegetational History of the British Isles* (Cambridge, 1970), 81–96; and "Newferry and the Boreal-Atlantic Transition," *New Phytologist* 98 (1984), 35–55.

11. Two summary papers are I. G. Simmons and J. B. Innes, "Late Mesolithic Land-Use and Its Impact in the English Uplands," *Biogeographical Monographs* 2 (1985), 7–17, and "Mid-Holocene Adaptations and Later Mesolithic Forest Disturbance in Northern England," *Journal of Archaeological Science* 14 (1987), 385–403.

12. The quotation I have in mind runs, "De foist time is happenstance, de second is coincidence, de toid time is enemy action." I regret I have been unable to verify the reference to this useful piece of probability theory.

13. I. G. Simmons, "Late Mesolithic Societies and the Environment of the Uplands of England and Wales," *Bulletin of the Institute of Archaeology, London* 16 (1979), 111–29; P. Mellars, "Fire Ecology, Animal Populations and Man: A Study of Some Ecological Relationships in Prehistory," *Proceedings of the Prehistoric Society* 42 (1976), 14–45.

14. R. Jacobi, J. H. Tallis, and P. Mellars, "The Southern Pennine Mesolithic and the Ecological Record," *Journal of Archaeological Science* 3 (1976), 307–20; I. G. Simmons and J. B. Innes, "Tree Remains in a North York Moors Peat Profile," *Nature, London* 294 (1981), 74–78.

15. See the discussion in R. Dennell, *European Economic Prehistory* (London, 1983).

16. I will, on request, send potential visitors a list of places from which to stand and stare.

Landschaft *and Linearity, by John R. Stilgoe*

1. On archetypes see C. G. Jung, *The Archetypes and the Collective Unconscious*, trans. R. F. C. Hull (Princeton: Princeton University Press, 1969), 1–36; Jung mentions spatial archetypes in *Mandala Symbolism*, trans. R. F. C. Hull (Princeton: Princeton University Press, 1969), 93–94.

2. For other definitions see Robert E. Dickinson, "Landscape and Society," *The Scottish Geographical Magazine* 55 (January 1939), 1–15; J. B. Jackson, "The Meaning of 'Landscape,'" *Kulturgeograft* 88 (1965), 47–50; Josef Schmithusen, "Was ist eine Landschaft?" *Erdkundliches Wisen* 9 (1964), 7–24; and Gabriele Schwarz, *Allgemeine Siedlungsgeographie* (Berlin: Walter de Gruyter, 1966), 162–220. Perhaps the most comprehensive analysis of the medieval *landschaft* is found in three works by Karl Siegfried Bader: *Das Mittelalterliche Dorf als Friedens—und Rechtsbereich* (Weimar: Bohlaus, 1957); *Dorfgenossenschaft und Dorfgemeinde* (Koln: Bohlau, 1962); and *Rechtsformen und Schichten der Liegenschaftsnutzung im Mittelalterlichen Dorf I* (Wein: Bohlaus, 1973).

3. Stephen Miller, "Politics and Amnesty International," *Commentary* 65 (March 1978), 58.

4. *Oxford English Dictionary* (New York: Oxford University Press, 1971), I, 1566–67.

5. Robert Coles, "Telic Reforms," *The New Yorker* 54 (March 13, 1978), 141.

6. *OED*, II, 3630–31.

7. John Conron, ed., *American Landscapes* (New York: Oxford University Press, 1971).

8. Henry James, *The American Scene* (New York: Harper and Brothers, 1970), v–vi; hereafter cited as James.

9. James, 38, 442, 50; in his novels James writes lovingly about pedestrian places; see *The Ambassadors* (1903; rpt. New York: W. W. Norton, 1964), 24.

10. James, 105, 98, 122.

11. James, 73; for another use of the metaphor see Hart Cranes *The Bridge* (Paris: Black Sun, 1930).

12. James, 46, 49, 385.

13. Timothy Dwight, *Travels in New England and New York* (Cambridge: Harvard University Press, 1969), 111, 221.

14. Frederick Law Olmsted, *A Journey in the Seaboard Slave States* (New York: Dix and Edwards, 1856); *A Journey Through Texas* (New York: Dix and Edwards, 1857); and *A Journey in the Back Country* (New York: Mason, 1861).

15. Emily Post, *By Motor to the Golden Gate* (New York: Appleton, 1917); Theodore Dreiser, *A Hoosier Holiday* (New York: John Lane, 1916); and George R. Stewart, *U.S. 40: Cross Section of the United States of America* (Boston: Houghton, 1953).

16. Adalbert Klaar, *Die Siedlungs-und hausformen des Wiener Waldes* (Stuttgart: J. Engelhorns, 1936).

17. Marc Bloch, *Land and Work in Medieval Europe*, trans. J. E. Anderson (Berkeley: University of California Press, 1967); G. G. Coulton, *The Medieval Panorama: The English Scene from Conquest to Reformation* (Cambridge, England: Cambridge University Press, 1938); Joan Thirsk, *The Agrarian History of England and Wales* (Cambridge: Cambridge University Press, 1967); Warren O. Ault, *Open-Field Farming in Medieval England* (New York: Barnes and Noble, 1972).

18. Jacob Grimm, *Deutsche Rechtsalterthumer*, eds. Andreas Heusler and Rudolf Hubner (Leipzig: Dietrich, 1899), I, 557–675.

19. Edward Chamberlayne, *Angliae Notitia: Or the Present State of England* (London, John Martyn, 1669), 76–77.

20. Jacob and Wilhelm Grimm, *German Folk Tales*, trans. Francis P. Magoun, Jr., and Alexander Krappe (Carbondale: Southern Illinois University Press, 1960).

21. Geoffrey Hindle, *A History of Roads* (Syracuse, N.J.: Citadel, 1971), 46–56; Hans Hitzer, *Die Strasse* (Muinchen: Callwey, 1971), 105–85; and Lewis Mumford, *The Culture of Cities* (1938; rpt. New York: Harcourt, 1970), 26, 80.

22. Emmerich Vattel, *The Law of Nations* (Dublin: Luke White, 1787), 80–82.

23. On the role of the city streets as meeting places see Pierre Lelievre, *La Vie des Cites de l'Antiquite a nous Jours* (Paris: Bourrelier, 1950), 11, and Richard Senett, *The Fall of Public Man* (New York: Knopf, 1977), 36 and *passim*.

24. See, for example, Nikolaus Lenau, "Der Postillon," for a description of the night-journeying: *Sammtliche Werke* (Stuttgart: Gottascher, 1855), I, 200–03.

25. Thomas De Quincey, *The English Mail Coach* (Boston: Ticknor, 1851).

26. Coleridge, *Poems* (Oxford: Oxford University Press, 1930), 29. See also Arnold Van Gennep, *The Rites of Passage*, trans. Monika B. Vizedom and Gabrielle L. Caffee (Chicago: University of Chicago Press, 1972), 15–25.

27. As examples, see the writings of Thomas Nelson Page, George W. Cable, Sarah Orne Jewette, George W. Sears, John Muir, and Dallas Lore Sharp.

28. Henry Olerich, *A Cityless and Countryless World* (New York: Arno, 1971). The following paragraphs are based on extensive study of approximately fifty-five utopian works, many of which are listed in Allyn Bailey Forbes, "The Literary Quest for Utopia, 1880–1900," *Social Forces* 6 (December 1927), 179–89.

29. Edward Bellamy, *Looking Backward* (Boston: Houghton, 1966); Bradford Peck, *The World of a Department Store* (Lewiston, Maine: The Author, 1900); Edgar Chambless, *Roadtown* (New York: Roadtown Press, 1910).

30. Olerich, 61.

31. Chambless, 53.

32. Thomas More, *Utopia*, ed. Edward Surtz (New Haven: Yale University Press, 1968), 61.

33. On the form of utopia see Louis Marin, *Utopiques: Jeux D'Espaces* (Paris: Editions de Minuit, 1973), and Wolfgang Biesterfeld, *Die Literarische Utopia* (Stuttgart: Metzler 1974), esp. 11.

34. See Douglas Fraser, *Village Planning in the Primitive World* (New York: Braziller, 1968); Joseph

Rykwert, *The Idea of a Town: The Anthropology of Urban Form in Rome, Italy, and the Ancient World* (Princeton: Princeton University Press, 1976); Mircea Eliade, *The Sacred and the Profane*, trans. Willard R. Trask (New York: Harcourt, 1959), 1–65; and Werner Muller, *Die Heilige Stadt* (Stuttgart: W. Kohlhammer, 1961).

35. Benton MacKaye, *New Exploration: A Philosophy of Regional Planning* (New York: Harcourt, 1928); Paul and Percival Goodman, *Communitas: Means of Livelihood and Ways of Life* (New York: Random, 1947): for a powerful visual linking of the archetypal *landschaft* with the crafted utopia, see Pare Lorentz's film *The City* (1939).

36. See Carl Steinitz, "Meaning and the Congruence of Urban Form and Activity," *American Institute of Planners Journal* (July 1968), 233–48; Kevin Lynch and Malcolm Rivkin, "A Walk Around the Block," *Landscape* 8 (Spring 1959), 24–34; and Kevin Lynch, *The Image of the City* (Cambridge, Mass.: MIT Press, 1960).

37. See, for example, Ursula K. Leguin, *The Wizard of Earthsea* (New York: Bantam, 1975); Herman Hesse, *Sidhartha, eine indische Dichtung* (Berlin: Fischer, 1922), which has been translated into English several times; Robert M. Pirsig, *Zen and the Art of Motorcycle Maintenance* (New York: Morrow, 1974); and most importantly, J. R. R. Tolkien, *The Hobbit, or There and Back Again* (New York: Ballantine Books, 1973). The dominant spatial feature in Tolkien's Middle Earth is the road.

38. Heiner Treinen, "Symbolische Ortsbezogenheit," *Kolner Zeitschrift fur Sociologie und Sozialpsychologie* 17 (1965): 73–97; Alexander Mitscherlich, *Die Unwirtlichkeit unserer Stadte* (Frankfurt am Main: Suhrkamp, 1965); and Hans Oswald, *Die Uberschatzte Stadt* (Olten: Walter, 1966).

Environmental Change in Colonial New Mexico, by Robert MacCameron

1. William Cronon, *Changes in the Land: Indians, Colonists and the Ecology of New England* (New York: Hill and Wang, 1983), vii.

2. Cronon, viii.

3. Richard White, *Land Use, Environment, and Social Change: The Shaping of Island County, Washington* (Seattle: University of Washington Press, 1980), 36. Neither Cronon nor White subscribes to the notion that Native Americans lived in any sort of perfect harmony or static relationship with nature prior to contact with Europeans. Cronon notes that the Indian practice in New England of burning forests to clear land for agriculture and to improve hunting "could sometimes go so far as to remove the forest altogether, with deleterious effects for trees and Indians alike." And White points out that the natives of Island County did not hesitate to alter natural systems when it was to their advantage to do so. For example, through the manipulation of the environment the Salish increased such desired plant species as bracken, nettles, and camas for use as food crops. Rather, most remarkable was the degree of environmental change that occurred in these far corners of English North America after contact. For both Cronon and White such change was both rapid and essentially linear.

4. Scholars generally agree, on the basis of archaeology and Spanish accounts, that pre-contact Pueblo settlements used some form of irrigation farming to grow maize, beans, squash, cotton and tobacco. While access to surface water removed certain restrictions and risks to agriculture, and, indeed, created surpluses allowing for greater social elaboration, studies now also indicate that environmental factors, some the result of using water control devices, constrained Pueblo farming and in the process effected changes in the land.

Irrigation farming, even in its crudest forms, likely sets in motion ecological chain reactions. Plants are introduced into habitats where they could not have sustained themselves previously; terracing changes the natural flow of streams; and man-made water diversions modify the natural vegetation, change the organic matter in the soil, and perhaps alter the migrating pattern of birds and animals. See Michael C. Meyer, *Water in the Hispanic Southwest: A Social and Legal History, 1550–1850* (Tucson: The University of Arizona Press, 1984), 19.

In addition, poor drainage and high evaporation can lead irrigation waters to deposit salts and other minerals that inhibit crop production. And as crop irrigation brings both fields and plants closer together the risk of crop loss due to diseases and insect pests is increased. See Linda S. Cordell, *Prehistory of the Southwest* (Orlando: Academic Press, Inc., 1984), 203–04.

Due to these factors, Cordell believes that good bottomland in the upper Rio Grande valley was very likely in increasingly short supply over the two centuries prior to the arrival of the Spanish. Studies of Pueblo land use and evidence of expansion, away from the river onto the plains, indicate that intensive efforts were being made to support a large population increase as Pueblos were abandoned and new communities founded. See Cordell "Prehistory: Eastern Anasazi," in *Handbook of North American Indians*, vol. 9, ed. Alfonso Ortiz (Washington, D.C.: Smithsonian Institution Press, 1979), 151. If this interpretation is correct, the carrying capacity (the maximum population that a particular environment can support indefinitely without leading to environmental degradation) along the Rio Grande itself may well have been reaching its limit at the time of Spanish contact.

5. A useful point of departure for understanding similarities and differences in frontier societies, and their relationship to the land, is the notion of an inclusive versus exclusive frontier. Spanish colonists frequently settled in areas of sedentary Indians, seeking Indian labor at the same time that they strove for Indian souls and mated with their women. Where there were no Indians there were no Spanish. Without a sharply defined racial barrier, this was a frontier of inclusion. In contrast, English colonists, while integrating economically with certain native populations to a degree, did not intermarry. Nor on any significant scale did they attempt to convert Indians to Protestantism. This then was a frontier of exclusion. See Alistair Hennessy, *The Frontier in Latin American History* (Albuquerque: University of New Mexico Press, 1978), 146–47; and Alfred Crosby, Jr., *The Columbian Exchange: The Biological and Cultural Consequences of 1492* (Westport, Conn.: Greenwood Press, 1972) and *Ecological Imperialism: The Biological Expansion of Europe 900–1900* (Cambridge: Cambridge University Press, 1986).

Add to this broad characterization the fact that the Province of New Mexico served as a defensive and missionary outpost over most of its colonial history. It never was an economic frontier on the order of New England, for example, a region which quickly became integrated into an international commercial trade network. In fact, Spain's loss of portions of its North American empire can be attributed to the advantages of England's expanding economic frontier, over Spain's defensive frontier, operating at long distance from centers of resources and population.

6. Richard I. Ford, "The New Pueblo Economy," in *When Cultures Meet: Remembering San Gabriel Del Yunge Oweenge* (Santa Fe: Sunstone Press, 1987), 73.

7. John R. Van Ness, "Hispanic Land Grants: Ecology and Subsistence in the Uplands of Northern New Mexico and Southern Colorado," in *Land, Water, and Culture: New Perspectives on Hispanic Land Grants*, eds. Charles L. Briggs and John R. Van Ness (Albuquerque: University of New Mexico Press, 1987), 204.

8. Hal Rothman, "Cultural and Environmental Change on the Pajarito Plateau," *New Mexico Historical Review* 64 (April 1989), 186. Rothman agrees fundamentally with Van Ness. He states that "American influence telescoped into a few years much more environmental and cultural change than Spanish practices had produced in nearly three hundred years." There were two reasons for these varying rates of change, according to Rothman. First, a marginal area such as New Mexico did not attract sufficient numbers of Europeans to effect dramatic environmental change; and, second, the "un-European," semiarid climate of New Mexico protected it from the "full brunt of the portmanteau biota of the Spanish" (188). Because many Old World plants, such as fruit trees, melons, and wheat, could exist only in proximity to water, more was needed than merely the presence of the Spanish and their descendants to "Europeanize" the plants and animals of New Mexico. What was required was the transformation of New Mexico into an economic frontier, creating opportunities to produce and transport commodities to market on a large scale.

9. Donald Worster, "Toward an Agroecological Perspective in History," *Journal of American History* 76 (March 1990), 1101.

10. Over the period of colonial rule, the Spanish conducted regular, and, by most accounts, fairly accurate censuses of the Province of New Mexico. (Because this study focuses upon the upper Rio Grande valley, the following population figures exclude the areas of El Paso and Zuni.) Only the estimates of Pueblo population at the time of the first Spanish settlement in 1598 and the Benavides counts of 1630 and 1635 have been the subject of any real debate. The figure for 1598 is generally agreed to be around 38,000, while Benavides estimated some 26,000 Pueblo Indians in 1630. The

sharp decline was attributable to disease, starvation owing to Spanish tribute and labor institutions, and flight to western Pueblos. Throughout the remainder of the seventeenth century and for most of the eighteenth century Pueblo population continued to decline. It was approximately 23,600 in 1680, the year of the Pueblo Revolt, 7,200 in 1706, 5,200 in 1752, 6,500 in 1776, 6,400 in 1805 and 1821. See Marc Simmons, "History of Pueblo-Spanish Relations to 1821," in *Handbook of North American Indians*, vol. 19, ed. Alfonso Ortiz (Washington, D.C.: Smithsonian Institution Press, 1979), 185.

11. Oakah L. Jones, Jr., *Los Paisanos: Spanish Settlers on the Northern Frontier of New Spain* (Norman: University of Oklahoma Press, 1979) 117–29.

12. Jones, 117–29.

13. Alvar W. Carlson, *The Spanish-American Homeland: Four Centuries in New Mexico's Rio Arriba* (Baltimore: The Johns Hopkins University Press, 1990), 9.

14. Marc Simmons, "The Rise of New Mexico Cattle Ranching," *El Palacio* 93 (Spring 1988), 7.

15. John O. Baxter, *Las Carneradas: Sheep Trade in New Mexico, 1700–1860* (Albuquerque: University of New Mexico Press, 1987), 90.

16. George P. Hammond and Agapito Rey, eds., *The Rediscovery of New Mexico, 1580–1594* (Albuquerque: University of New Mexico Press, 1966) 89, 230.

17. Simmons, "New Mexico Cattle Ranching," 7.

18. Baxter, 24.

19. Baxter, 92.

20. Marc Simmons, "The Chacon Economic Report of 1803," *New Mexico Historical Review* 60 (January 1985), 87.

21. David J. Weber, *The Taos Trappers: The Fur Trade in the Far Southwest, 1540–1846* (Norman: University of Oklahoma Press, 1971), 10.

22. Peter Bakewell, "Ecological Effects of Silver Mining in Colonial Spanish America," paper presented at the American Historical Association annual meeting, December, 1985.

23. Hester Jones, "Uses of Wood by the Spanish Colonists in New Mexico," *New Mexico Historical Review* 7 (July 1932).

24. Hammond and Rey, 221, 230.

25. Josiah Gregg, *Commerce of the Prairies* (Norman: University of Oklahoma Press, 1954), 113.

26. W. W. H. Davis, *El Gringo: New Mexico and Her People* (Lincoln: University of Nebraska Press, 1982), 356.

27. Simmons, "New Mexico's Colonial Agriculture," *El Palacio* 89 (Spring 1983), 9.

28. Edward W. Smith, *Adobe Bricks in New Mexico* (Socorro: New Mexico Bureau of Mines and Mineral Resources, 1982), 1113.

29. Malcolm Ebright, "New Mexican Land Grants: The Legal Background," in *Land, Water, and Culture: New Perspectives on Hispanic Land Grants*, 23.

30. Carlson, 31, 69–70.

31. D. W. Meinig, *The Shaping of America: A Geographical Perspective on 500 Years of History*, vol. 1 (New Haven: Yale University Press, 1986), 240.

32. John Van Ness, "Hispanic Land Grants," 193–94.

33. Nancy Hunter Warren, "The Irrigation Ditch (Photo Essay)," *El Palicio* 86 (Spring 1980), 28. See also Daniel Tyler, "Dating the Caño Ditch: Detective Work in the Pojoaque Valley," *New Mexico Historical Review* 61 (January 1986) and Stanley Crawford, *Mayordomo: Chronicle of an Acequia in Northern New Mexico* (Albuquerque: University of New Mexico Press, 1988).

34. Marc Simmons, "Spanish Irrigation Practices in New Mexico," *New Mexico Historical Review* 47 (April 1972), 145.

35. Arthur Goss, "The Value of Rio Grande Water for the Purpose of Irrigation," New Mexico College of Agriculture and the Mechanic Arts, Agricultural Experiment Station, *Bulletin* (November 1893), 34.

36. Frank E. Wozniak, "Irrigation in the Rio Grande Valley, New Mexico: A Study of the Development of Irrigation Systems Before 1945" (Santa Fe: The New Mexico Historic Preservation Division, 1987), 38, photocopied.

37. Wozniak, 25.

38. Ford, "The New Pueblo Economy," 74.

39. Ford, 86.

40. Vorsila L. Bohrer, "The Prehistoric and Historic Role of the Cool-Season Grasses in the Southwest," *Economic Botany* 29 (July-September 1975), 203.

41. Ford, 87.

42. Christopher Vecsey, "American Indian Environmental Religions," in *American Indian Environments: Ecological Issues in Native American History*, eds. Christopher Vecsey and Robert W. Venables (Syracuse: Syracuse University Press, 1980), 10.

43. Frances Leon Swadesh, "Structure of Spanish-Indian Relations in New Mexico," in The *Survival of Spanish American Villages*, ed. Paul Kutsche (Colorado Springs: Research Committee, Colorado College, 1979), 53–61.

44. Swadesh, 61.

45. Yi-Fu Tuan, Cyril E. Everard, and Jerold G. Widdison, *The Climate of New Mexico* (Santa Fe: State Planning Office, 1969), 158.

46. Harold C. Fritts, "Tree-Ring Evidence for Climatic Changes in Western North America," *Monthly Weather Review* 93 (July 1965), 421, 430–31.

47. Fluctuations in the climate of colonial New Mexico can be viewed as well in the broader context of the climatic phenomenon known as the Little Ice Age. Between 1430 and 1850, the Northern Hemisphere's climate was allegedly cooler than periods either before or after. Therefore, the upper Rio Grande valley, over this time, may have experienced, on average, larger snow cover, enhanced freezing, and, therefore, shorter growing seasons. On the other hand, cooler and more moist summers may have produced excellent harvests. More importantly, however, climate conditions during the Little Ice Age were far from stable, and there were complex spatial patterns of warming and cooling throughout the period. To draw causal relations between the LIA and long-term environmental change in the upper Rio Grande valley is therefore risky at best. See T. M. L. Wrigley et al., eds., *Climate and History: Studies in Past Climates and Their Impact on Man* (Cambridge: Cambridge University Press, 1981), 17.

48. Ronald U. Cooke and Richard W. Reeves, *Arroyos and Environmental Change in the American South-West* (Oxford: Clarendon Press, 1976), 16.

49. John R. Kummel and Melvin I. Dyer, "Consumers in Agroecosystems: A Landscape Perspective," in *Agricultural Ecosystems: Unifying Concepts*, eds. Richard Lowrance, Benjamin R. Stinner, and Garfield J. Hause (New York: John Wiley & Sons, 1984), 65.

50. Dan Scurlock, "The Rio Grande Bosque: Ever Changing," *New Mexico Historical Review* 63 (April 1988), 135.

51. Spanish Archives of New Mexico, I, 1118.

From Conservation to Environment, by Samuel P. Hays

1. For this theme see Samuel P. Hays, *Conservation and the Gospel of Efficiency* (Cambridge, 1958).

2. A somewhat larger ecological context for analysis of the soil conservation movement is Donald Worster, *Nature's Economy* (San Francisco, 1977), 189–253.

3. For the Taylor Grazing Act and its implementation see Phillip O. Foss, *Politics and Grass: The Administration of Grazing on the Public Domain* (Seattle, 1960); Marian Clawson, *The Bureau of Land Management* (New York, 1971); William Voigt, Jr., *Public Grazing Lands: Use and Misuse by Industry and Government* (New Brunswick, N.J., 1976).

4. See, especially, James B. Trefethen, *An American Crusade for Wildlife* (New York, 1975), esp. 243–55.

5. The University of Michigan School of Forestry, for example, became the School of Natural Resources when Samuel Dana succeeded Filibert Roth, a protégé of Gifford Pinchot, as Dean. One of Dana's major innovations was to bring wildlife more fully into the curriculum. Interview with Carl Holcomb, student at the School in the early 1930s and editor of its student magazine, *Michigan Forests*, in 1934.

6. *Forests and Waters* was one of the titles of the magazine published by the American Forestry Association (now *American Forests*) in the early twentieth century. It was also the name of the ad-

ministering agency for Pennsylvania resources, the Department of Forests and Waters until 1970 when it was changed to the Department of Environmental Resources.

7. The National Parks Association added this new title to its publication in 1970. At the same time it revised its own name from National Parks Association to National Parks and Conservation Association.

8. A brief statement of this competition between the Park Service and the Forest Service, and a detailed analysis of the Olympic case is Ben W. Twight, "The Tenacity of Value Commitment: The Forest Service and the Olympic National Park," Ph.D. thesis, University of Washington, Seattle, 1971.

9. See, for example, Ralph Borsodi, *Flight from the City: An Experiment in Creative Living on the Land* (New York and London, 1933), and a publication, edited by Mildred Loomis, which grew out of Borsodi's inspiration, *The Green Revolution*, 1962–, published in the 1960s at the School of Living, Freeland, Maryland.

10. For a brief description of this social context see Samuel P. Hays, "The Structure of Environmental Politics Since World War II," in *Journal of Social History* 14–4 (Summer 1981), 719–38.

11. See William Ashworth, Hells Canyon, *The Deepest Gorge on Earth* (New York, 1977); interview with Brock Evans (May 1980) who at the time of the Hells Canyon decision was Pacific Northwest representative of the Sierra Club.

12. These developments are worked out more fully in Samuel P. Hays, "The Role of Forests in American History," paper presented at a conference on the Future of the American Forests, sponsored by the Conservation Foundation, Seattle, Washington, April 1979. See also Hays, "Gifford Pinchot and the American Conservation Movement," in Carroll W. Purcell, Jr., ed., *Technology in America: A History of Individuals and Ideas* (Cambridge, 1981), 151–62.

13. This analysis of the relationship between the U.S. Forest Service and the wilderness movement is drawn from a variety of sources, including Twight (n. 8); James Gilligan, "The Development of Policy and Administration of Forest Service Primitive and Wilderness Areas in the Western United States" (2v), Ph.D. thesis, University of Michigan, 1953; articles in the *Living Wilderness* and the *Sierra Bulletin* from the mid-1930s onward; accounts of wilderness politics in state publications such as "Wild Oregon," "Wild Washington," "The Wilderness Record" (California); and the Sierra Club *National News Report*.

14. For accounts of these divergent views see articles in *Forest Planning*, Forest Planning Clearinghouse, Eugene, Oregon, which began publication in April 1980.

15. For a continuing treatment of issues arising from this program see the annual *Proceedings* of the National Watershed Congress, first held in 1953 and annually thereafter.

16. The issues in stream channelization can be followed in Committee on Government Operations, U.S. House of Representatives, *Stream Channelization* (4 parts), 92nd Congress, 1st Session (Washington, D.C., 1971).

17. For a survey of nongame wildlife issues, see Wildlife Management Institute, "Current Investments, Projected Needs & Potential New Sources of Income for Nongame Fish & Wildlife Programs in the United States" (Washington, D.C., 1975).

18. Wildlife Management Institute, "The North American Wildlife Policy, 1973," which includes a copy of the "American Game Policy, 1930" (Washington, D.C., n.d.).

19. For two recent compilations of value changes over the past several decades see Joseph Veroff, Elizabeth Douvan, and Richard A. Kulka, *The Inner American: A Self-Portrait From 1957 to 1976* (New York, 1981); Daniel Yankelovich, *New Rules: Searching for Self-Fulfillment in a World Turned Upside Down* (New York, 1981).

20. See Arnold Mitchell, "Social Change: Implications of Trends in Values and Lifestyles," VALS Report No. 3, Stanford Research International, Menlo Park, 1979; and John Naisbitt, "The New Economic and Political Order of the 1980s," speech given to the Foresight Group, Stockholm Sweden, April 17, 1980, available from the Center for Policy Process, Washington, D.C. For periodic coverage of work on value changes consult *Leading Edge Bulletin: Frontiers of Social Transformation*, published by Interface Press, Los Angeles, California.

21. See, for example, Mary Keys Watson, "Behavioral and Environmental Aspects of Recreational Land Sales," Ph.D. thesis, Department of Geography, Pennsylvania State University, 1975.

22. A considerable number of public opinion polls by the Gallup, Harris, and Roper organiza-

tions indicate the range of expression of environmental values. There are also numerous polls on specialized environmental subjects which reflect environmental values. Among them are: Stephen R. Kellert, "American Attitudes, Knowledge and Behaviors Toward Wildlife and Natural Habitats," study funded by the U.S. Fish and Wildlife Service, of which three of four phases were completed as of the end of 1980. The titles of each of phases I, II, and III were: "Public Attitudes Toward Critical Wildlife and Natural Habitat Issues," "Activity of the American Public Relating to Animals," and "Knowledge, Affection and Basic Attitudes Toward Animals in American Society." See also the Gallup Organization, "National Opinions Concerning the California Desert Conservation Area," study conducted for the Bureau of Land Management (Princeton, 1978); and Opinion Research Corporation, "The Public's Participation in Outdoor Activities and Attitudes Toward National Wilderness Areas," prepared for the American Forest Institute (Princeton, 1977).

23. This sectional analysis is derived from tabulation of environmental votes in the U.S. House of Representatives, 1970–77, originally prepared by the League of Conservation Voters, Washington, D.C., for each congressional session.

24. One type of evidence which reflects these growing interests is the "field guide," which grew rapidly in extent and circulation in the 1960s and 1970s. The most traditional format was that represented by the Peterson guide series which identified birds, plants, and animals. But there were an increasing number of hiking guides which included considerable information about the natural environment through which one hiked. Often each new site around which public natural environment interest arose led to a guide which enabled people to "find their way" and to appreciate what they saw. In 1980 the Sierra Club began to publish a new series of regional "naturalist guides" which provided similar assistance to seeking out a wide range of natural environmental areas.

25. One of the major expressions of this interest was nature photography. This cannot be pinned down quantitatively, but not wholly irrelevant was the rise in photography as a whole in American society. The 1980 edition of the survey of American participation in the arts indicated that the number of Americans engaged in photographic pursuits rose from 19 percent in 1975 to 44 percent in 1980. Westerners tended to be more active (56 percent participation in 1980) than those in other parts of the country. See American Council for the Arts, *Americans and the Arts* (New York, 1981), 37.

26. Two documents which reflect this emphasis at a governmental level are U.S. Senate, Select Committee on Nutrition and Human Needs, *Dietary Goals for the United States* (Washington, D.C.: GPO, 1977); and U.S. Department of Health, Education, and Welfare, Public Health Service, *Healthy People: The Surgeon General's Report on Health Promotion and Disease Prevention* (Washington, D.C.: GPO, 1979). See also, as a representative more popular statement, *Environmental Science and Technology*, April 1970, 275–77, interview with Dr. Paul Kotin, director, National Institute for Environmental Health Sciences, including the statement, "Now people are interested not merely in not being very sick but in being very well."

27. The economic role of the health food industry can be followed in its trade publication, *Whole Foods*, published in Santa Ana, California, beginning in January 1978. For data on the level of business see "First Annual Report on the Industry," in *Whole Foods: Natural Food Guide* (And/Or Press, Berkeley, California, 1979), 268–74.

28. In late 1981 a new magazine, *American Health*, was announced, subtitled "Fitness of Body and Mind." The initial direct-mail test to establish the existence of potential readers brought a 7.2 percent response when 5 percent is considered to be very good. The initial subscription offer will go to readers of a variety of publications, such as *Runners' World, Psychology Today,* and other science, health, class, food, and self-help magazines, which, taken as a whole, reflect the varied dimensions of the value changes associated with new attitudes toward health (*New York Times*, November 23, 1981).

29. Some of these issues have now become "classic," the subject of book-length writing. See, for example, Allan R. Talbot, *Power Along the Hudson: The Storm King Case and the Birth of Environmentalism* (New York, 1972); Barry M. Casper and Paul David Wellstone, *Powerline: The First Battle of America's Energy War* (New York, 1980); Paul Brodeur, *Expendable Americans* (New York, 1973).

30. For an analysis in this vein see William E. Shands and Robert G. Healy, *The Lands Nobody Wanted* (The Conservation Foundation, Washington, D.C., 1977).

31. A classic expression of the environmental view in this debate is Amory B. Lovins, *Soft Energy*

Paths: Toward a Durable Peace (Cambridge, Mass., 1977). An analysis of the values implicit in these contrasting positions is included in Avarham Shama and Ken Jacobs, *Social Values and Solar Energy Policy: The Policy Makers and the Advocates,* Solar Energy Research Institute, October 1979, SERI-RR-51-329.

32. See, for example, Susan Jay Kleinberg, "Technology's Step-Daughters: The Impact of Industrialization Upon Working Class Women, Pittsburgh, 1865–1890," University of Pittsburgh, Ph.D. thesis, 1973.

33. The most celebrated book in this debate was Donella H. Meadows and Dennis L. Meadows, *The Limits to Growth* (New York, 1972). See also a reply by H. S. D. Cole, et al., eds., *Models of Doom: A Critique of the Limits to Growth* (New York, 1973). For one important item in the energy debate see Robert Stobaugh and Daniel Yergen, *Energy Futures* (New York, 1979).

34. See Roger W. Sant, et al., *Eight Great Energy Myths: The Least-Cost Energy Strategy—1978–2000,* Energy Productivity Report No. 4, The Energy Productivity Center, Mellon Institute, Arlington, Va., 1981.

35. See items in n. 29; see also Hays, "The Structure of Environmental Politics Since World War II," n. 10.

36. See, for example, Craig R. Humphrey and Frederick R. Buttel, *Environment, Energy and Society* (Belmont, Calif., 1982), which describes Rachel Carson as an "important catalyst for the environmental movement" and the Santa Barbara oil spill as a "pivotal event" (7, 122). This book, the most comprehensive account yet from the perspective of "environmental sociology," gives heavy emphasis to the campus student movement of 1969–1970 as the source of environmental concern.

37. The Sierra Club, for example, grew from 7,000 in 1952 to 70,000 in 1969, and the Wilderness Society from 12,000 in 1960 to 54,063 in 1970.

38. The relative significance of widely shared social and economic changes on the one hand, and dramatic events on the other is a major set of alternatives in many historical analyses. The environmental scene is an especially striking case of the way in which preoccupation with the more publicized events has obscured the more fundamental changes.

39. For hearings leading up to the appointment of the NORRRC see United States Senate, Committee on Interior and Insular Affairs, 85th Congress, 1st Session, Hearing, "Outdoor Recreation Resources Commission," May 15, 1957 (Washington, D.C.: GPO, 1957). See accounts of the outdoor recreation situation, which reflect varied responses to it, in *American Forests,* December 1960, 58–59, and November 1961, 40–41 and 55–56; and in *Living Wilderness* in issues through 1959 to 1962, for example, Winter–Spring 1962, 3–9.

40. For the American Wilderness Alliance see its publications, *Wild America* (1979–present) and "On the Wild Side" (1979–present). See also *Proceedings, 1980 Western Wilderness and Rivers Conference,* Denver, Colorado, November 21–22, 1980, distributed and apparently published by the Alliance.

41. The work of the Nature Conservancy can best be followed in its quarterly publication, *Nature Conservancy News* (Arlington, Va.), which began publication in 1951.

42. For items on the air quality issue see John C. Esposito, *Vanishing Air* (New York, 1970) and Richard J. Tobin, *The Social Gamble* (Lexington, Mass., 1979). On water quality see David Zwick and Marcy Benstock, *Water Wasteland* (New York, 1971) and Harvey Lieber, *Federalism and Clean Waters* (Lexington, Mass., 1975).

43. The pesticide controversy has produced several "tracts for the times," among them Frank Graham, Jr., *Since Silent Spring* (Boston 1970); Rita Gray Beatty, *The DDT Myth: Triumph of the Amateurs* (New York, 1973); Georg Claus and Karen Bolander, *Ecological Sanity* (New York, 1977); and Robert van den Bosch, *The Pesticide Conspiracy* (New York, 1978).

44. For the early stages of this issue see Committee on Merchant Marine and Fisheries, Subcommittee on Fisheries and Wildlife Conservation, 89th Congress, 2nd Session, Hearing, "Estuarine and Wetlands Legislation," June 16, 22–23, 1966. One can follow the issue as it developed leading up to the Coastal Zone Management Act of 1972. That Act was an important example of how a major environmental thrust, a proposal for a system of national estuarine areas to be managed by the National Park Service, much akin to the newly emerging concept of seashores and lakeshores, was al-

most completely turned back. It appeared in the 1972 Act in the very limited form of "estuarine research areas."

45. One of the most significant backgrounds to the 1969 National Environmental Policy Act lay in the concern of the Fish and Wildlife Service about the failure of federal agencies to consider the impacts of development on fish and wildlife habitat, and especially their failure to "consult" with the agency under the Fish and Wildlife Coordination Act. One such issue was the dredge and fill practices of the U.S. Army Corps of Engineers who refused to consider "impacts" other than those on the maintenance of navigation channels. To rectify this problem, Rep. John Dingell of Michigan included in a proposed estuarine area act a section which would require the Fish and Wildlife Service to approve each permit granted by the Corps. See Committee on Merchant Marine and Fisheries, Subcommittee on Fisheries and Wildlife Conservation, 90th Congress, 1st Session, Hearing, "Estuarine Areas," (Washington, D.C.: GPO, 1967); see especially testimony of Alfred B. Fitt, U.S. Army Corps of Engineers, 119–207. One of the major thrusts leading up to NEPA took place in the Sub-committee on Fisheries and Wildlife under the leadership of Rep. Dingell. See its hearing, "Environmental Quality," 91st Congress, 1st Session, on HR 6750, a bill designed to amend the Fish and Wildlife Coordination Act (Washington, D.C.: GOP, 1969). It should be emphasized that in its origins NEPA was an inter-agency review and not a public review process. It constituted a far more diluted response to the problem of inter-agency review than did the Dingell proposal for a "Veto" or "dual permit" procedure, since it gave agencies the authority only to comment and not to veto the actions of other agencies. Only under modifications by the Nixon administration and the courts did NEPA become an instrument of public review.

46. Toxic chemical cases were numerous and have generated a considerable amount of writing. See, for example, Ralph Nader, Ronald Brownstein, and John Richard, eds., *Who's Poisoning America: Corporate Polluters and Their Victims in the Chemical Age* (San Francisco, 1981); Michael H. Brown, *Laying Waste: The Poisoning of America by Toxic Chemicals* (New York, 1979).

47. For some more recent events in these affairs see "Exposure," newsletter published by the Environmental Action Foundation (Washington, D.C.), February 1980–; and "The Waste Paper," published by the Sierra Club, Spring 1980– (Buffalo, New York).

48. A brief, concise statement of environmental energy perspectives is in Gerald O. Barney, ed., *The Unfinished Agenda* (New York, 1977), 50–68. The most eloquent speaker for the environmental energy view was Amory Lovins (fn. 31).

49. A source which provides one of the most comprehensive views of this concern is the series of "whole earth" catalog publications. These include Stewart Brand, ed., *The Whole Earth Catalog* (Menlo Park, Calif., 1968); *The Last Whole Earth Catalog* (1971); *The Whole Earth Epilog* (1974); and *The Next Whole Earth Catalog* (1980). The spirit of personal autonomy is expressed by a statement introducing the 1980 edition: "So far remotely done power and glory—as via government, big business, formal education, church—has succeeded to the point where gross defects obscure actual gains. In response to this dilemma and to these gains a realm of intimate, personal power is developing—the power of individuals to conduct their own education, find their own inspiration, shape their own environment, and share the adventure with whoever is interested. Tools that aid this process are sought and promoted by The Next Whole Earth Catalog" (2). See also *Mother Earth News* (n. 9) for a view of the range of facets in this perspective.

50. On many and varied occasions there were expressions of cooperation and joint action. The most extensive occurred in 1981 when the Global Tomorrow Coalition was formed, by the end of the year comprising 53 environmental organizations, most of them formed during the environmental era. See its publication, *Interaction* (Washington, D.C.), the first issue of which appeared November/December 1981.

51. See a varied set of newspaper clippings and articles, author file; also interviews with national leaders of the Sierra Club and the National Parks and Conservation Association concerning their varied efforts to maintain working liaison with Secretary of the Interior James Watt.

52. See report, "Protecting the Environment: A Statement of Philosophy," drawn up by the 14-member Task Force on the Environment, co-chaired by Dan W. Lufkin and Henry L. Diamond; see also accounts of the Task Force, list of its members, and a summary of its report in *Environment Re-*

porter, October 17, 1980, 812, and January 30, 1981, 1855–56. The Task Force report was rejected and its personnel replaced as lead administration advisers on environmental affairs by another group, quite divorced from earlier environmental activity, headed by Norman Livermore, a former administrator in California during the governorship of Ronald Reagan. See *Environment Reporter,* January 12, 1980, 1226. For a brief account of this transition in advisers see *Wilderness Report,* December 1980.

53. See The Harris Survey, "Substantial Majorities Indicate Support for Clean Air and Clean Water Acts," June 11, 1981, in the form of a news release.

54. From October 1, 1980, to October 1, 1981, Sierra Club membership increased by 35 percent, and in October 1981 went over the 250,000 mark. Organizations less politically oriented grew in both membership and financial contributions, but less rapidly.

55. See *Washington Post,* November 15, 1981, Section L-1 for a brief account of the "green vote" in the 1981 elections; see also clippings (author file) from New Jersey newspapers concerning participation in electoral politics there during the summer primaries.

56. This conclusion differs markedly from the views of environmental sociologists; see, for example, Humphrey and Buttel, *op. cit.,* 123–27, who speak of "rise and fall" rather than "persistent evolution" of environmental affairs. Their analysis seems to rest not on an examination of environmental values as social and political phenomena, but on their judgment as to the balance of political forces involved in a few selected "environmental problems."

57. The following analysis is rarely made explicit in contemporary writings, but rests more on my own judgment about tendencies and implications inherent in environmental activity.

58. A recent nonideological analysis which assumes this approach is Jerry A. Kurtzweg and Christina Nelson Griffin, "Economic Development and Air Quality: Complementary Goals for Local Governments," in *Journal of the Air Pollution Control Association,* 31–11, November 1981, 1155–62.

59. Perhaps it is not coincidental that the environmental movement was distinctively weak, compared with regional levels of education and urbanization, in the old "factory belt" of the North—as measured by votes in the U.S. House of Representatives on environmental issues, and as concluded from an analysis of environmental affairs within those states.

60. Hence, support from the corporate business community for "natural areas" programs, for example on the part of the Nature Conservancy, but opposition to wilderness which involved larger tracts of land on which conflicts with development were far more likely to occur.

61. The close relationship between production efficiency and environmental efficiency is described in Michael G. Royston, *Pollution Prevention Pays* (Pergamon Press, New York, 1979).

62. This setting for innovation was defined especially in the water quality program in which "technology standards" were adopted in the 1972 Clean Water Act and in which as a result the Environmental Protection Agency was required to analyze existing technologies to decide which was "the average of the best."

63. For a view on one aspect of this problem see Hyman G. Rickover, "Getting the Job Done Right," *New York Times,* November 25, 1981, op. ed. page.

64. Nicholas A. Ashford and George R. Heaton, "The Effects of Health and Environmental Regulation on Technological Change in the Chemical Industry: Theory and Evidence," in Christopher T. Hill, ed., *Federal Regulation and Chemical Innovation* (American Chemical Society, Washington, D.C., 1979), 45–66.

65. The classic case with respect to TVA was the controversy over the construction of Tellico Dam in the Little Tennessee River. But this was only one of many such issues. These can be followed in the monthly newsletter of the Tennessee Citizens for Wilderness Planning.

66. The "OSHA/Environment Network" organized to defend both occupational and community environmental protection programs during the early years of the Reagan administration extended earlier more informal cooperation into a more formal organization. See author clipping file, and miscellaneous documents on activities of the Network during 1981 (author file).

67. This is reflected in party and ideological analyses of environmental support. See "The Public Speaks Again: A New Environmental Survey," *Resources,* Resources for the Future, Washington, D.C., No. 60, September–November 1978.

68. This view seems to be implicit in a wide range of environmental issues, especially in the nation's countryside and wildlands.

69. See the activities of the Environmental Policy Center, Washington, D.C., with respect to appropriations for the construction of dams; this was led throughout the 1970s by Brent Blackwelder of the Center. See also "Alternative Budget Proposals for the Environment, Fiscal Years 1981 & 1982," drawn up by 9 national environmental organizations to suggest ways in which the Reagan administration could reduce federal expenditures.

70. A useful account of issues involving the Corps of Engineers is Daniel A. Mazmanian and Jeanne Nienaber, *Can Organizations Change: Environmental Protection, Citizen Participation, and the Corps of Engineers* (The Brookings Institution, Washington, D. C., 1979).

71. For these value changes see items in nn. 19 and 20. Naisbitt, especially, provides a useful analysis by distinguishing between "leading sectors" and "lagging sectors" of value change; he sorts out states which are in the one or the other category. It may not be without significance that most of the "leading" states with respect to environmental value change also seem to be states which have high percentages of women serving as state legislators. For this data see reports compiled since 1974 on women in state legislators by the National Women's Education Fund, Washington, D.C.

72. For an expression of this "sense of place," see a view about the role of the watershed as a place: ". . . watershed consciousness focuses on place. . . . The journey of this perspective is right out your window—the immediate valleys and hills that surround you, that channel rain and snowmelt into your nearest creeks and rivers and lakes. It is a first excursion of thought into the place you live." See Stewart Brand, ed., *The Next Whole Catalog* (Menlo Park, Calif., 1980), "Streaming Wisdom," 64–67.

The Evolution of Public Environmental Policy, by Richard H. K. Vietor

1. *Sierra Club v. Ruckelshaus,* 355 F. Supp. 253, 256 (D.D.C. 1972).

2. Carl Bagge reprinted in *Coal Mining and Processing,* December 1975.

3. Department of Health, Education, and Welfare, *Proceedings: National Conference on Air Pollution,* November 18–20, 1958. Index of Conference Participants.

4. U.S. Congress, Senate, Public Works Committee, Hearings on Air Pollution Control, September 9–11, 1963 (88th Cong., 1st ses.).

5. Ibid., 120–21.

6. *Sierra Club: An Introduction* (Sierra Club pamphlet, n.d.), 1.

7. Charles O. Jones, "Speculative Augmentation in Federal Air Pollution Policy-Making," *Journal of Politics,* vol. 36 (1974), 445.

8. Edmund Muskie quoted in *Coal Age,* August, 1967, 28.

9. U.S. Congress, Senate, Subcommittee on Air and Water Pollution, Air Quality Act of 1967—Hearings on S. 780 (90th Cong., 1st ses.), 2025, 2157–63.

10. 42 U.S.C. Sec. 1857–18571 (Supp. V, 1970).

11. Supra note 1.

12. U.S. Congress, Senate, Public Works Committee, *Senate Report No. 403* (90th Cong., 1st ses.), 1967, 2.

13. National Coal Policy Conference, Inc., *A Guide to the Air Quality Act of 1967* (Washington, D.C., NCPC, Inc., 1968), 11.

14. Ibid., 18.

15. Sidney Edleman, Chief, Environmental Health Branch (HEW), "Draft Policy Statement on Guidelines for Review of State Ambient Air Quality Standards," May 21, 1968 (in files of Don Walters, Environmental Protection Agency, Research Triangle Park, North Carolina), 2.

16. Ibid., 8.

17. Ibid., 9.

18. Ibid., 10.

19. Ibid.

20. "Draft of Guidelines, October 10, 1968," fn. 9 supra, 3.

21. "Draft of Guidelines, October 29 1968," fn. 9 supra, 1, 5.

22. Memorandum, Dean S. Mathews, Deputy Chief, Abatement Program, December 30, 1968, fn. 9 supra.

23. Marginal notations on "Draft of Guidelines, March 28, 1969," from Bureau of Abatement and Control, fn. 9 supra.

24. "Draft of Guidelines, February 19, 1969," fn. 9 supra.

25. John T. Middleton, Commissioner, NAPCA, "Air Conservation in 1969: Who will make the decisions?" delivered at the air pollution briefing in Phoenix, Arizona, February 4, 1969, fn. 9 supra.

26. John T. Middleton, Commissioner, NAPCA, *Public Policy and Air Pollution Control*, presented at the Penjerdel Regional Conference, Swarthmore, Pennsylvania, June 11, 1969 (Wash., D.C., NAPCA, 1969), 4–5.

27. U.S. Department of Health, Education, and Welfare, NAPCA, *Guidelines for the Development of Air Quality Standards and Implementation Plans* (Washington, D.C., 1969).

28. Environmental Protection Agency, Air and Water Programs Office, Records: Contract Records, file CP-1 (Federal Records Center, Suitland, Md., accession no. 412-73-5, Box 17, AMC file), correspondence from J. A. Overton, president of the American Mining Congress, to L. A. DuBridges, President's science advisor, July 22, 1969, 4–5.

29. EPA, Records (acc. no. 412-73-5), Box 17, AMC file, letter from AMC to Governors, July, 1969, 6.

30. Gallup Poll, conducted May 2–3, 1970, in *New York Times,* August 30, 1970.

31. John C. Esposito, *Vanishing Air* (New York: Grossman Publishers, 1970).

32. Charles Jones, "Speculative Augmentation," 449–53.

33. U.S. Congress, House of Representatives, Interstate and Foreign Commerce Committee, Hearings on Air Pollution Control and Solid Waste Recycling, March 5, 16–20, April 14, 1970 (91st Cong., 2nd ses.).

34. Quoted in James Miller, "Air Pollution" in J. Rathelsburger, ed., *Nixon and the Environment* (New York: Village Books, 1972), 14.

35. Pub. L. 91-604, 84 Stat. 1676, 42 U.S.C. Sec. 1857–18571.

36. Ibid.

37. House Air Pollution Hearings, 1970, 297.

38. U. S. Congress, Senate, Public Works Committee, Hearings on Air Pollution (91st Cong., 2nd ses.), 1970, 132–33, 159.

39. U. S. Congress, Senate, Public Works Committee, *Senate Report No. 1196,* (91st Cong., 2nd ses.), 1970, 11.

40. U.S. Congress, Senate, Public Works Committee, Hearings on Nondegradation Policy of the Clean Air Act, July 24, 1973 (93rd Cong., 1st ses.), 279–81.

41. 36 *Fed. Reg.* 1502, January 30, 1971.

42. Natural Resources Defense Council, "Comments," March 15, 1971 (in Environmental Protection Agency, Records Office, "1971 Ambient Air Standards File," Wash., D.C.), 9.

43. 36 *Fed. Reg.* 8187, April 30, 1971.

44. *Sierra Club v. Ruckelshaus,* 344 F. Supp. 253 (D.D.C. 1972), *aff'd,* 4 ERC 1815 (D.C. Cir. 1972), *aff'd,* 41 W.S.L.W. 4825 (U.S., June 11, 1973).

45. 36 *Fed. Reg.* 6680, April 7, 1971.

46. "Log and analysis of comments on 1971 proposed implementation plan guidelines," EPA, Records Office, Wash., D.C.

47. Memorandum of John Middleton, reprinted in U.S. Congress, Senate, Subcommittee on Air and Water Pollution, *Implementation of the Clean Air Act Hearings,* February 1972 (92nd Cong., 2nd ses.), 47–48.

48. Supra notes 34 and 47. See also, Henry J. Steck, "Private Influence on Environmental Policy," *Environmental Law* 5 (1975) 241–282.

49. U.S. Congress, House, Subcommittee on Public Health and Environment, *Clean Air Act Oversight* (92nd Cong., 1st and 2nd ses.), January 26–28, 1972, 479.

50. Ibid. See also, EPA Records (acc. no. 412-73-5), Box 17, file CR-2, memorandum from J. Middleton to E. Tuerk, July 19, 1971.

51. House Oversight Hearings, 1972, 529.

52. Supra note 50.

53. *Senate Clean Air Act Hearings*, 1972, 43–55.

54. Edward Tuerk, Office of Program Management Operations, Environmental Protection Agency, interview with author at EPA offices, Wash., D.C., May 13, 1974.

55. Supra note 50.

56. Executive Order 11523, in U.S. Congress, Senate, Subcommittee on Intergovernmental Relations, Hearings on Advisory Committees (91st Cong., 2nd ses.), December 8, 10, 17, 1970, 502.

57. Minutes Summaries, Meeting of National Industrial Pollution Control Council, May 6, 1971 (Records Section, Department of Commerce Library, Washington, D.C.).

58. National Industrial Pollution Control Council, Administrative Files, Box No. 9, folder, "Memos," and folder, "Permit Program—Army Corps of Eng." (Freedom of Information Access at Main Commerce Bldg., Wash., D.C.).

59. Senate Oversight Hearings, 1972, 3.

60. Supra note 1.

61. Supra note 44.

62. *New York Times,* June 17, 1973, sec. 4, 4.

63. *Pittsburgh Post-Gazette,* June 12, 1973.

64. National Coal Association, *56th Annual Meeting, June, 1973* (transcript at NCA offices, Wash., D.C.).

65. *Senate Hearings on Nondegradation,* 1973, 5–44, 67–96.

66. 38 *Fed. Reg.* 18986, July 16, 1973.

67. Environmental Protection Agency, *Technical Support Document—EPA Regulations for Preventing the Significant Deterioration of Air Quality* (EPA-450/2-75-001), January 25, 10–11.

68. Ibid., 14.

69. Ibid., 15.

70. 39 *Fed. Reg.* 42513, December 5, 1974.

71. *EPA Significant Deterioration Support Document,* 4–5.

72. Supra note 70.

73. U.S. Congress, Senate, Committee on Public Works, Subcommittee on Environmental Pollution, *Implementation of the Clean Air Act—1975,* March 19–20, April 21–13, 1975 (94th Cong., 1st ses.), 895–97.

74. Ibid., 21–23, 1368–72.

75. Ibid.

76. Pub. L. 91-604, Sec. 111(a)(1).

77. Environmental Protection Agency, Transcript: Public Hearings and Conference on Status of Compliance with Sulfur Oxide Emission Regulations by Power Plants, October 18–November 2, 1973 (Washington, D.C.: EPA Records Office).

78. Sulfur Oxide Control Technology Assessment Panel (SOCTAP), *Final Report on Projected Utilization of Stack Gas Cleaning Systems* (Washington, D.C.: EPA, 1973).

79. *EPA, Capital Improvements Study,* fn. 78 supra, 439–62. Also, for example, J. D. Geist, Vice-President, New Mexico Public Service Co., testified on pollution control costs at Four Corners Project. Power plant costs, merely to meet requirements of secondary standards, were $125/kilowatt, above and beyond the $100/kilowatt cost of generating the power. That was for cleaning low-sulfur coal 90 percent, less than strict no-significant deterioration limits would require, 871–73.

80. *Senate Clean Air Act Hearings,* 1975, 895–97.

81. *Senate Hearings on Nondegradation,* 1973, 67–96. Also, statement by Frank Zarb, Administrator, Federal Energy Agency, in *Senate Clean Air Act Hearings,* 1975, 338–41.

82. Federal Energy Administration, "Draft Environmental Impact Statement: Energy Independence Act of 1975 and Related Tax Proposals (DES 75-2)," March, 1975, fn. 73 supra, 664.

83. Ibid.

84. Dennis Meadows et al., *The Limits to Growth* (New York: New American Library, 1972).

85. Malcolm Scully, "Growth or No-Growth? The Debate Intensifies," *Chronicle of Higher Education,* November 3, 1975.

86. Shell Oil Company, *Shell Shareholder News,* November 1975, 2.

87. Ibid.

88. U.S. Congress, Senate, Committee on Environment and Public Works, "Clean Air Amendments of 1977," *Report No. 95-127* (95th Cong., 1st ses.), May 1977.

Reconstructing Environmentalism, by Robert Gottlieb

1. Alston's speech is from the Our Vision of the Future panel session, People of Color Environmental Leadership Summit, October 26, 1991; see also "Reshaping the Environmental Movement," in Louis Head and Valerie Taliman, *Voces Unidas,* Southwest Organizing Project, vol. 1, no. 4, 9; Interviews with Dana Alston, Michael Fischer, John Adams.

2. There are to be sure notable exceptions to this approach, such as William Cronon, whose masterful *Nature's Metropolis: Chicago and the Great West* (New York: W.W. Norton, 1991) provides a compelling argument about the intricate relationships between the metropolis, the natural environment, rural areas, and resource flows; Martin Melosi, whose writings on the urban environment and urban environmental movements (including, most recently, his essay "The Place of the City in Environmental History," *Environmental History Review,* Spring 1993; see also, *Garbage in the Cities: 1880–1980* (College Station: Texas A&M University Press, 1981); and his edited volume *Pollution and Reform in American Cities, 1870–1930* (Austin: University of Texas Press, 1980) have helped situate a new type of environmental history; and Joel Tarr, whose writings on public health issues and the urban environment (see, for example, his essay in the *American Journal of Public Health,* "Industrial Wastes and Public Health: Some Historical Notes, Part 1, 1876–1932," September 1975) have provided a crucial link for environmental historians.

3. Marshall's comments are from "Wilderness as Minority Right," an August 27, 1928, article for the *Service Bulletin* of the Forest Service, cited in James Glover, *A Wilderness Original: The Life of Bob Marshall* (Seattle: The Mountaineers, 1986), 96.

4. The "region which contains no permanent inhabitants" quote is from "The Problem of the Wilderness," Robert Marshall, *Scientific Monthly* 30 (February 1930): 148; the "vast lonely expanse" quote is from *Arctic Village,* Robert Marshall, H. Smith & R. Haas, N.Y., 1933, 198; see also George Marshall, "Bob Marshall and the Alaska Arctic Wilderness," *Living Wilderness,* Autumn 1970, Vol. 34, No. 111, 29–32

5. The "people can not live" quote is from "Recreational Limitations to Silviculture in the Adirondacks," Robert Marshall, *Journal of Forestry* 23, No. 2 (February 1925): 176. In this article, Marshall argued for continued protection of virgin forest areas in the Adirondacks, both for its proximity to the New York metropolitan area as a recreational resource, and as a source of inspiration in its undeveloped state. "The finest formal forest, the most magnificently artificially grown woods," Marshall wrote for his audience of professional foresters, "can not compare with the grandeur of primeval woodland. In these days of over-civilization, it is not mere sentimentalism which makes the virgin forest such a genuine delight" (173).

6. In a Forest Service publication, the agency promoted forest service lands as places where there were "few rules and few crowds" and where you could "bring your family, have a picnic, gather wood for your campfire (but be careful with it), hike in the mountains—or take a genuine wilderness trip into some of the ruggedest country on the continent." See *Forest Outings,* Russell Lord, ed., U.S. Forest Service, Washington, D.C., 1940, cited in *A Wilderness Original,* 95.

7. Cited in the *Nation,,* December 20, 1933, 696. The quotes from *The People's Forests* (Harrison Smith and Robert Haas, N.Y., 1933) can be found on 123, 211.

8. The quote, "there shall be no roads," is from a draft of a text for the Forest Service that Marshall prepared and is reproduced as "Protection at Last for Wilderness," in *Living Wilderness,* July 1940, 3.

9. The Bauer/Marshall correspondence is described in *A Wilderness Original,* 218–19.

10. Murie sought to directly associate Marshall with this elite posture in his essay, "Wilderness Is for Those Who Appreciate," *Living Wilderness,* July 1940, 5. On the redbaiting of Bob Marshall, see, Benjamin Stolberg, "Muddled Millions: Capitalist Angels of Left-Wing Propaganda," *Saturday Evening Post,* February 15, 1941, 9; "High Federal Aides Are Linked to Reds at House Hearing,"

New York Times, August 18, 1938; "WPA Union Called Communist Plan," *New York Times*, April 18, 1939.

11. For an outline of Robert Marshall's will, see "We Want No Straddlers," Stephen Fox, *Wilderness*, Winter 1984, 10, and *A Living Wilderness*, 272; see also the letter to the editor, written by his close friend Gardner Jackson after Marshall's death, in *Nation*, Vol. 149, December 2, 1939, 635.

12. The "I chose medicine" quote is from *Exploring the Dangerous Trades: The Autobiography of Alice Hamilton, M.D.* (Boston: Little Brown, 1943), 38.

13. The "ideal place from which to observe" quote is from Barbara Sicherman, *Alice Hamilton: A Life in Letters* (Cambridge: Harvard University Press, 1984), 4. The typhoid epidemic issue is also discussed in Wilma R. Slaight, *Alice Hamilton: First Lady of Industrial Medicine*, Ph.D. diss., Case Western University, 1974, 24–27.

14. The "here was a subject" quote is from *Exploring the Dangerous Trades*, 115.

15. The "wash hands or scrub nails" quote is from *Exploring the Dangerous Trades*, 122. The concept of industrial lead poisoning "inevitability" is discussed in Alice Hamilton, *The White Lead Industry in the United States, With an Appendix on the Lead-Oxide Industry, Bulletin of the Bureau of Labor*, No. 95, Washington, D.C., 1912, 190.

16. "It seemed natural and right" is a quote from *Exploring the Dangerous Trades*, 269; the duty "to the producer" quote is from Hamilton, *Industrial Poisons in the United States* (New York: The MacMillan Company, 1925), 541.

17. The "laboratory material" quote is from *Exploring the Dangerous Trades*, 294; see also Alice Hamilton and Gertrude Seymour, "The New Public Health," *Survey*, Vol. XXXVIII, No. 3, April 21, 1917, 59–62; Hamilton, "The Scope of the Problem of Industrial Hygiene," *Public Health Reports*, October 20, 1922, Vol. 37, No. 42, 2604–08.

18. Alice Hamilton, "What Price Safety? Tetra-ethyl Lead Reveals a Flaw in our Defenses," *Survey Midmonthly*, June 15, 1925, Vol. LIV, No. 6. 333.

19. Harvard established three informal conditions for Hamilton's appointment: no use of the Harvard Club, no football tickets, and no participation in the commencement procession. Women were only admitted to the medical school after 1945. See, "Alice Hamilton: First Lady of Industrial Medicine", 135; see also *Alice Hamilton: A Life in Letters*, 209–18.

20. Rachel Carson, *Silent Spring* (New York: Fawcett Crest Edition, 1964).

21. Carson's National Book Award acceptance Speech is reprinted in Paul Brooks, *The House of Life: Rachel Carson at Work* (Boston: Houghton Mifflin, 1972), 127–29.

22. The question of pesticide research and Carson's argument about "purchased expertise" is discussed in *Since Silent Spring*, 55–68.

23. Loren Eisley, cited in "Using a Plague to Fight a Plague," *Saturday Review*, September 29, 1962, 18.

24. *Silent Spring*, 262; for a discussion of Carson the writer, see Carol Gartner, *Rachel Carson* (New York: Frederick Ungar, 1983), and Paul Brooks, *Speaking for Nature: How Literary Naturalists from Henry Thoreau to Rachel Carson Have Shaped America* (Boston: Houghton Mifflin, 1980), as well as Brooks' *The House of Life*, which also includes excerpts of Carson's writings.

25. The "thanks to a woman" quote is from Edwin Diamond, "The Myth of the Pesticide Menace," *Saturday Evening Post*, September 28, 1963, 16–18; the "priestess of nature" quote is from Clarence Cottam, "A Noisy Reaction to Silent Spring," *Sierra Club Bulletin*, January 1963, 4.

26. See, Letters in the *Sierra Club Bulletin*, March 1963, 18, and the April–May 1963 issue as well. Despite the acknowledgment of controversy, the feedback generated by *Silent Spring* was far less than the "heated debate" that had gone on for several months concerning "which climbing classification system [the Club] should use—NCCS [National Climbing Classification System] or decimals." This was reflected by more than two pages of letters in the June 1963 issue of the Bulletin (8–9, 12).

27. David Brower, *For Earth's Sake*, 215. An article in *National Wildlife*, the publication of the National Wildlife Federation, suggested that Carson's book "might have gone too far" in its effort to "shock the public" about pesticide dangers. The NWF reviewers argued that by frightening the public "into believing there is not legitimate use of chemicals" Carson could cause "unneeded restrictions which would hamstring future research and progress. It might mean that we would never have an-

other DDT, the chemical miracle that rescued millions of lives from hunger and disease." See, R. G. Lynch and Cliff Ganschow, "Pesticides: Man's Blessing or Curse?" *National Wildlife*, February-March 1963, 10–17.

28. Carson's statement is cited in "Silent Spring—III," *New Yorker*, June 30, 1962, 67.

29. See the interview with Thomas Kimball (1991).

Searching for a "Sink" for an Industrial Waste, by Joel A. Tarr

1. For elaboration of this concept, see, Joel A. Tarr, "The Search for the Ultimate Sink: Urban Air, Land, and Water Pollution in Historical Perspective," J. Kirkpatrick Flack, ed., *Records of the Columbia Historical Society of Washington, D.C.,* in vol. 41 (Charlottesville, Va., 1984), 1–29.

2. Arthur C. Bining, *Pennsylvania Iron Manufacture in the Eighteenth Century* (Harrisburg, Pa., 1973), 19–38; Michael Williams, *Americans and Their Forests: A Historical Geography* (New York, 1989), 104–10; and Paul F. Paskoff, *Industrial Evolution: Organization, Structure, and Growth of the Pennsylvania Iron Industry, 1750–1860* (Baltimore, Md., 1983), 1–38, 91–105. Some charcoal kilns, developed shortly before the Civil War, attempted to capture wood chemicals; charcoal retorts, developed in the 1870s and 1880s, were specifically designed to capture volatile wood chemicals. See, R. H. Schallenberg and D. A. Ault, "Raw Materials Supply and Technological Change in the American Charcoal Iron Industry," *Technology and Culture* 18 (July 1977): 453–54.

3. Peter Temin, *Iron and Steel in Nineteenth-Century America: An Economic Inquiry* (Cambridge, Mass., 1964), 82–83; Paskoff (n. 2 above), 18.

4. Williams (n. 2 above), 147–50, 337–44. Charcoal production, however, was much less important as a cause of deforestation than clearing for agricultural purposes.

5. Bining (n. 2 above), 59.

6. Schallenberg and Ault (n. 2 above), 436–66.

7. Alfred D. Chandler, Jr., "Anthracite Coal and the Beginnings of the Industrial Revolution in the United States," *Business History Review* 46 (Summer 1972): 151–64. Chandler maintains that this substitution occurred not because of demand for new energy sources but rather because anthracite operators acquired the means to develop and make available the supply of anthracite.

8. Paskoff (n. 2 above), 101–02; Temin (n. 3 above), 58–62.

9. The total production of charcoal iron rose until 1890 but the proportion made with the fuel diminished sharply as a fraction of total production. See Temin (n. 3 above), 266–69; and Williams (n. 2 above), 337–44.

10. For a discussion of the anthracite region, see Donald L. Miller and Richard E. Sharpless, *The Kingdom of Coal: Work, Enterprise, and Ethnic Communities in the Mine Fields* (Philadelphia, Pa., 1985).

11. John Fulton, *Coke: A Treatise on the Manufacture of Coke and Other Prepared Fuels and the Savings of By-Products* (Scranton, Pa., 1906), 145.

12. John Fulton, "A Report on Methods of Coking," in *Second Geological Survey of Pennsylvania: 1875* (Harrisburg, Pa., 1876), 122–23; Raymond Foss Bacon and William A. Hamor, *American Fuels* in vol. 1 (New York, 1922, 2 vols.), 6–9; and Frederick Moore Binder, *Coal Age Empire: Pennsylvania Coal and Its Utilization to 1860* (Harrisburg, Pa., 1974), 118–19.

13. Temin (n. 3 above), 268–69; Kenneth Warren, *The American Steel Industry, 1850–1970: A Geographical Interpretation* (Oxford, England, 1973), 77–80, 110, 200–06; and Chandler, "Anthracite Coal," 141–81.

14. Ibid., 122–23.

15. Fulton (n. 13, above), 146–57.

16. The process of drawing coke by hand was immensely difficult. Not only was the work heavy but the laborers were also exposed to the heat of the ovens and dust and fumes. Workers willing to perform these difficult tasks were difficult to find and attempts were made to develop mechanical coke drawers as well as mechanical levelers and watering devices. See, Walter W. Macfarren, "Coke Drawing Machines," in *Proceedings of the Engineers Society of Western Pennsylvania* 23 (November 1907): 451–509; and F. C. Keighly, "The Connellsville Coke Region," *Engineering Magazine* 20 (October 1900): 39.

17. The yield of high-class Connellsville coal coked in beehive ovens in 1875 was 63 percent. Franklin Platt, *Special Report on the Coke Manufacture of the Youghiogheny River Valley. Second Geological Survey of Pennsylvania: 1875,* 63; in 1911 the U.S. Geological Survey reported the average yield nationally for beehive ovens was 64.7 percent. See, F. H. Wagner, *Coal and Coke* (New York, 1916), 296–99 and Fulton (n. 13 above), 147–48. Higher-quality coke could be produced by increasing the length of the coking process to 72 hours.

18. H. N. Eavenson, *The First Century and a Quarter of the American Coal Industry* (Baltimore, Md., 1942), 380–84, 579–83; David Demarest and Eugene D. Levy, "A Relict Industrial Landscape," *Landscape* 29 (1986): 30–32; and Muriel E. Sheppard, *Cloud By Day: The Story of Coal and Coke and People* (Uniontown, Pa., 1947), 22–24.

19. William L. Affelder, "Jones & Laughlin's Coke Plant," *Mines and Minerals* 29 (December 1908): 195–99.

20. W. A. Buckhout, "The Effect of Smoke and Gas Upon Vegetation," *Annual Report of the Pennsylvania State Department of Agriculture* (Harrisburg, Pa., 1900), 180–82. One observer of the beehive coke region wrote that the ammonia "thrown into the atmosphere . . . by coke ovens . . . no doubt add[s] materially to the fertility of our land; for nature never wastes anything, and it is in all probability precipitated." See, John W. Boileau, *Coal Fields of Southwestern Pennsylvania: Washington & Greene Counties* (Pittsburgh: Privately printed, 1907), 56.

21. See, Shephard (n. 18 above). Some accounts describe the whole Connellsville region as frequently blanketed with fumes emitted from beehive coke ovens. See, for instance, William A. Metcalf, "On Smoke," *Proceedings of the Engineer's Society of Western Pennsylvania* 8 (1892): 29–30. Miners and coke oven workers often planted gardens close to the ovens but were careful to place them downwind of the ovens. The fumes from the ovens trapped heat in the locality, leading to a longer growing season. For pictures of "prize winning" gardens, as well as of beehive coke oven plants, see John K. Gates, *The Beehive Coke Years: A Pictorial History of Those Times* (Uniontown, Pa., 1990).

22. *Joseph Lentz v. Carnegie Bros. & Co.,* 145 *Pa.* 613 (1891). Fumes and other emissions from the beehive ovens almost certainly affected the health of the workers at the ovens and possibly the health of nearby residents but the author has been unable to find a study documenting this. Industrial hygiene, which would have dealt with worker's health, was in its infancy at the time and attention focused on accidents rather than long-term chronic effects of industrial work. See, for instance, Sara J. Davenport, *Bibliography of Bureau of Mines Publications Dealing with Health and Safety in the Mineral and Allied Industries 1910–46, U.S. Bureau of Mines Technical Paper 705* (Washington, D.C., 1948), 23–24.

23. For an interesting interpretation of common law rulings on pollution cases in the nineteenth and early twentieth centuries, see Christine Rosen, "A Litigious Approach to Pollution Regulation: 1840–1906" (unpublished paper, School of Business, University of California, Berkeley, 1990).

24. In the Sanderson Cases, the Pennsylvania Supreme Court held that injured parties could not collect damages because of harm done to their property from mine acid drainage because mine acid was a result of the "natural and necessary use" of property. See, *Sanderson v. Pennsylvania Coal Co.,* 86 *Pa.* 401 (1878) and 102 *Pa.* 370 (1883); *Pennsylvania Coal Co. v. Sanderson,* 94 *Pa.* 302 (1880) and 113 *Pa.* 126, 146 (1886).

25. See, *Brown et al. v. Torrence,* 7 *Pa. State Reports* 186 (1880); *Adam Robb v. Carnegie Bros. & Co.,* 145 *Pa.* 324 (1891); and *Joseph Lentz v. Carnegie Bros. & Co.,* 145 *Pa.* 612 (1891); and *Campbell v. Bessemer Coke Company,* 23 *Pa. Superior Ct.* 374 (1903).

26. Christine Rosen, "Cost Benefit Analysis in Pollution Nuisance Law 1840–1904" (unpublished paper, University of California at Berkeley, 1990).

27. See, for instance, *Adam Robb v. Carnegie Bros. & Co.,* 145 *Pa.* 324 (1891).

28. J. F. Slagle, *A Digest of the Acts of Assembly and a Code of the Ordinances of the City of Pittsburgh* (Pittsburgh, Pa., 1869), 269; and Hiram Schock (comp. and ed.), *Digest of the General Ordinances and Laws of the City of Pittsburgh to March 1, 1938* (Pittsburgh, Pa., 1938), 384–85.

29. Affelder (n. 19 above), 197. No mention has been found of other attempts to control pollution from beehive ovens although at the Continental No. 1 Plant of the H. C. Frick Coke Company, located near Uniontown, Pennsylvania, a system was developed that used a tall stack to burn by-products

and generate steam. See, also, Charles Catlett, "Increasing the Coke Yield from Bee-Hive Ovens," *Cassier's Magazine* 24 (October 1903): 534–42.

30. Mellon Institute, *Some Engineering Phases of the Smoke Problem,* Bulletin no. 6, Mellon Institute Smoke Investigation (Pittsburgh, Pa., 1914), 68–69. The report notes that in these coke ovens, the gases were "conveyed by long passages to the stacks where they are discharged." Since most bee-hive coke ovens vented through the roof, the use of the stacks suggests that the J & L ovens were included in the observation.

31. In 1904, however, in the case of *Sullivan v. Jones & Laughlin Steel Company,* the Pennsylvania Supreme Court sustained an injunction issued by a lower court that enjoined J & L Steel from operating its blast furnaces in a manner that caused substantial damage to nearby residential areas. The suit had been engendered by J & L's use of fine Mesabi ore that produced severe dustfall problems. See, *Sullivan, Appellant, v. Jones & Laughlin Steel Company,* 208 *Penn. State Reports* 540 (1904). See, also, "Decision of the Supreme Court of Pennsylvania on the Use of Mesabi Ores," *The Bulletin of the American Iron and Steel Association* (July 10, 1904), 100–01. Full compliance with the injunction, however, was difficult to achieve given the nature of the technology, and the courts held in 1908 that the modifications installed by J & L constituted adequate compliance with the decree in spite of evidence that the plaintiffs were still suffering injury. See, *Sullivan v. Jones & Laughlin Steel Co.,* 222 *Pa* 72 (1908).

32. Sam H. Schurr and Bruce C. Netschert, *Energy in the American Economy, 1850–1975* (Baltimore, Md., 1960), 98, n. 75.

33. Fulton (n. 11 above), 311–12; Catlett (n. 29 above), 541; and Bacon and Hamor (n. 14 above), vol. 1, 133–34.

34. For contemporary discussions of beehive coke oven wastes see Catlett, "Increasing the Coke Yield From Bee-Hive Ovens," 534–42; William Gilbert Irwin, "Coke-Making in the United States," *Cassier's Magazine* 19 (January 1901): 205; Heinrich J. Freyn, "The Wastefulness of Coke Ovens," *Scientific American Supplement No. 1998* (April 1914): 246–47; and Floyd W. Parsons, "Everybody's Business: Needless Coal Wastes," *Saturday Evening Post* (March 13, 1920): 36–38.

35. M. Camp and C. B. Francis, *The Making, Shaping and Treating of Steel* (Pittsburgh, Pa., 1919, 4th ed.), 101–16.

36. Camp and Francis (n. 35 above), 101–02; and F. H. Wagner, (n. 19 above), 303–06.

37. Camp and Francis (n. 35 above), 116–41.

38. See, W. W. Davis, "The Semet-Solvay By-Product Oven," *Journal of the Engineers' Society of Pennsylvania* 2 (October 1910): 401. For the conservation movement, see Samuel P. Hays, *Conservation and the Gospel of Efficiency* (Cambridge, Mass., 1959).

39. William H. Blauvelt, "The By-Product Coke Oven," *Engineering News* (August 27, 1908), 60: 218–23; and Aaron J. Ihde, *The Development of Modern Chemistry* (New York, 1964), 454–61.

40. C. A. Meissner, "The Modern By-Product Coke Oven," in James T. McCleary, ed., *Year Book of the American Iron and Steel Institute* (New York, 1913), 118–78.

41. Aaron J. Hyde, *The Development of Modern Chemistry* (New York, 1964), 614–17, 671–94.

42. Parsons (n. 36 above), 38.

43. Wilbert G. Fritz and Theodore A. Veenstra, *Regional Shifts in the Bituminous Coal Industry With Special Reference to Pennsylvania* (Pittsburgh, Pa., 1935), 116–20.

44. Beehive ovens continued to be constructed in situations where the expense of a by-product plant could not be justified. See, Warren (n. 13 above), 112–15; and Eavenson (n. 18 above), 582–85. During World War II, many beehive ovens were fired-up to meet wartime demands. See, G. S. Scott, J. A. Kelley, E. L. Fish, and L. D. Schmidt, *Modern Beehive Coke-Oven Practice, Preliminary Report,* U.S. Bureau of Mines R. I. 3738 (Washington, D.C., December 1943).

45. Richard G. Luthy, "Problems of Water Reuse in Coke Production," in E. Joe Middlebrooks, ed., *Water Reuse* (Ann Arbor, Mich., 1982), 501–20.

46. Phenols is the broad term applied to the mono-hydroxy derivatives of the benzene ring, and includes phenol, cresols, and xylenols. Coke oven waste includes a combination of all the phenols. See, Ohio River Valley Water Sanitation Commission, *Phenol Wastes: Treatment by Chemical Oxidation* (Cincinnati, Ohio, 1951), 9.

47. R. D. Leitch, "Stream Pollution by Wastes from By-Product Coke Ovens: A Review, With Special Reference to Methods of Disposal," *Public Health Reports* 40 (September 25, 1925): 2022.

48. Stuart Galishoff, "Triumph and Failure: The American Response to the Urban Water Supply Problem, 1860–1923," in Martin V. Melosi, ed., *Pollution and Reform in American Cities, 1870–1930* (Austin, Tex., 1980), 50–54.

49. H. P. Bohmann, "Find Causes of Obnoxious Tastes in Milwaukee Water," *Engineering News-Record* 82 (January 23, 1919): 181–82. H. R. Crohurst, a sanitary engineer with the USPHS, estimated in 1924 that 1 part of phenol waste to 75–100 million parts of chlorinated water would bring detectable tastes and odors. See, E. S. Tisdale, "Cooperative State Control of Phenol Wastes on the Ohio River Watershed," *Journal of American Water Works Association* 18 (1927): 575; and Wellington Donaldson, "Industrial Wastes in Relation to Water Supplies," *American Journal of Public Health* 11 (March 1921): 195. Interestingly, very high concentrations of chlorine would neutralize the phenol. See, Ohio River Valley Water Sanitation Commission, *Phenol Wastes*, 9.

50. "Discussion of Report of Committee No. 6 on Industrial Wastes in Relation to Water Supply," *JAWWA* 12 (1924): 411–15; and F. Holman Waring, "Results Obtained in Phenolic Wastes Disposal Under the Ohio River Basin Interstate Stream Conservation Agreement," *AJPH* 19 (1929): 758–70.

51. "Progress Report of Committee on Industrial Wastes in Relation to Water Supply," *JAWWA* 10 (1923): 415–30; 16 (1926): 302–03.

52. Waring (n . 50 above), 758–70.

53. "Progress Report on Recent Developments in the Field of Industrial Wastes in Relation to Water Supply," *JAWWA* 16 (1926): 310–11.

54. In 1912, the city of New Castle, Pennsylvania, sued the Carnegie Steel Company of Farrel, Pennsylvania, to stop it from polluting New Castle's water supply with phenolic wastes from its by-product coke oven. The company settled the case out of court and agreed to prevent further difficulties. See, F. Holman Waring, "Results Obtained in Phenolic Wastes Disposal under the Ohio River Basin Interstate Stream Conservation Agreement," *AJPH* 19 (1929): 759–61.

55. The Youghiogheny River waters were plagued by extreme hardness, acidity, and turbidity. See, E. C. Trax, "A New Raw Water Supply for the City of McKeesport, Pennsylvania," *Journal of the American Water Works Association* 3 (1916): 947–53.

56. See, Camp and Francis (n. 35 above), 103–04, for a description of the Clairton by-product coke plant.

57. *McKeesport v. Carnegie Steel Company*, 66 Pittsburgh Legal Journal 695 (1918).

58. 66 *Pittsburgh Legal Journal* 696.

59. E. B. Besselievre, "Statutory Regulation of Stream Pollution and the Common Law," *Transactions of the American Institute of Chemical Engineers* 16 (1924): 217–29; and John H. Fertig, "The Legal Aspects of the Stream Pollution Problem," *AJPH* 16 (August 1926): 782–88.

60. Joel A. Tarr, "Industrial Wastes and Public Health: Some Historical Notes, Part 1, 1876–1932," *AJPH* 75 (September 1985): 1059–67.

61. W. L. Stevenson, "Pennsylvania Sanitary Water Board," *Engineering News-Record* 91 (October 25, 1923): 684–85; and Almon L. Fales, "Progress in the Control of Pollution by Industrial Wastes," *AJPH* 18 (1928): 715–21. The Board established three classes of streams: those that were relatively pure; those in which pollution needed to be controlled; and those that were already so polluted that they could not be used for public water supplies or recreational purposes, and therefore were not worthy of pollution removal. Some conservationists criticized the Pennsylvania law as focusing too strongly on drinking water quality and not paying enough attention to larger environmental questions. See, Tarr, "Industrial Wastes and Public Health," 1062.

62. Philip P. Wells, Deputy Attorney General, to Charles E. Miner, M.D., Secretary of Health, April 12, 1924, printed in 72 *Pittsburgh Legal Journal* 648; and "Discussion of Report of Committee No. 6 on Industrial Wastes in Relation to Water Supply," *JAWWA* 12 (1924): 411, comments of W. L. Stevenson.

63. John E. Monger, "Administrative Phases of Stream Pollution Control," *AJPH* 16 (August 1926): 790–91. Existing plants, however, were exempt from the requirement.

64. Ellis Hawley, "Herbert Hoover, the Commerce Secreatariat, and the Vision of an Associative State, 1921–28," *Journal of American History* 61 (1974): 116–40.

65. Monger (n. 63 above), 16: 792–94.

66. W. L. Stevenson, "The Sanitary Conservation of Streams by Cooperation," *Transactions of the American Institute of Chemical Engineers* 27 (1931): 13–21. The agreement also required the firms to notify state authorities immediately in case of an accident that threatened water supplies.

67. John E. Monger to Homer S. Cummings, November 6, 1923, U.S. Public Health Service Archives, Record Group 90, National Archives, Washington, D.C. Cooperation was sometimes slow in forthcoming, and the threat of legal action was always in the background. See, "Discussion of Report of Committee No. 6 on Industrial Wastes in Regard to Water Supply," *JAWWA* 12 (May 22, 1924): 411–12, comments of J. W. Ellms.

68. Tisdale (n. 49 above), 18: 577–78.

69. Monger to Cummings, November 6, 1923.

70. Tarr (n. 60 above), 75: 1064.

71. See a series of letters from various state departments of health to Surgeon General Homer S. Cummings in USPHS Records, Record Group 90, National Archives, Washington, D.C.

72. John E. Monger to Surgeon General Homer S. Cummings, February 7, 1923, H. H. Streeter to Cummings, December 24, 1923, and Wade Frost to Cummings, November 19, 1923, U.S. Public Health Service Records, Record Group 90, National Archives, Washington, D.C.

73. Tisdale (n. 49 above), 18: 256–57. The state health officials officially requested that the USPHS conduct an investigation of the health effects of phenols and that the Bureau of Mines investigate its "industrial" aspects. Only the Bureau of Mines study was published. See, Leitch (n. 47 above), 40: 2021–26. Leitch was a chemical engineer employed by the Bureau of Mines.

74. See, Tisdale (n. 49 above), 256–57.

75. Waring (n. 50 above), 764–66. The agreement was eventually signed by the states of Kentucky, New York, Maryland, Illinois, Indiana, Tennessee, North Carolina, and Virginia. This compact was an early example of a single-purpose basinwide agreement. For a discussion of such agreements, see Roscoe C. Martin, et al., *River Basin Administration and the Delaware* (Syracuse, N.Y., 1960), 227–37.

76. For the growing importance of engineers within the field of public health and their conflict with physicians, see Joel A. Tarr, et al., "Disputes Over Water Quality Policy: Professional Cultures in Conflict," *AJPH* 70 (April 1980): 427–35, and Joel A. Tarr et al., "Water and Wastes: A Retrospective Assessment of Wastewater Technology in the United States, 1800–1932," *Technology & Culture* 25 (April 1984): 246–50.

77. Willard W. Hodge, "Waste Disposal Problems in the Coal Mining Industry," in Willem Rudolfs, ed., *Industrial Wastes: Their Disposal and Treatment* (New York, 1953), 312–418; and Luthy (n. 45 above), 501–20. The quenching ratio was about 500 gallons of water to quench the coke produced by one ton of coal. Other methods such as biological treatment were also used but primarily for discharge to sewers.

78. Monger to Cummings, August 6, 1923, and enclosures.

79. Waring (n. 50 above), 761.

80. Ibid., 761–63.

81. Leitch (n. 47 above), 40: 2021–26. See, also, Luthy (n. 45 above), 511.

82. Sanitary Engineering Division, Chicago Board of Health, "Water Contamination With Phenol Wastes," *Annual Reports, 1926–30* (Chicago, 1930), 470–76, 492–504; Arthur E. Gorman, "Pollution of Lake Michigan: Survey of Sources of Pollution," *Civil Engineering* 3 (1933): 519–22; and Craig E. Colten, *Industrial Wastes in the Calumet Areas, 1869–1970* (Champaign, Ill., September 1985).

83. Norman F. Prince, chemist and engineer from the Rochester Gas and Electric Corporation, in "Discussion," *JAWWA* 18 (1927): 582–83. See, also, F. W. Sperr, Jr., "Disposal of Phenol Wastes from By-Product Coke Plants," *AJPH* 19 (1929): 907.

84. See, for instance, "Discussion," *JAWWA* 18:584.

85. Mellon Institute (n. 30 above), 72. By-product ovens did produce fumes and steam but not necessarily heavy smoke pollution.

86. In 1969 Allegheny County, Pennsylvania, the home of U.S. Steel's Clairton coke works, approved new Air Pollution Rules and Regulations requiring, "The water utilized for the quenching of coke, prior to use as a quenching agent, shall be of a quality as may be discharged into the nearest stream or river, in accordance with the Acts of the Commonwealth of Pennsylvania." Compliance with this act would have required water completely free of phenols, an almost impossible task, and the code was amended in 1970 to require that 99 percent of the phenol be removed. Since that time, the coke industry and the regulatory authorities have engaged in constant conflict and negotiation about control of coke plant emissions to the environment, with the primary concern focused on health damages to workers and county residents. For a discussion of these acts and of subsequent events, see Charles O. Jones, *Clean Air: The Policies and Politics of Pollution Control* (Pittsburgh, Pa., 1975).

87. In 1929, for instance, a case was filed in Allegheny County by a group of plaintiffs trying to secure compensation for personal and property damage from coke oven chimney wastes. The case was not disposed of until 1940, having been to the U.S. Supreme Court, the Pennsylvania Supreme Court, and to the U.S. Court of Appeals a number times on procedural grounds. The merits of the case were never dealt with and the plaintiffs secured no relief. See, J. Philip Bromberg, *Clean Air Act Handbook* (Rockville, Md., 1985, 2nd. ed.), 16.

88. "Enforcement through persuasion and cooperation" was also characteristic of enforcement of the Allegheny County Air Pollution statutes from 1949 through the 1960s. See ibid., 94.

89. In 1924 the congress had passed the Oil Pollution Act, but it only applied to coastal waters and made no reference to water supplies.

90. Quoted in Tisdale (n. 49 above), 580.

91. The preferred vehicle for dealing with interstate river pollution problems was the Ohio River Valley Sanitation Commission (ORSANCO), an interstate compact founded by authority of Congress in 1936. See, Edward J. Cleary, *The ORSANCO Story: Water Quality Management in the Ohio Valley under an Interstate Compact* (Baltimore, Md., 1967), 249–83.

Personal Boundaries in the Urban Environment, by Raymond W. Smilor

1. "The Noise Nuisance," *Current Literature* 29 (November 1900), 508.

2. Floyd W. Parsons, "The Devils of Din," *Saturday Evening Post* 203 (November 8, 1930), 16–17.

3. E. L. Godkin, "Noises," *Nation* 56 (June 15, 1893), 433.

4. "Noise That Protects," *Scribner's Magazine* 46 (October 1909), 506–07.

5. "A Crusade for Quiet," *Outlook* 102 (September 28, 1912), 157–59.

6. William A. Lloyd, "Noise as a Nuisance," *University of Pennsylvania Law Review and American Law Register* 82 (April 1934), 567–82.

7. Horace Gay Wood, *A Practical Treatise on the Law of Nuisances in Their Varied Forms: Including Remedies Therefor at Law and in Equity* (Albany, N.Y.: John D. Parsons, Jr., 1875), 583.

8. Appeal of the Ladies' Decorative Art Club, 13 Atl. 537 (Pa. 1888).

9. Joseph A. Joyce and Howard C. Joyce, *Treatise on the Law Governing Nuisances* (Albany, N.Y.: Matthew Bender and Co., 1906), 228. See also Wood, *A Practical Treatise*, 586–87.

10. *Butterfield v. Klaber*, 52 How. Prac. (N.Y.) 255, per Sanford, J.

11. Wood, *A Practical Treatise*, 599.

12. Lloyd, "Noise as a Nuisance," 571. See also Wood, *A Practical Treatise*, 598–99.

13. George A. Spater, "Noise and the Law," *Michigan Law Review* 63 (June 1965), 1373–1410. This article also appears in James L. Hildebrand, *Noise Pollution and the Law* (Buffalo, New York: William S. Hein & Co., Inc., 1970), 22–59.

14. Henry Hazlitt, "In Dispraise of Noise," *Century* 120 (January 1930), 4–6.

15. H. M. Johnson, "Noise—A Social Problem," *Harper* 159 (October 1929), 561–71.

16. Wood, *A Practical Treatise*, 584.

17. Lloyd, "Noise as a Nuisance," 569–70.

18. *Stevens v. Rockport Granite Co.*, 216 Mass. 486, 104 N.E. 373 (1914).

19. *Boston Advertiser*, February 27, 1886.

20. For a discussion of the various cases see Joyce, Treatise, 216–37, and Lloyd, "Noise as a Nuisance," 569–82.

21. Edward S. Morse, "The Steam Whistle: A Menace to Public Health," address read before the Massachusetts Association of Boards of Health at Boston, January 27, 1905.

22. "Whist!" *Living Age* 198 (September 2, 1893), 522–24.

23. *New York Times*, July 19, 1895.

24. Philip G. Hubert, Jr., "For the Suppression of City Noises," *North American Review* 159 (November 1894), 633–35.

25. Imogen B. Oakley, "Public Health Versus the Noise Nuisance," *National Municipal Review* 4 (April 1915), 231–37.

26. *New Imperial and Windsor Hotel Co. v. Johnson*, I Ir.R. 327, 336 (1912).

27. *Richards v. Washington Terminal Co.*, 233 U.S. 546 (1914).

28. For a further discussion of the case see Spater, "Noise and the Law," 30–45.

29. These same difficulties exist today. See Stuart F. Levin, "Law and the Municipal Ecology: Part Two: Noise Pollution," *National Institute of Municipal Law Officers*, Washington, D.C., 1970, 55–87, reproduced by the Environmental Protection Agency.

30. James J. Putnam, "Discussion," Transactions of the Fifteenth International Congress on Hygiene and Demography, Washington, D.C., September 23–28, 1912, 540–41.

31. Cyril M. Harris, *Handbook of Noise Control* (New York: McGraw-Hill Book Company, Inc., 1957), Section 39, 2–12.

32. For a complete listing of the municipal ordinances against noise that follow see *Anti-Noise Ordinances of Various Cities*, Compiled for the Committee on Health of the Chicago City Council, Dr. Willis O. Nance, Chairman, by the Municipal Reference Library, Chicago, 1913, 1–36. To ample noise abatement legislation, this report drew together laws from 31 cities from every section of the country.

33. *Goodrich v. Busse et al.*, 247 Ill. 500, 93 N.E. 292, 139 Am. St. Rep. 335 (1910).

34. *Chicago Tribune*, July 27, 1911.

35. *Chicago Tribune*, July 25–29, 1911.

36. "Hooting Nuisance," *Living Age* 270 (August 19, 1911), 508–10.

37. *New York Times*, December 23, 1906.

38. "Noises and Their Suppression," *Journal of the American Medical Association* 49 (July 27, 1907), 232.

39. *Statutes of the United States of America* Passed at the Second Session of the Fifty-Ninth Congress, 1906–1907), v. 35, Part 1, Chap. 892, 881. See also Congressional Record, 59th Congress, 2d Session, January 11–31, 1907, 1093.

40. *House Reports*, v. 1, No. 5625, 59th Congress, 2d Session, 1906–1907.

41. *New York Times*, December 8, 1906.

42. *New York Times*, July 17, 1928.

43. These problems with enforcement persisted. See Harris *Handbook of Noise Control*, Section 39, 4, 12.

44. "Anti-Noise Policeman," *Outlook* 107 (June 27, 1914), 438–39. See also William T. Watson, M.D., "Baltimore's Anti-Noise Crusade," *National Municipal Review* 3 (July 1914), 385–89.

45. Dr. Willis O. Nance, "The Noise Problem in Chicago," address read to the City Club of Chicago, June 17, 1913; appeared in The City Club Bulletin, 6 (July 23, 1913), 229–38.

46. *New York Times*, September 20, 1913.

47. Willis O. Nance, M.D., "Gains Against the Nuisances: II. Noise and Public Health," *National Municipal Review* 11 (October 1922), 326–32.

48. "What We Are Doing and How," radio talk by Lewis H. Brown, Chairman, over Station WEAF, January 28, 1930; appeared in *City Noise, Noise Abatement Commission, Department of Health, City of New York* (New York: The Academy Press, 1930), 259–65.

Equity, Eco-racism and Environmental History, by Martin V. Melosi

Portions of this chapter are derived from some of my earlier studies, including *Coping with Abundance: Energy and Environment in Industrial America* (New York, 1985), 296–97; "Public History and the Environment," *Public Historian* 15 (Fall 1993): 11–13; "The Origins and Development of Environmental History," Linda Moore, et al., *Instructors Resource Manual: America's History* (New York, 1993), E39–E43.

1. In "The Significance of the Frontier in American History" (1893), Frederick Jackson Turner argued that the West was the seedbed of democracy, individualism, prosperity, and inventiveness. Walter Prescott Webb made the environment central to his study of the West in such books as *The Great Plains* (1931). And James C. Malin—regarded by many as the father of American environmental history—presented a more sophisticated analysis of the western environment in *The Grassland of North America* (1947), stressing the complex interrelationship between humans and the natural world. But neither Webb nor Malin looked at nature strictly on its own terms. Like Turner, Webb wanted to prove that the West was unique or exceptional, whereas Malin sought to justify his belief in the right of the individual to exploit the environment for economic ends.

2. For some general treatments of the modern environmental movement, see Samuel P. Hays, *Beauty, Health, and Permanence: Environmental Politics in the United States, 1955–1985* (Cambridge, 1987); Robert Paehlke, *Environmentalism and the Future of Progressive Politics* (New Haven, 1989); and Robert Gottlieb, *Forcing the Spring: The Transformation of the Environmental Movement* (Washington, D.C., 1993).

3. See Donald Worster, "Doing Environmental History," in Donald Worster, ed., *The Ends of the Earth* (New York, 1988), 290–91; Lawrence Rakestraw, "Conservation History: An Assessment," *Pacific Historical Review* 41 (1972): 276; John Opie, "Environmental History: Pitfalls and Opportunities," *Environmental Review* 7 (1983): 10; Barbara Leibhardt, "Interpretation and Causal Analysis: Theories in Environmental History," *Environmental Review* 12 (Spring 1988): 23–24.

4. For example, Roderick Nash's *Wilderness and the American Mind* (New Haven, 1967) moved beyond the notion of wilderness as the anthropocentric "American frontier" to uncover broader, aesthetic significance in human contact with the natural world. Samuel P. Hays, provided a model for later work on conservation and reform with his seminal *Conservation and the Gospel of Efficiency* (Cambridge, 1959).

5. Donald Worster's *Nature's Economy: A History of Ecological Ideas* promoted an ecological interpretation of history in some sense like James C. Malin had done several decades earlier, but Worster's book was not published until 1977. See also Richard White, "American Environmental History: The Development of a New Historical Field," *Pacific Historical Review* 54 (August 1985): 297–335.

6. See Donald Worster, "Ecology of Order and Chaos," *Environmental History Review* 14 (Spring/Summer 1990): 1–18.

7. See Carolyn Merchant, *The Death of Nature: Women, Ecology, and the Scientific Revolution* (San Francisco, 1980).

8. See Martin V. Melosi, "The Place of the City in Environmental History," *Environmental History Review* 17 (Spring 1993): 1–23.

9. For example, see Richard White, *The Roots of Dependency: Subsistence, Environment, and Social Change among the Choctaws, Pawnees, and Navahoes* (Lincoln, 1983); J. Donald Hughes, *American Indian Ecology* (El Paso, 1983); William Cronon, *Changes in the Land: Indians, Colonists, and the Ecology of New England* (New York, 1983); Calvin Martin, *Keepers of the Game: Indian-Animal Relationships in the Fur Trade* (Berkeley, 1978).

10. On the diversity of the modern environmental movement, see Robert Gottlieb, "Reconstructing Environmentalism: Complex Movements, Diverse Roots," *Environmental History Review* 17 (Winter 1993): 2–3.

11. The 1987 Women in Toxics Organizing conference, held in Arlington, Virginia, played an important part in promoting the central role of women in the antitoxics efforts. See Gottlieb, *Forcing the Spring*, 203, 207–12; Andrew Szasz, *Ecopopulism: Toxic Waste and the Movement for Environmental Justice* (Minneapolis, 1994), 90, 150–53.

12. Szasz, *Ecopopulism*, 5.

13. Lois Marie Gibbs, "Celebrating Ten Years of Triumph," *Everyone's Backyard* 11 (February 1993): 2.

14. Szasz, *Ecopopulism*, 6, 69–72. See also "The Grassroots Movement for Environmental Justice," *Everyone's Backyard* 11 (February 1993): 3.

15. Robert D. Bullard, ed., *Confronting Environmental Racism: Voices from the Grassroots* (Boston, 1993), 9.

16. Cynthia Hamilton, "Coping with Industrial Exploitation," ibid., 63.

17. Robert D. Bullard, *Dumping in Dixie: Race, Class, and Environmental Quality* (Boulder, Colo., 1994; second ed.), xiii.

18. Stella M. Capek, "The 'Environmental Justice' Frame: A Conceptual Discussion and an Application," *Social Problems* 40 (February 1993): 8.

19. Bunyan Bryant and Paul Mohai, eds., *Race and the Incidence of Environmental Hazards: A Time for Discourse* (Boulder, Colo., 1992), 1–2.

20. Gottlieb, "Reconstructing Environmentalism: Complex Movements, Diverse Roots," 16–17. See also Charles T. Rubin, *The Green Crusade: Rethinking the Roots of Environmentalism* (New York, 1994), 182–87.

21. Quoted in Karl Grossman, "The People of Color Environmental Summit," Robert D. Bullard, ed., *Unequal Protection: Environmental Justice and Communities of Color* (San Francisco, 1994), 272.

22. The Sierra Club, National Wildlife Federation, Audubon Society, Environmental Defense Fund, Environmental Policy Institute, Friends of the Earth, Greenpeace, and so forth.

23. Bryant and Mohai, eds., *Race and the Incidence of Environmental Hazards*, 6. See also Marc Mowrey and Tim Redmond, *Not in Our Backyard: The People and Events That Shaped America's Modern Environmental Movement* (New York, 1993), 413–37.

24. See Dana A. Alston, ed., *We Speak for Ourselves: Social Justice, Race and Environment* (Washington, D.C., 1990), 3; "From the Front Lines of the Movement for Environmental Justice," *Social Policy* 22 (Spring 1992): 12; Robert D. Bullard, "Anatomy of Environmental Racism and the Environmental Justice Movement," in Bullard, ed., *Confronting Environmental Racism*, 22–23; Pat Bryant, "Toxics and Racial Justice," *Social Policy* 20 (Summer 1989): 51.

25. Karl Grossman, "Environmental Racism," *Crisis* 98 (April 1991): 15.

26. See Janet Kreger, "Ecology and Black Student Opinion," *Journal of Environmental Education* 4 (Spring 1973): 30–34; Carl Anthony, "Why African Americans Should Be Environmentalists," *Earth Island Journal* (Winter 1990): 43–44; Paul Mohai, "Black Environmentalism," *Social Science Quarterly* 71 (December 1990): 744; Frederick H. Buttel and William L. Flinn, "Social Class and Mass Environmental Beliefs: A Reconsideration," *Environment and Behavior* 10 (September 1978): 433–50.

27. Alston, ed., *We Speak for Ourselves*, 3.

28. Henry Vance Davis, "The Environmental Voting Record of the Congressional Black Caucus," Bryant and Mohai, eds., *Race and the Incidence of Environmental Hazards*, 55–63; "Do Environmentalists Care About Poor People?" 52; "Beyond White Environmentalism," *Environmental Action* (January/February, 1990): 19, 27.

29. Dorceta Taylor, "Can the Environmental Movement Attract and Maintain the Support of Minorities?" Bryant and Mohai, *Race and the Incidence of Environmental Hazards*, 38.

30. Such indicators as monetary (e.g., donations to environmental causes), political (e.g., attendance at meetings), legal (e.g., participating in litigation), educational (e.g., attending courses, workshops), nature (e.g., visits to national parks), wildlife (e.g., hunting/fishing), membership affiliation, and so forth may not unveil levels of black interest in environmental issues.

31. Dorceta E. Taylor, "Blacks and the Environment: Toward an Explanation of the Concern and Action Gap Between Blacks and Whites," *Environment and Behavior* 21 (March 1989): 175–98.

32. Barbara Deutsch Lynch, "The Garden and the Sea: U.S. Latino Environmental Discourse and Mainstream Environmentalism," *Social Problems* 40 (February 1993): 108–18. See also "Beyond White Environmentalism," 24–27.

33. The CRJ was founded in 1963 after the assassination of black activist Medgar Evers, church bombings in Birmingham, Alabama, and other anti–civil rights activities. Also note that prior to Chavis, others, such as sociologist Robert Bullard, have been exploring the issue of environmental racism since the late 1970s.

34. See Charles Lee, "Toxic Waste and Race in the United States," Bryant and Mohai, eds., *Race and the Incidence of Environmental Hazards*, 10–16, 22–27; Rosemari Mealy, "Charles Lee on Environmental Racism," *We Speak for Ourselves*, 8; Paul Mohai and Bunyan Bryant, "Environmental Racism: Reviewing the Evidence," ibid., 163–69; Grossman, "Environmental Racism," 16–17; Grossman,

"From Toxic Racism to Environmental Justice," *E: The Environmental Magazine* 3 (May/June, 1992): 30–32; Dick Russell, "Environmental Racism," *Amicus Journal* 11 (Spring 1989): 22–25; Bryant, "Toxics and Racial Justice," 49–50.

35. Bullard, ed., *Confronting Environmental Racism*, 3. See also Grossman, "Environmental Racism," 31.

36. Bullard, *Unequal Protection*, xvi.

37. Memorandum, William Clinton, February 11, 1994; "Not in My Backyard," *Human Rights* 20 (Fall 1993): 27–28; Bryant and Mohai, eds., *Race and the Incidence of Environmental Hazards*, 5; Grossman "The People of Color Environmental Summit," 287; Bullard, "Conclusion: Environmentalism with Justice," Bullard, ed., *Confronting Environmental Racism*, 195.

38. Michel Gelobter, "Toward a Model of 'Environmental Discrimination,'" Bryant and Mohai, eds., *Race and the Incidence of Environmental Hazards*, 64–73. For a contrary view, see Mohai and Bryant, "Environmental Racism: Reviewing the Evidence," 164.

39. Leonard P. Gianessi, Henry M. Peskin, Edward Wolff, "The Distributional Effects of Uniform Air Pollution Policy in the United States," *Quarterly Journal of Economics* 93 (May 1979): 281.

40. Ibid., 281–96.

41. Gelobter, "Toward a Model of 'Environmental Discrimination,'" 72.

42. Vicki Been, "What's Fairness Got to Do with It? Environmental Justice and the Siting of Locally Undesirable Land Uses," *Cornell Law Review* 78 (September 1992): 1014–15. See also Rachel D. Godsil, "Remedying Environmental Racism," *Michigan Law Review* 90 (November 1991): 394; Matthew Rees, "Black and Green," *New Republic* 206 (March 2, 1992): 15–16; Grossman, "From Toxic Racism to Environmental Justice," 35.

43. "Beyond White Environmentalism," 21. See also Russell, "Environmental Racism," 25–30.

44. Alice L. Brown, "Environmental Justice: New Civil Rights Frontier," *Trial* 29 (July 1993), 48, 51–52. See also Bullard, *Dumping in Dixie*, xiii–xiv; "'Environmental Racism': It Could Be a Messy Fight," *Business Week* (May 20, 1991): 116.

45. Brown, "Environmental Justice," 52; Been, "What's Fairness Got to Do with It?" 1084–85; Godsil, "Remedying Environmental Racism," 420–26. See also C. Miller, "Efficiency, Equity and Pollution: The Case of Radioactive Waste," *Environment and Planning* 19 (1987): 913–18.

46. Grossman, "Environmental Racism," 15.

47. See Martin V. Melosi, *Coping with Abundance: Energy and Environment in Industrial America* (New York, 1985), 296–97. See also Samuel P. Hays, *Beauty, Health, and Permanence: Environmental Politics in the United States, 1955–1985* (New York, 1987).

48. In the late nineteenth and early twentieth centuries, urban environmentalists formed many local groups to confront the impacts of industrialization and urbanization, although they did not come to dominate the emerging new environmental movement, nor did they give attention to matters of race and class. See Martin V. Melosi, "Environmental Reform in the Industrial Cities: The Civic Response to Pollution in the Progressive Era," Kendall E. Bailes, *Environmental History: Critical Issues in Comparative Perspective* (Lanham, Md., 1985), 494–515.

49. Grossman, "From Toxic Racism to Environmental Justice," 34–35.

50. See Peter S. Wenz, *Environmental Justice* (New York, 1988); Brent A. Singer, "An Extension of Rawls' Theory of Justice to Environmental Ethics," *Environmental Ethics* 10 (Fall 1988): 217–31.

51. Allan Schnaiberg, "Redistributive Goals versus Distributive Politics: Social Equity Limits in Environmental and Appropriate Technology Movements," *Sociological Inquiry* 53 (Spring 1983): 214. See also John A. Hird, "Environmental Policy and Equity" The Case of Superfund," *Journal of Policy Analysis and Management* 12 (1993): 323–35; Peter Nijkamp, "Equity and Efficiency in Environmental Policy Analysis: Separability Versus Inseparability," Allan Schnaiberg, et al., *Distributional Conflicts in Environmental-Resource Policy* (New York, 1986), 61–73.

52. Gottlieb, *Forcing the Spring*, 235–40.

53. Clayton R. Koppes, "Efficiency, Equity, Esthetics: Shifting Themes in American Conservation," in Donald Worster, *The Ends of the Earth: Perspectives on Modern Environmental History* (New York, 1988), 233–36, 251. I have argued elsewhere that "wise use" was not a tool for equity but a "happy compromise" for government officials who began to realize that they faced a potential contradic-

tion in promoting economic growth, on the one hand, and providing stewardship over the public lands, on the other. The wise use concept provided a middle ground to support sustained economic growth. In either case, however, efficiency, not equity or esthetics, dominated Progressive Era America and beyond. See Martin V. Melosi, "Energy and Environment in the United States: The Era of Fossil Fuels," *Environmental Review* 11 (Fall 1987): 167–68.

54. Andrew Hurley, *Environmental Inequalities: Class, Race, and Industrial Pollution in Gary, Indiana, 1945–1980* (Chapel Hill, 1995), xiii–iv.

55. See Alfred Crosby, *Ecological Imperialism: The Biological Expansion of Europe, 900–1900* (Cambridge, 1986)

56. Barbara Rose Johnston, ed., *Who Pays the Price? The Sociocultural Context of Environmental Crisis* (Washington, D.C., 1994), 234–35.

Rice, Water, and Power, by Mart A. Stewart

1. David Doar, *Rice and Rice Planting in the South Carolina Low Country* (Charleston, S.C.: The Charleston Museum, 1970), 8.

2. For the "hydraulic civilizations" thesis, see Karl Wittfogel, *Oriental Despotism: A Comparative Study of Total Power* (New Haven: Yale University Press, 1957). For the "specific qualities of water" and the basic obligations of those who would manipulate it, see 15–27. Works that consider "hydraulic cultures" and irrigation communities, as well as the specific features of the political economy of irrigation organizations, and that at least nod in Wittfogel's direction, include Karl W. Butzer, *Early Hydraulic Civilization in Egypt: A Study in Cultural Ecology* (Chicago: University of Chicago Press, 1976); most of the essays and papers in E. Walter Coward, Jr., ed., *Irrigation and Agricultural Development in Asia: Perspectives from the Social Sciences* (Ithaca: Cornell University Press, 1980), Theodore E. Downing and Mcguire Gibson, eds., *Irrigation's Impact on Society* in Anthropological Papers of the University of Arizona, No. 25 (Tucson: University of Arizona Press, 1974), Julian H. Steward, Robert Adams, Donald Collier, Angel Palerm, Karl A. Wittfogel, and Ralph L. Beals, *Irrigation Civilizations: A Comparative Study* in Social Science Monographs I (Westport: Greenwood Press, 1981), and Richard B. Woodbury, ed., *Civilization in Desert Lands* in Anthropological Papers of the University of Utah, No. 62 (University of Utah Press, 1970); Clifford Geertz, "The Wet and the Dry: Traditional Irrigation in Bali and Morocco," *Human Ecology* 1 (March 1972), 23–40; Thomas F. Glick, *Irrigation and Society in Medieval Valencia* (Cambridge: Belknap Press, 1970); Robert C. Hunt and Eva Hunt, "Canal Irrigation and Local Social Organization," *Current Anthropology* 17 (September 1976), 389–98; E. R. Leach, "Hydraulic Society in Ceylon," *Past and Present* 15 (April 1959), 2–26; Arthur Maass and Raymond L. Anderson, . . . *and the Desert Shall Rejoice: Conflict, Growth, and Justice in Arid Environments* (Cambridge: MIT Press, 1978).

Donald Worster creates a taxonomy for many of these that is structured around a thoughtful reconsideration of Wittfogel's work, in *Rivers of Empire: Water, Aridity, and the Growth of the American West* (New York: Pantheon, 1985). See especially 17–60, and 342–46. Worster's discussion of Wittfogel has been, in turn, drawn into the eye of the interpretive whirlpool of irrigation studies, in a review essay of several studies of local irrigation systems in the West: Gerald D. Nash, "Eden or Armageddon? Recent Studies of Water in the Twentieth-Century West," *Journal of Forest History* 33 (October 1989), 197–201.

3. See Henri Lefebvre, "Space: Social Product and Use Value," in *Critical Sociology: European Perspectives*, ed. and trans. J. W. Freiburg (New York: Irvington Publishers, 1979); Rhys Isaac, *The Transformation of Virginia, 1740–1790* (Chapel Hill: University of North Carolina Press, 1982), 18–19; D. W. Meinig, "Introduction," in Meinig, ed., *The Interpretation of Ordinary Landscapes: Geographical Essays* (New York: Oxford University Press, 1979), 6; J. B. Jackson, "The Vernacular Landscape," in Edmund C. Penning-Rowsell and David Lowenthal, eds., *Landscape Meanings and Values* (London: Allen and Unwin, 1986), 67–77; John Stilgoe, *Common Landscapes of America, 1580–1845* (New Haven: Yale University Press, 1982). See also Denis Cosgrove, *Social Formation and Symbolic Landscape* (London: Croom Helm, 1984), 41–63, for a general discussion of the effects of capitalist production on shaping the landscape. Charles E. Orser's discussion of plantation space and power relations, though from

the exceedingly hard, material vantage point of a historical archaeologist, is especially relevant to a discussion of plantation landscapes: "Toward a Theory of Power for Historical Archaeology: Plantations and Space," in *The Recovery of Meaning: Historical Archaeology in the Eastern United States*, ed. by Mark P. Leone and Parker B. Potter, Jr. (Washington, D.C.: Smithsonian Institution Press, 1988), 313–43, esp. 316–21.

4. The best description of the geography can be found in Sam B. Hilliard, "The Tidewater Rice Plantation: An Ingenious Adaptation to Nature," *Geoscience and Man: Coastal Resources* 12 (June 1975), 58, 61. A revised and expanded version is: Hilliard, "Antebellum Tidewater Rice Culture in South Carolina and Georgia," in James R. Gibson, ed., *European Settlement and Development in North America: Essays on Geographical Change in Honour and Memory of Andrew Hill Clark* (Toronto: University of Toronto Press, 1978), 91–105. See also Douglas C. Wilms, "The Development of Rice Culture in 18th Century Georgia," *Southeastern Geographer* 12 (1972), 45–57. On the "perfection" of the system, see R. F. W. Allston, "Essay on Sea Coast Crops," *DeBow's Review* 16 (June 1854), 612–13, and Hilliard, "The Tidewater Rice Plantation," 58, 61.

5. Peter H. Wood, *Black Majority: Negroes in Colonial South Carolina* (New York: W. W. Norton, 1974), 63–91; Todd Savitt, *Medicine and Slavery: The Diseases and Health Care of Blacks in Antebellum Virginia* (Urbana: University of Chicago Press, 1978), 17–32.

6. Wilms, "The Development of Rice Culture," 55.

7. R. W. F. Allston, "Rice," *DeBow's Review* 4 (December 1847), 506; Hilliard, "Tidewater Rice Plantations," 64.

8. Hilliard, "Tidewater Rice Plantations," 64; T. D. Mathews, F. W. Stapor, Jr., C. R. Richter, J. V. Miglarese, M. E. McKenzie, and L. A. Barclay, eds., *Ecological Characterization of the Sea Island Coastal Region of South Carolina and Georgia*, Vol. I: *Physical Features of the Characterization Area*, U.S. Fish and Wildlife Service, Office of Biological Services (Washington, D.C., 1980), 80.

9. Lewis Cecil Gray, *History of Agriculture in the Southern United States to 1860* (Washington, D.C.: Carnegie Institution, 1933), vol. 2, 726; James C. Darby, "On the Embanking and Preparation of Marsh Land, for the Cultivation of Rice," *Southern Agriculturist* 2 (January 1829), 23–28; Hilliard, "Tidewater Rice Plantations," 59–61; Allston, "Rice," 506–11; Allston, "Sea Coast Crops," 605.

10. Hugh Starnes, "The Rice-Fields of Carolina," *The Southern Bivouac* 2 (November 1886), 335.

11. The most succinct definition, explaining the difference between a task labor system and a gang labor system, as Philip Morgan has pointed out, is in Gray, *Agriculture in the Southern U.S.*, vol. 1, 550–51. Also see Morgan, "Work and Culture: The Task System and the World of Lowcountry Blacks, 1700 to 1880," *William and Mary Quarterly* 39 (October 1982), 564–83.

12. Fernand Braudel, *The Structures of Everyday Life: The Limits of the Possible*, trans. Sian Reynolds (New York: Harper & Row, 1979), 147; Daniel C. Littlefield, *Rice and Slaves: Ethnicity and the Slave Trade in Colonial South Carolina* (Baton Rouge: Louisiana State University Press, 1981), 84–92.

13. Roswell King to Pierce Butler, March 28, 1807, Butler Family Papers (hereinafter BFP), Historical Society of Pennsylvania (hereinafter HSP).

14. Mart A. Stewart, "Land Use and Landscapes: Environment and Social Change in Coastal Georgia," Ph.D. dissertation, Emory University, 1988, 317–30, 357–66, 383–90. Tension between the ideal of a well-regulated community and the disorder of human relationships was common on all plantations, large and small, throughout the antebellum South. See James Oakes, *The Ruling Race: A History of American Slaveholders* (New York: Vintage Books, 1982), 153–91. The unique demands of hydraulic agriculture made this tension more problematic on lowcountry rice plantations, however.

15. Roswell King to Pierce Butler, August 2, 1818, July 23, 1808, BFP, HSP.

16. Stewart, "Land Use and Landscapes," 308–09.

17. Roswell King to Pierce Butler, February 18, 1809, August 17, 1808, BFP, HSP; Orrin Sage Wrightman and Margaret Davis Cate, *Early Days of Coastal Georgia* (St. Simons Island: Fort Frederica Association, 1955), 163; King to Butler, July 8, 1803, February 18, 1809, BFP, HSP; Pierce Butler to William Page, November 15, 1797, Pierce Butler Letterbook, 1787–1822, HSP. Many of the ex-slaves who made claims against the federal government through the Southern Claims Commission after the Civil War listed and described the kinds of property they had and the crops and animals they raised. For a summary of claims that yield small details of this nature, see Mart Stewart, "What Na-

ture Suffers to Grow": Life, Labor, and Landscape on the Georgia Coastal Plain, 1680–1920 (University of Georgia Press, forthcoming). See also Louis Manigault to Charles Manigault, December 28, 1852, and William Capers, Sr., to Louis Manigault, October 3, 1860, in James M. Clifton, ed., *Life and Labor on Argyle Island: Letters and Documents of a Savannah River Rice Plantation, 1833–1867* (Savannah: Beehive Press, 1978), 133, 307.

Additional information on the importance of benne to coastal black Georgians can be found in Savannah Unit, Georgia Writers' Project, Works Projects Administration, *Drums and Shadows: Survival Studies Among the Georgia Coastal Negroes* (Athens: University of Georgia Press, 1986), 71, 178.

See Philip D. Morgan, "The Ownership of Property by Slaves in the Mid-Nineteenth Century Low Country," *Journal of Southern History* 49 (August 1983), 399–422, for a summary of property ownership by slaves in Liberty County.

18. Elizabeth J. Reitz, Tyson Gibbs, and Ted A. Rathbun, "Archaeological Evidence for Subsistence on Coastal Plantations," in *The Archaeology of Slavery and Plantation Life,* ed. Theresa A. Singleton (New York: Academic Press, Inc., 1985), 183–85; Roswell King to Pierce Butler, August 13, 1808, May 21, 1820, June 14, 1813, June 5, 1814, September 17, 1815, BFP, HSP; William Hampton, William R. Adams, Carolyn Rock, and Janis Kearney-Williams, "Foodways on the Plantations at Kings Bay: Hunting, Fishing, and Raising Food," in William Hampton, ed., *Historical Archaeology of Plantations at Kings Bay, Camden Country, Georgia,* Report of Investigations 5 (Gainesville: University of Florida Department of Anthropology, 1987), 225–27, 231, 233, 241–42, 244–76.

19. Tidewater slaves had more time to exploit a richer environment than slaves on interior plantations. See Peter Kolchin, "Reevaluating the Antebellum Slave Community: A Comparative Perspective," *Journal of American History* 70 (December 1983), 591–92; Singleton, "Archaeology of Slavery," 185; Reitz, et al., "Subsistence on Coastal Plantations," 185; John Solomon Otto, *Cannon's Point Plantation, 1794–1860: Living Conditions and Status Patterns in the Old South* (New York: Academic Press, 1984), 164.

20. An Act for ordering and governing Slaves within this Province, May 10, 1770, in Howell Cobb, comp., *General and Public Statutes of the State of Georgia* (New York: Edward O. Jenkins, 1859), 596; Roswell King to Pierce Butler, July 20, 1817, BFP, HSP.

21. Roswell King, Jr., to Thomas Butler, August 9, 1828, BFP, HSP.

22. Roswell King, Jr., to Thomas Butler, August 9, September 27, 1835, BFP, HSP. See Stewart, "Land Use and Landscapes," 390–93, for a fuller discussion of this point.

23. See, for example: Glynn County Inferior Court Minutes, March 11, 22, 1852, Book D, 1814–1870, 191–206, Microfilm 82/70, Georgia Department of Archives and History.

24. Afro-American animal tales were common in lowcountry Georgia and South Carolina. See, for example, Georgia Writers Project, *Drums and Shadows,* 79, 110–11, 160–61, 171. An older collection, with no notes on informants, is Charles C. Jones, Jr., *Negro Myths from the Georgia Coast Told in the Vernacular* (New York: Houghton, Mifflin and Co., 1888). Patricia Jones-Jackson describes the discernment of distinct features of particular animals that sea island storytellers bring to their tales: *When Roots Die: Endangered Traditions on the Sea Islands* (Athens: University of Georgia Press, 1987), 16–17, 171n–172n. For interpretive essays on the tales and their cultural meanings, see Charles Joyner, *Down by the Riverside: A South Carolina Slave Community* (Chicago: University of Illinois Press, 1984), 172–95, and Lawrence Levine, *Black Culture and Black Consciousness: Afro-American Folk Thought from Slavery to Freedom* (New York: Oxford University Press, 1977), 81–135. For discussions of the close ties of extended plantation Afro-American families with particular locales, see Herbert G. Gutman, *The Black Family in Slavery Freedom, 1750–1925* (New York: Pantheon Books, 1976), 208–11; Patricia Guthrie, "Catching Sense: The Meaning of Plantation Membership Among Blacks on St. Helena Island, South Carolina," Ph.D. dissertation, University of Rochester, 1977, 114–29. For the meaning of a "sense of place," and the kind of landscapes that embody it, see John Brinkerhoff Jackson, *Discovering the Vernacular Landscape* (New Haven: Yale University Press, 1984), 55.

25. Eric Foner, *Nothing But Freedom: Emancipation and Its Legacy* (Baton Rouge: Louisiana State University Press, 1983), 82–108. For Georgia, see "Visit to 'Gowrie' and 'East Hermitage' Plantations, Savannah River," March 22, 1867, in Clifton, ed., *Life and Labor,* 363; Frances Butler Leigh, *Ten Years on a Georgia Plantation* (London: Richard Bentley & Son, 1883), 26, 54–56, 66, 268, 156. A variety of

examples of concessions that Georgia freedmen were able to wrest from planters are outlined in Paul Cimbala, "The Freedmen's Bureau, the Freedmen and Sherman's Grant in Reconstruction Georgia, 1865–1867," *Journal of Southern History* 55 (November 1989), 629–30.

26. Frances Butler to P. C. Hollis, March 10, 1868, Wister Family Papers (herinafter WFP), HSP; Statement of Crop of 1879, WFP, HSP; James Couper to P. C. Hollis, October 17, 1880, BFP, HSP; Inventory of Property on Butlers Island, December 10, 1884, BFP, HSP; Frances Butler Leigh to P. C. Hollis, January 20, 1886, WFP, HSP; Inventory of Implements at Butlers Island, February 14, 1881, BFP, HSP; Leigh, *Ten Years*, 270–71; Mary Granger, ed., *Savannah River Plantations* (Savannah: Georgia Historical Society, 1947), 20–21, 22, 24, 250, 253.

27. Wilms, "The Development of Rice Culture," 56; Foner, *Nothing But Freedom*, 109; John Scott Strickland, "Traditional Culture and Moral Economy: Social and Economic Change in the South Carolina Low Country, 1865–1910," in Steven Hahn and Jonathan Prude, *The Countryside in the Age of Capitalist Transformation* (Chapel Hill: University of North Carolina Press, 1985), 166–68; Frances Butler Leigh to P. C. Hollis, December 11, 1886, WFP, HSP; Pete Daniel, *Breaking the Land: The Transformation of Cotton, Tobacco, and Rice Cultures Since 1880* (Chicago: University of Chicago Press, 1985), 38–50.

28. Leigh, *Ten Years*, 207–08; James Couper to P. C. Hollis, March 11, April 7, November 10, 1879, BFP, HSP; Plantation Expenses, 1880, 1881, and Eleven Months of 1882, BFP, HSP. On the reluctance of South Carolina lowcountry freedmen to do ditching and banking and the effects of this on rice planting, see Amory Austin, *Rice: Its Cultivation, Production, and Distribution in the United States and Foreign Countries*, U.S. Department of Agriculture, Division of Statistics, Report No. 6 (Washington, D.C.: Government Printing Office, 1893), 78–79; John Scott Strickland, "'No More Mud Work': The Struggle for the Control of Labor and Production, in Low Country South Carolina, 1863–1880," in Walter J. Fraser, Jr., and Winfred B. Moore, Jr., eds., *The Southern Enigma: Essays on Race, Class, and Folk Culture* (Westport, Conn.: Greenwood Press, 1983), 51.

29. James Couper to P.C. Hollis, April 10, August 26, 1880, BFP, HSP.

30. The persistence of the rice economy here appears to reflect Karl Wittfogel's observation about the inertia of hydraulic regimes. See Wittfogel, *Oriental Despotism*, 421–23, for his conclusions about the "staying-power" of hydraulic regimes. Marvin Harris also discusses this idea from another perspective, in a chapter on the "hydraulic trap," in *Cannibals and Kings: The Origins of Cultures* (New York: Random House, 1977), 153–64. He uses the "hydraulic civilization" example to argue that the choices humans make at certain historical moments about strategies of resource exploitation carry their own force, once the strategies are carried out, and become major commitments that are hard to get out of when conditions change. Peter Coclanis, however, credits this inertia in South Carolina to the effects of the structural domination of the lowcountry economy for two centuries by a plantation agriculture geared toward the production of a commodity for the international market, to the constraints of the swampy lowcountry environment, and to the absence of viable economic alternatives: "Rice . . . constituted the best hope for sustained economic progress that the area ever had." See *The Shadow of a Dream: Economic Life and Death in the South Carolina Low Country, 1670–1920* (New York: Oxford University Press, 1989), 142, 132–57.

31. Two devastating hurricanes hit the coast in 1893, and were especially destructive to the Savannah River plantations. Other hurricanes caused further damage in 1894, 1898, 1906, 1910, and 1911: Lawrence S. Rowland, "'Alone on the River': The Rise and Fall of the Savannah River Rice Plantations of St. Peters Parish, South Carolina," *South Carolina Historical Magazine* 88 (July 1987), 149; James M. Clifton, "Twilight Comes to the Rice Kingdom: Postbellum Rice Culture on the South Atlantic Coast," *Georgia Historical Quarterly* 62 (Summer 1978), 151.

32. Frances Butler Leigh identified the pattern black movement would take, in an observation about the hands on the Butlers Island plantation shortly after the War: "Many of them left, not to work for anyone else, but to settle on their own properties in the pine woods." See *Ten Years*, 156. Also see Paul Cimbala, "The Freedmen and Sherman's Grant," especially 529–632; and *Savannah Morning News*, January 30, 1880. Land ownership by coastal blacks increased dramatically between 1870 and 1900: Eric Foner, *Nothing But Freedom*, 108–09; Albert Smith Colby, "Down Freedom's Road: The Contours of Race, Class, and Property Crime in Black-Belt Georgia, 1866–1910," Ph.D. dissertation,

University of Georgia, 1982, 133–48; Steven W. Engerrand, "'Now Scratch or Die': The Genesis of Capitalistic Agricultural Labor in Georgia, 1865–1880," Ph.D. dissertation, University of Georgia, 1981, 179. In Georgia, for example, black landholding in coastal Glynn County tripled between 1880 and 1900 (from 5,862 to 14,761 acres), and increased from 8,521 to 11,691 in Bryan County: W. E. Burghardt Du Bois, "The Negro Landholder in Georgia," in Carroll D. Wright, ed., *Bulletin of the Department of Labor* 35 (July 1901), 692–93, 695, 698–99, 721–22, 735, 739–41.

33. The economy of lowcountry Afro-Americans occupied a peripheral position to the larger Georgia, South Carolina, and American economy, but their participation in this larger economy, especially by way of the expanding coastal plain lumber industry, was crucial. Many of the lumbermen and raftsmen were black, as were turpentine workers in the area. See the *Darien Timber Gazette,* July 18, 1874; Carlton A. Morrison, "Raftsmen of the Altamaha," M.A. thesis, University of Georgia, 1970, 38, 42; and Thomas F. Armstrong, "The Transformation of Work: Turpentine Workers in Coastal Georgia, 1865–1901," *Labor History* 25 (Fall 1984), 523–28. The quotation is from David D. Long and James E. Ferguson, "Soil Survey of Glynn County, Georgia," *Field Operations of the Bureau of Soils, 1911* (Washington, D.C.: Government Printing Office, 1914), 598–99.

"Dammed at Both Ends and Cursed in the Middle," by Brian Donahue

1. William Brewster, *Concord River* (Cambridge, Mass., 1937); Ludlow Griscom, *The Birds of Concord* (Cambridge, Mass., 1949); Ann Zwinger and Edwin Way Teale, *A Conscious Stillness: Two Naturalists on Thoreau's Rivers* (New York, 1982); Richard K. Walton, *Birds of the Sudbury River Valley—An Historical Perspective* (Lincoln, MA, 1984).

2. D. L. Child, "Memorial," *Report of the Joint Special Committee upon the Subject of the Flowage of Meadows on Concord and Sudbury Rivers* (Boston, 1860), xcvii.

3. Morton J. Horwitz, *The Transformation of American Law, 1780–1860* (Cambridge, Mass., 1977).

4. Gary Kulik, "Dams, Fish and Farmers: Defense of Public Rights in Eighteenth-Century Rhode Island," in Steven Hahn and Jonathan Prude, eds., *The Countryside in the Age of Capitalist Transformation* (Chapel Hill, N.C., 1985).

5. The ecological conditions that created these grassy meadows are now altered: beavers and Indians have vanished, the river level has changed several times, new plant species have arrived, even the chemistry of the water has changed. For a discussion of the ecology of the meadows past and present see Richard J. Eaton, *A Flora of Concord* (Cambridge, Mass., 1974), 18–27.

6. The small percentage of land devoted to English hay in 1791 can be seen in Table 1 (see n. 44). It was even lower earlier in the eighteenth century. For example, only 459 acres of English hay were reported in Concord in 1749, compared to 722 in 1791, and 2206 in 1850. See Brian Donahue, "The Forests and Fields of Concord: An Ecological History" in David Hackett Fischer, ed., *Concord: The Social History of a New England Town, 1750–1850* (Waltham, Mass., 1983). See also Jared Eliot, *Essays Upon Field Husbandry in New England and Other Papers, 1748–1762* (New York, 1934); and Howard S. Russell, *A Long, Deep Furrow: Three Centuries of Farming in New England* (Hanover, N.H., 1976).

7. Early problems with the meadows are discussed in Lemuel Shattuck, *A History of the Town of Concord* (Boston and Concord, 1835), 14–16. Petitions to the Legislature for relief are quoted in Shattuck and in *Report upon Flowage,* Appendix G. The labor of transforming the meadows is largely unrecorded, but evidence of it can be found in deeds which often mention ditches as boundaries for meadow lots. These ditches are still dimly visible from the air today.

8. Grass species are mentioned in farmers' testimony in *Report upon Flowage;* see also Eaton, *Flora of Concord;* and Edwin Way Teale's discussion in Zwinger and Teale, *A Conscious Stillness,* 130–35.

9. D. L. Child, "Memorial," *Report upon Flowage,* xcii.

10. See figures on acreage and yield in Table 2:1 in Donahue, "Forests and Fields of Concord," 34.

11. *Report upon Flowage,* xcii.

12. Ibid. Testimony of John B. Wright.

13. Bradford Touvey and Francis H. Allen, eds., *Journal of Henry D. Thoreau* (New York, 1962), vol. VI, 422–24.

14. Christopher Roberts, *The Middlesex Canal, 1793–1860* (Boston, 1938).

15. *New England Farmer*, XIV, 1862, 254, 274, 280.

16. *Report upon Flowage*, farmers' testimony, *passim*. Also see *Journal of Henry Thoreau*, vol. XI, 122; vol. V, 381.

17. "Unjust Operation of Law," *New England Farmer*, 1857, 271. Internal evidence suggests this article was written either by Col. David Heard or perhaps David L. Child.

18. *Report upon Flowage*, 27.

19. "The Proprietors of Sudbury Meadows vs. The Proprietors of the Middlesex Canal," *Massachusetts Reports*, vol. 40, 1839, 36–53; "Unjust Operation of Law," 271.

20. "William Heard vs. Proprietors of the Middlesex Canal," *Massachusetts Reports*, vol. 46, 1842, 81–88. See also "Stevens vs. Proprietors of the Middlesex Canal," *Massachusetts Reports*, vol. 12, 1815, 409, for the precedent upon which this decision was based.

21. Roberts, *Middlesex Canal*.

22. "David Heard vs. Charles P. Talbot and another," *Massachusetts Reports*, vol. 73, 1856, 113–21.

23. "Unjust Operation of Law," 271.

24. *Report of Experiments and Observations on the Concord and Sudbury Rivers in the Year 1861 by Commissioners* (Boston, 1862).

25. *Journal of Henry Thoreau*, vol. XII, 247.

26. Helen Fitch Emery, *The Puritan Village Evolves: A History of Wayland, Massachusetts* (Wayland, 1981).

27. *Report upon Flowage*, xciii.

28. Caleb Eddy to Levi Bartlett, Boston, January 8, 1835. Middlesex Canal Records, Box 31, Folder 22.

29. *Report of the Cochituate Water Board on the Petition of the Proprietors of Meadows on the Sudbury River* (Boston, 1859). Middlesex Canal Records, Box 2, Folder 36. See also "Unjust Operation of Law," 271.

30. This is the way Thoreau heard the phrase as it made the rounds, *Journal of Henry Thoreau*, vol. XIII, 149. The *New England Farmer*, XI, 1859, reported it as "a dam at both ends, and a curse between them." Both sources attribute it to Col. Heard.

31. *Report upon Flowage*, xcx.

32. Ibid., Statement of Respondents.

33. Ibid., 56.

34. *New England Farmer* XII, 1860, 260.

35. *Decision of the Supreme Court of Massachusetts upon the Constitutionality of the Act of 1860 to Remove the Dam across the Concord River at Billerica* (as reported in "Monthly Law Reporter") (Boston, 1861), 36.

36. *Report of Experiments and Observations on the Concord and Sudbury Rivers in the Year 1861 by Commissioners* (Boston, 1862).

37. *Report of the Joint Committee of the Legislature, March 27, 1862* (Boston, 1862), 10.

38. *Report upon Flowage*, testimony of Nathan Barrett, John Sherman, and Charles Cutting. See also Lawrence Eaton Richardson, *Concord River* (Barre, Mass., 1964).

39. John Tuttle, Tilly Buttrick, Nathan Barrett, and Peter Wheeler to the Directors of Middlesex Canal, Concord, Massachusetts, August 24, 1807. Middlesex Canal Records, Box 1, Folder 14.

40. Robert A. Gross, *The Minutemen and Their World* (New York, 1976). See also James Kimenker, "The Concord Farmer: An Economic History, 1750–1850," in Fischer, *Concord: Social History*.

41. On the shift to commercial agriculture see Winifred B. Rothenberg, "A Price Index for Rural Massachusetts, 1750–1855," *Journal of Economic History* 39, 1979; Rothenberg, "The Emergence of Farm Labor Markets and the Transformation of the Rural Economy: Massachusetts, 1750–1855," *Journal of Economic History* 48, 1988; Robert A. Gross, "Culture and Cultivation: Agriculture and Society in Thoreau's Concord," *Journal of American History* 69, 1982; and Kimenker, "The Concord Farmer." See also Donahue, "Forests and Fields of Concord," for further agroecological analysis of the transition.

42. See Sarah F. MacMahon, "A Comfortable Subsistence: A History of Diet in New England, 1630–1850" (Ph.D. Thesis, Brandeis University, 1982) on the shift to wheat flour.

43. Tillage land in Concord peaked at 1487 acres, or 20 percent of all improved acreage in 1771. It fell to 1064, or 13 percent by 1791, and maintained this level throughout the first half of the nineteenth century, being 1068 acres in 1850. See Table 1 (n. 44) and Donahue, "Forests and Fields of Concord."

44. TABLE 1.: Land-use Acreages, By Percentage: Three Concord River Valley Towns

	Tillage	English Hay	Fresh Meadow	Pasture	Woods	Unimproved
SUDBURY						
1791	12	06	15	21	46	
1831	08	08	13	27	12	32
1840	07	09	13	28	11	32
1850	09	12	13	25	11	30
1860	10	17	13	25	26	07
WAYLAD						
1791	09	09	15	32	34	
1831	08	12	17	27	10	26
1840	07	12	15	30	11	25
1850	09	14	14	30	10	23
1860	11	16	14	29	23	05
CONCORD						
1791	09	06	15	35	36	
1801	09	07	17	29	28	10
1811	09	08	17	24	27	14
1821	09	09	17	30	25	11
1831	08	09	15	29	15	24
1840	09	11	14	28	14	24
1850	08	16	11	27	11	28
1860	08	16	13	26	14	23
1865	08	17	10	34	25	05
1885	10	26	23	34	05	
1895	13	33	20	29	03	
1905	10	27	13	48	02	

Sources: Mass. Tax Valuations, 1791–1860; Mass. State Census, 1865–1905.

45. See Henry Colman, *Fourth Report of the Agriculture of Massachusetts—Counties of Franklin and Middlesex* (Boston, 1841), 197–98.

46. Colman, *Fourth Report,* 355.

47. Records of the Concord Farmers' Club, 1856, 1857, Concord Free Public Library.

48. Henry D. Thoreau, "Walking," *The Natural History Essays* (Salt Lake City, 1980), 118.

49. Concord Farmers' Club Records, 1860.

50. Ralph Waldo Emerson, "Address to Middlesex Agricultural Society," *Agriculture of Massachusetts* (Boston, 1858), 17.

51. *New England Farmer,* 1857, 288. See also Colman, *Fourth Report,* 375.

52. *Argument of Hon. Henry F. French of Boston, March 12, 1862, before the Joint Committee of the Leg. of Mass., on the Petition for the Repeal of the Act of 1860 . . .* (Boston, 1862).

53. *Report upon Flowage,* farmers' testimony, *passim.*

54. See Brian Donahue, "Skinning the Land: Economic Growth and the Ecology of Farming in Nineteenth Century Massachusetts" (Unpub., 1984) for a more extended discussion of these agroe-

cological woes; and Donahue, "Forests and Fields of Concord" for a detailed look at several farms in Concord.

55. This is evident from examination of individual farms in Concord. See Donahue, "Forests and Fields."

56. David H. Miller, *Water at the Surface of the Earth: An Introduction to Ecosystem Hydrodynamics* (New York, 1977); J. W. Hornbeck, R. S. Pierce, and C. A. Federer, "Streamflow Changes after Forest Clearings in New England," *Water Resource Research* 6, 1970.

57. Donahue, "Skinning the Land." From a low point of about 10 percent, forest cover in the valley came up to about 25–35 percent between 1850 and 1870, and remained at roughly this level through 1900.

58. James H. Patric and Ernest M. Gould, "Shifting Land Use and the Effects on River Flow in Massachusetts," *Journal AWWA*, 1976.

59. Ibid. Interestingly, Sudbury River flow has increased again since 1950, presumably because a lot of the watershed is now paved.

60. Even the heavens may have turned against the farmers: a preliminary study by William Baron suggests rainfall may have been on the increase in the 1840s, 1850s and 1860s after an abnormally dry regime from the 1760s to 1820s. William R. Baron, "Historical Climates of the Northeastern United States—Seventeenth through Nineteenth Centuries," in George P. Nichols, ed., *Holocene Human Ecology in Northeastern North America* (New York, 1988).

Irrigation, Water Rights, and the Betrayal of Indian Allotment, by Donald J. Pisani

1. Alvin M. Josephy, Jr., "Here in Nevada a Terrible Crime," *American Heritage*, 21 (June 1970), 93–100. See also Martha C. Knack and Omer C. Stewart, *As Long as the River Shall Run: An Ethnohistory of Pyramid Lake Reservation* (Berkeley, 1984); Lowell Smith and Pamela Deuel, "The California-Nevada Interstate Water Compact: A Great Betrayal," *Cry California*, 7 (Winter 1971–1972), 24–35; and Donald J. Pisani, "The Strange Death of the California-Nevada Compact: A Study in Interstate Water Negotiations," *Pacific Historical Review* 37 (November 1978), 637–58.

2. *Winters v. United States*, 207 U.S. 577 (1908).

3. *Arizona v. California*, 373 U.S. 546 (1963). The Paiutes pressed suit against the Truckee-Carson Irrigation District in 1973 to define their Winters doctrine rights. They hoped for sufficient water to stabilize the level of the lake and create a new fishing industry. Their case found its way to the Supreme Court in 1983 only to be rejected on grounds that the Supreme Court shows far less sympathy for Indian claims today than it did in the 1960s and early 1970s, and its decision may have far-reaching implications for the claims of other tribes in the arid West. See Knack and Stewart, *As Long as the River Shall Run*, 351–58.

4. Francis Paul Prucha's *Indian-White Relations in the United States: A Bibliography of Works Published, 1975–1980* (Lincoln, 1982), 66–68, included a special section on water rights with forty-seven citations. However, virtually all were articles published in legal journals. The one notable exception, discussed later in this paper, was Norris Hundley's superb piece in the October 1978, issue of the *Western Historical Quarterly*.

5. The best broad surveys of Indian land policy in the late nineteenth and early twentieth centuries are J. P. Kinney, *A Continent Lost—A Civilization Won: Indian Land Tenure in America* (Baltimore, 1937); Frederick E. Hoxie, *A Final Promise: The Campaign to Assimilate the Indians, 1880–1920* (Lincoln, 1984); Leonard A. Carlson, *Indians, Bureaucrats, and Land: The Dawes Act and the Decline of Indian Farming* (Westport, Conn., 1981); and D. S. Otis, *The Dawes Act and the Allotment of Indian Lands* (Norman, Oklahoma, 1973). The three most useful surveys of federal reclamation pay almost no attention to the effect of government irrigation projects on Indian water rights. See Michael Robinson, *Water for the West: The Bureau of Reclamation, 1902–1977* (Chicago, 1979); Paul Gates's chapter in his magisterial survey, *History of Public Land Law Development* (Washington, D.C., 1968), 635–98; and William Warne, *The Bureau of Reclamation* (New York, 1973).

6. This article is meant to suggest to historians of Indians and natural resources an important

and neglected field. A broad history of the federal trusteeship of Indian resources—including hydroelectric power sites, timber, coal, oil, and uranium, as well as water—is badly needed.

7. Robert G. Dunbar, *Forging New Rights in Western Waters* (Lincoln, 1983), 3. Ethnohistorians have greatly expanded our understanding of the deep cultural differences revealed in ways Euro-Americans and Native Americans thought about time, religion, the land, labor, and technology. However, we must be careful not to fall into the trap of concluding that cultural differences alone prevented Indians from adapting to settled agriculture. Not only did some tribes practice irrigation long before contact with the Spanish, but many showed a remarkable capacity to absorb new tools and values from other Indians as well as non-Indians. There are many reasons Indians questioned the value of irrigation. On Montana reservations, for example, the growing season was too short and the soil incapable of producing those high-value crops—such as beans, cotton, and melons—raised in the American southwest. Stock raising promised a greater return with less labor. On other reservations, Indians lacked the implements and knowledge to farm. In still other cases, they feared that improving land would invite further white greed.

8. The classic statement on the value of irrigation is in William Ellsworth Smythe, *The Conquest of Arid America* (New York, 1900). See also Donald J. Pisani, *From the Family Farm to Agribusiness: The Irrigation Crusade in California and the West, 1850–1931* (Berkeley, 1984).

9. Porter J. Preston and Charles A. Engle, "Report of Advisors on Irrigation on Indian Reservations," in *Hearings before a Subcommittee of the Committee on Indian Affairs*, Senate, July 8, 10, 11, 12, and 17, 1929 (Washington, D.C., 1930), 2217, 2229.

10. *U.S. Statutes at Large*, XXVI, 1011, 1039.

11. Commissioner of Indian Affairs, *Annual Report, 1893* (Washington, D.C., 1893), 50, hereafter report year only is cited; "Message from the President of the United States, Transmitting Certain Reports upon the Condition of the Navajo Indian Country," S. Ex. Doc. 68, 52 Cong., 2 sess., serial 3056, 1893.

12. Commissioner of Indian Affairs, *Annual Report* (1891), 50–52; *Annual Report* (1893), 47–50; *Annual Report* (1894), 24–26; *Annual Report* (1895), 28–32; and *Annual Report* (1900), 70, 267.

13. Commissioner of Indian Affairs, *Annual Report* (1895), 26.

14. Commissioner of Indian Affairs, *Annual Report* (1900), 59.

15. The commissioner observed in 1898 that on the Crow Reservation a $6 payment made to the Indians twice a year for land ceded to the government had been spent on frivolous items while money earned in wages "they regard much more highly and expend with much more care and discretion." Commissioner of Indian Affairs, *Annual Report* (1898), 49. See also *Annual Report* (1897), 29, 33; and *Annual Report* (1906), 14–15.

16. Commissioner of Indian Affairs, *Annual Report* (1895), 28; *Annual Report* (1900), 58; and *Annual Report* (1901), 63; "Irrigation Projects on Indian Reservations, Etc.," H. Doc. 1268, 63 Cong., 3 sess., serial 6891, 1914; and "Indian Irrigation Projects, 1919," H. Doc. 387, 66 Cong., 2 sess., serial 7769, 1919, 10.

17. Ray P. Teele, *Irrigation in the United States* (New York, 1915), 218; Commissioner of Indian Affairs, *Annual Report* (1914), 38; *Annual Report* (1915), 218; Commissioner of Indian Affairs, *Annual Report* (1914), 38; *Annual Report* (1915), 44; *Annual Report* (1916), 42; *Annual Report* (1919), 38; and *Annual Report* (1920), 23.

18. *U.S. Statutes at Large*, XXXIII, 224; Department of Interior, *Thirteenth Annual Report of the Reclamation Service, 1913–1914* (Washington, D.C., 1915), 73–80. In fairness to the Reclamation Service, government engineers expected the members of Indian families to pool their individual allotments so that in most cases farms would be much larger than five acres. However, because a white husband and wife could each file on a farm, the Indians still received much less land.

19. Commissioner of Indian Affairs, *Annual Report* (1906), 148–50; and *Annual Report* (1909), 42; "Truckee-Carson Irrigation Project, Etc.," H. Doc. 211, 59 Cong., 2 sess., serial 5152, 1907, 3; and *Forestry and Irrigation, 13* (June 1907), 288–89; *Churchill County Standard* (Fallon, Nevada), October 20 and December 1, 1906, and May 18, 1907. The policy of permitting Indians to sell part of their allotments before the twenty-five-year trust period was extended to all Indian land within the proposed Reclamation Service project boundaries by *U.S. Statutes at Large*, XXXIV, 327.

20. *U.S. Statutes at Large*, XXXIII, 302–05, 225, 360; Department of Interior, *Third Annual Report of the Reclamation Service, 1903–1904* (Washington, D.C., 1905), 83, 328–29.

21. *U.S. Statutes at Large*, XXXIV, 53. Congress had authorized allotment of Yakima lands in December, 1904 (*U.S. Statutes at Large*, XXXIII, 595). At that time proceeds from the sale of surplus land were used to pay for irrigation works constructed by the Indian Bureau, not the Reclamation Service. The 1906 statute permitted the Reclamation Service to take over the Indian Bureau's canal system. This and related legislation added arable land to the public domain.

22. Commissioner of Indian Affairs, *Annual Report* (1906), 89; and *Annual Report* (1907), 50–51; Department of Interior, *Ninth Annual Report of the Reclamation Service, 1909–1910* (Washington, D.C., 1911), 32; and *Fifteenth Annual Report of the Reclamation Service, 1915–1916* (Washington, D.C., 1916), 547. On the three Montana reservations, the Reclamation Service planned to irrigate 423,000 acres. See the *Eighth Annual Report of the Reclamation Service, 1908–1909* (Washington, D.C., 1910), 90–96.

23. Commissioner of Indian Affairs, *Annual Report* (1905), 7.

24. Commissioner of Indian Affairs, *Annual Report* (1896), 30; and *Annual Report* (1900), 63. In 1895, a contract was signed with the company, which promised to provide sufficient water to irrigate about 120,000 acres. However, the company's "ditch" was little more than rough embankments of sandy, porous soil. It quickly washed out. The commissioner reported that Fort Hall Indians branded the scheme a "deception and fraud" (61).

25. Department of Interior, *First Annual Report of the Reclamation Service from June 17 to December 1, 1902* (Washington, D.C., 1903), 81–82; Commissioner of Indian Affairs, *Annual Report* (1900), 59; *Annual Report* (1904), 7–21; *Annual Report* (1906), 87; *Annual Report* (1916), 37, 43; and *Annual Report* (1919), 41. For a good summary of the violation of Pima water and the Indian reaction, see "Letters and Petition with Reference to Conserving the Rights of the Pima Indians, of Arizona, to the Lands of their Reservation and the necessary Water Supply for Irrigation," 62 Cong., 1 sess., U.S. Congress, House Committee Print H-3840, 1911.

26. For an overview of the Reclamation Service's position on state as opposed to federal water rights see Donald J. Pisani, "State vs. Nation: Federal Reclamation and Water Rights in the Progressive Era," *Pacific Historical Review*, (August 1982), 265–82.

27. Department of Interior, *Third Annual Report of the Reclamation Service, 1903–1904* (Washington, D.C., 1905), 268; and Commissioner of Indian Affairs, *Annual Report* (1906), 83.

28. Commissioner of Indian Affairs, *Annual Report* (1906), 83; *Annual Report* (1907); *Annual Report* (1908), 31–32; and *Annual Report* (1913); "Irrigable Lands on Uintah Indian Reservation, Utah," S. Doc. 414, 66 Cong., 3 sess., serial 7794, 1921, 4.

29. *Winters v. United States*, 207 U.S. 564 (1908); Norris Hundley, Jr., "The Dark and Bloody Ground of Indian Water Rights: Confusion Elevated to Principle," *Western Historical Quarterly* 9 (October 1978), 455–82; and Hundley, "The 'Winters' Decision and Indian Water Rights: A Mystery Reexamined," *Western Historical Quarterly* 13 (January 1982), 17–42.

30. *U.S. Statutes at Large*, XXV, 113–33. For summaries of the conditions that led to the suit see 1–13 of "Brief of the Appellee in United States Circuit Court of Appeals for the Ninth Circuit," in RG60, Records of the Department of Justice, box 221, file 58730, National Archives; and *Winters v. United States*, 143 Fed. Rep. 740 at 740–42 (1906).

31. Hundley, "The 'Winters' Decision and Indian Water Rights," 39; U.S. Supreme Court, "Brief for the United States in *Winters v. United States*," *Records and Briefs*, v. 207, Library of Congress Law Library; *Winters v. United States*, 207 U.S. 564 (1908).

32. *Kansas v. Colorado*, 206 U.S. 46 (1907).

33. "Memorandum: Conflicting Attitude Dept. Justice on Irrigation Matters," December 8, 1905, in Record Group 115, Records of the Bureau of Reclamation, General Administrative and Project Records, 1902–1919, "(762) Legal Discussions—General," National Archives. See also the undated position paper prepared by the Indian Bureau, "Memorandum Relative to Cases Involving Water Rights of Indians in Montana and Washington," in box 231, "Legal Discussions—General, thru Dec. 31, 1907," and Pisani, "State vs. Nation," 278.

34. Attorney General to the President, December 18, 1905, box 221, file 58730, Record Group 60, Records of the Department of Justice (NA/RG60), National Archives.

35. Ethan Allen Hitchcock to the President, January 5, 1906, box 221, file 5870, NA/RG60.

36. F. E. Leupp to secretary of the Interior, April 3, 1906, box 221, file 58730, NA/RG60.

37. In this case, "cooperation" meant that the Wyoming State Engineer had extended the time for "final Proof" (i.e. actually using water) on Wind River Reservation water claims for several years. See Commissioner of Indian Affairs, *Annual Report* (1911), 16.

38. For example, on June 8, 1912, Yakima Indians urged Congressman John H. Stephens, head of the House Committee on Indian Affairs, to reject a bill that would have reduced the water available to Indians on the Ahtanum River in Washington. They used the river long before the Reclamation Service opened that section of the Yakima Project. Even though the stream ran through the reservation, the secretary of the Interior had, according to the Indians, granted 75 percent of the water to white farmers. "Memorial of the Yakima Tribe of Indians," H. Doc. 1304, 62 Cong., 3 sess., serial 6500, 1913, 3. For the Supreme Court decision see *United States v. Powers*, 305 U.S. 533 (1939).

39. House Joint Resolution 250 (Stephens), *Congressional Record*, 62 Cong., 2 sess., House, February 22, 1912, 2344.

40. "Water Rights of the Indians, Etc.," H. Doc. 1274, 63 Cong., 3 sess., serial 6888, 1914, 3–4.

41. *Congressional Record*, 63 Cong., 2 sess., Senate, June 17, 1914, 10587-600; June 18, 1914, 10652–57; June 20, 1914, 10768–89; June 23, 1914, 10937–46; and June 24, 1914, 11019–36.

42. Ibid., 10595, 10769–70, 10936.

43. The Reclamation Service feared that selling improved Indian land for fair market value would discourage potential white settlers who had to pay construction costs, and the price of setting up a farm and breaking up the land, in addition to the charge for the land.

44. This was not a special concession. The Reclamation Act of 1902 provided interest-free loans, and when the Reclamation Service took over irrigation on the Flathead Reservation, it simply reaffirmed and extended this substantial inducement to take up government land.

45. *Congressional Record*, 63 Cong., 2 sess., June 24, 1914, 11019–34.

46. Ibid., 11023.

47. Ibid., 11034.

48. Ibid., 10773–74, for Robinson's comments.

49. Ibid., 11024, 11027.

50. Ibid., 11028.

51. On June 24, 1914, the Senate approved a compromise amendment limiting federal irrigation expenditures on the Flathead Reservation to $100,000 unless the attorney general certified that the Indians "are protected and confirmed in their water rights." This amendment failed to win endorsement in the House. That most senators had little understanding of or interest in Indian water rights was revealed in the vote: 29 for, 20 against, and 47 not voting. Ibid., 11019, 11036.

52. "Report of Commission to Investigate Irrigation Projects on Indian Lands," H. Doc. 1215, 63 Cong., 3 sess., serial 6888 (Washington, D.C., 1914), 38–40; and Commissioner of Indian Affairs, *Annual Report* (1914), 39.

53. Commissioner of Indian Affairs, *Annual Report* (1915), 47. In 1915, the Reclamation Service claimed to be able to irrigate nearly 89,000 acres on those three reservations, but less than 4,000 acres were cultivated, most by whites. Department of Interior, *Fourteenth Annual Report of the Reclamation Service, 1914–1915* (Washington, D.C., 1915), 117, 122, 128.

54. For examples of the persistence of the water rights problem see "Irrigable Lands on Uintah Indian Reservation, Utah," S. Doc. 414, 66 Cong., 3 sess., serial 7794, 1921, 4; S. Rep. 706, 67 Cong., 2 sess., serial 7951, 1922; H. Rep. 624, 67 Cong., 2 sess., serial 7955, 1922; and "For Settlement of Water Rights in the Toole County Irrigation District, Montana, Affecting Indians of the Blackfeet Indian Reservation," S. Rep. 1073, 67 Cong., 4 sess., serial 8155, 1923.

55. Department of Interior, *Nineteenth Annual Report of the Reclamation Service, 1919–1920* (Washington, D.C., 1920), 439–40. Agricultural development on the Flathead and Fort Peck reservations followed a similar course. See the statistics on 447 and 456 of the same report.

56. Porter J. Preston and Charles A. Engle, "Report of Advisors on Irrigation on Indian Reservations," in Senate, *Hearings before a Subcommittee of the Committee on Indian Affairs*, July 8, 10, 11, 12, and 17, 1929, Congressional print S545-2-B (Washington, D.C., 1930), 2210–61. Preston and Engle

were chosen to conduct the study presumably because they would take a more detached view than engineers in Interior. For more general surveys of reservation conditions see "Survey of Conditions of Indians in the United States," U.S. Senate, *Hearings before the Committee on Indian Affairs*, 70 Cong., 1 sess., January 10 and 13, 1928 (Washington, D.C., 1928); and *Hearings before a Subcommittee of the Committee on Indian Affairs*, Senate, September 12–14, 1932 (Washington, D.C., 1934).

57. Preston and Engle, "Report of Advisors on Irrigation on Indian Reservations," 2217–20.

58. In most western states, water rights did not attach directly to particular parcels of land. They could be acquired by tenants as well as landlords.

59. Commissioner of Indian Affairs, *Annual Report* (1914), 29. For good overviews of the leasing system, but with little attention to irrigation and water rights, see Otis, *The Dawes Act and the Allotment of Indian Lands*, 98–123; and Kinney, *A Continent Lost—A Civilization Won*, 214–48. See also Hoxie, *A Final Promise*, 168; and Carlson, *Indians, Bureaucrats, and Land*, 136–38.

60. Preston and Engle, "Report of Advisors on Irrigation on Indian Reservations," 2213, 2237–38; and "Extension of Time for Payment of Charges Due on Indian Irrigation Projects," S. Rep. 586, 71 Cong., 2 sess., serial 9186, 1930, 1. See also H. Rep. 943, 72 Cong., 1 sess., serial 9492, 1932; H. Rep. 1372, 72 Cong., 1 sess., serial 9493, 1932; S. Rep. 1197, 72 Cong., 2 sess., serial 9647, 1933; and S. Rep. 908, 73 Cong., 2 sess., serial 9770, 1934.

61. Donald J. Pisani, "Reclamation and Social Engineering in the Progressive Era," *Agricultural History* 57 (January 1983), 46–63.

62. Hundley, "The Dark and Bloody Ground of Indian Water Rights," 480.

63. Carlson, *Indians, Bureaucrats and Land*, offers substantial statistical evidence to support his case. See in particular the appendices, 181–206.

Biotic Change in Nineteenth-Century New Zealand, by Alfred W. Crosby

1. Peter Kalm, *Travels into North America* (Barre, Mass.: Imprint Society, 1972), 24.

2. Joseph Banks, *Endeavor Journal of Joseph Banks, 1768–1771* (Sydney: Angus & Robertson, 1962), 11, 8.

3. James Cook, *The Journals of Captain James Cook on his Voyages of Discovery, 1: The Voyage of the Endeavor, 1768–1771* (Cambridge: The Hakluyt Society, 1955), I, 276–78; Joseph D. Hooker, "Note on the Replacement of Species in the Colonies and Elsewhere," *Natural History Review*, 1864, 126.

4. Harrison M. Wright, *New Zealand, 1769–1840: Early Years of Western Contact* (Cambridge, Mass.: Harvard University Press), 36–37; D. Ian Pool, *The Maori Population of New Zealand, 1769–1971* (Aukland: Aukland University Press), 235.

5. Charles Darwin, *The Origins of Species* (New York: New American Library, 1958), 332; W. T. L. Travers, "Of the Changes Effected in the Natural Features of a New Country by the Introduction of Civilized Races," *Transactions and Proceedings of the New Zealand Institute*, II, 299–330.

6. Elmer D. Merrill, *The Botany of Cook's Voyage* (Waltham, Mass.: Chronica Botanica, 1954), 227; Charles Darwin, *The Voyage of the Beagle* (Garden City, N.Y.: Doubleday, 1962), 426; Alexander Bathgate, *Colonial Experiences*, (Glasgow: James MacLehose, 1874), 229; T. F. Cheeseman, "The Naturalized Plants of Aukland Provincial District," *Transactions and Proceedings of the New Zealand Institute*, XV, 269–70.

7. Hooker, 126.

8. Wright, 62; Peter Buck, "Medicine Amongst the Maoris in Ancient and Modern Times" (Ph.D. diss., Alexander Turnbull Library, Wellington, New Zealand, n.d.), 82–83; W. H. Goldie, "Moaori Medical Lore," *Transactions and Proceedings of the New Zealand Institute*, XXXVII, 84.

9. Augustus Earle, *Narrative of a Residence in New Zealand* (London: Oxford University Press, 1966), 178; Arthur W. Thomson, *The Story of New Zealand: Past and Present as Savage and Civilized* (London: John Murray, 1859), 214–16.

10. N. L. Edson, "Mortality from Tuberculosis in the Maori Race," *New Zealand Medical Journal* XLII, 2, February 1943, 102, 105; F. D. Fenton, *Observations on the State of the Aboriginal Inhabitants of New Zealand* (Aukland: W. C. Wilson, 1859), 21, 29.

11. A. H. McLintock, ed., *An Encyclopedia of New Zealand* (Wellington: R. E. Owen, 1966), II, 823–24; C. E. Adams, "A Comparison of the General Mortality in New Zealand, in Victoria and New

South Wales, and in England," *Transactions and Proceedings of the New Zealand Institute*, XXXI, 661.

12. Travers, 312–13.

13. John Clayton, *The Reverend John Clayton, a Person with a Scientific Mind. His Writings and Other Related Papers* (Charlottesville, Va.: University Press of Virginia, 1965), 138; Elton, 100.

Australian Nature, European Culture, by Thomas R. Dunlap

1. Geoffrey Bolton, *Spoils and Spoilers* (Sydney: George Allen and Unwin, 1981). Derek Whitelock, *Conquest to Conservation* (Adelaide: Wakefield, 1985). William J. Lines, *Taming the Great South Land* (Sydney: George Allen and Unwin, 1991).

2. Lines, *Taming the Great South Land*, 278.

3. Richard White, "American Environment History: The Development of a New Historical Field," *Pacific Historical Review* 54 (August 1985), 335.

4. Areas where the aboriginal populations remained a significant part of the population or where Anglo legal and social institutions were not dominant are somewhat different cases.

5. Cliff Ollier, "Physical Geography and Climate of Australia," in Australian Bureau of Statistics, *Year Book Australia, 1988* (Canberra: Commonwealth Government Printer, 1988), 202.

6. Ollier, "Physical Geography," 207–14. On ecological constraints and climate see Graeme Caughley, "Ecological Relationships," in Caughley, Neil Shepherd, and Jeff Short (editors), *Kangaroos: Their Ecology and Management in the Sheep Rangelands of Australia* (Melbourne: Cambridge University Press, 1987), 159–87. Caughley suggested (author's interview, June 1990) that equilibrium models of ecosystems were inappropriate and misleading in the chaotic Australian system. His viewpoint is not universal, but it is at least defensible.

7. A. L. Burt made this point in commenting on the political development of frontier societies, "If Turner Had Looked at Canada, Australia, and New Zealand When He Wrote about the West," in Walker D. Wyman and Clifton B. Kroeber (editors), *The Frontier in Perspective* (Madison: University of Wisconsin Press, 1965), 59–77.

8. Ollier, "Physical Geography," 202–56. Population density map after page 256; P. Laut, "Changing Patterns of Land Use in Australia," 547–56, in Australian Bureau of Statistics, *Year Book Australia, 1988*. Canada is demographically comparable, its population clustered in a few strips along the border with the United States, but agriculture is possible in much more of the land.

9. Quoted in Alfred Crosby, *Ecological Imperialism* (New York: Cambridge University Press, 1986), 7. A longer development of the European picture of Australian nature is in F. G. Clarke, *The Land of Contrarities* (Clayton: Melbourne University Press, 1977), Chapter Nine.

10. Bureau of Flora and Fauna, *Fauna of Australia*, Volume 1A (Canberra: Australian Government Publishing Service, 1987) provides an overview. On the vegetation and its differences see Stephen Pyne, *Burning Bush* (New York: Henry Holt, 1991), 15–67.

11. Bernard William Smith, *European Vision and the South Pacific, 1768–1850* (London: Oxford University Press, 1960) deals with the general process, as, from a somewhat different stand, does Paul Carter, *The Road to Botany Bay* (Chicago: University of Chicago Press, 1987). On European art, see Bernard William Smith, *Australian Painting, 1788–1970* (Melbourne: Oxford University Press, 1971), 82–85. On the kangaroo, R. M. Younger, *Kangaroo Images Through the Ages* (Melbourne: Hutchinson, 1988). The Mitchell Library, Sydney, New South Wales, has an invaluable collection of early white Australian views of the land.

12. Quoted in Ross Gibson, *The Diminishing Paradise* (Sydney: Angus and Robertson, 1984), 50. Watling is a favorite, appearing also in Bernard William Smith, "The Artist's Vision of Australia," Smith, *The Antipodean Manifesto* (Melbourne: Oxford University Press, 1976), and Alec H. Chisholm (editor), *Land of Wonder* (Sydney: Angus and Robertson, 1964).

13. On art and the landscape, Smith, "The Artist's Vision of Australia." Chisholm, *Land of Wonder*, provides a good sampling of Australian nature writing.

14. White's work, first published in 1788, was one of the premier nature books in English and, as one commentator noted: "The early cheap editions were carried all over the world by expatriate Englishmen, sick for the dells and hillocks of home." Richard Mabey, introduction to Penguin edition (New York: Penguin, 1977), viii.

15. Donald Fleming, "Science in Australia, Canada, and the United States: Some Comparative Remarks," *Actes du Dixieme Congres International d'Histoire des Sciences* (Paris: Hermann, 1962), 183. On science and national identity in North America see Suzanne Zeller, *Inventing Canada* (Toronto: University of Toronto Press, 1987).

16. P. J. Stanbury, "The Discovery of the Australian Fauna and the Establishment of Collections," in Bureau of Flora and Fauna, *Fauna of Australia*, Volume 1A, 202–26. Sybil Jack, "Cultural Transmission: Science and Society to 1850," and Ian Inkster and Jan Todd, "Support for the Scientific Enterprise, 1850–1900" in R.W.Home (editor), *Australian Science in the Making* (Melbourne: Cambridge University Press, 1988), 45–66 and 102–32. For a comparative perspective on the museums, see Susan Sheets-Pyenson, *Cathedrals of Science* (Kingston: McGill-Queens University, 1988).

17. Fleming, "Science in Australia, Canada, and the United States" on the problems of visiting scientists. The standard model of scientific development, now heavily criticized, is in George Basalla, "The Spread of Western Science," *Science* 156, 611–22. For its application to Australia see R. W. Home's introduction to Home, *Australian Science in the Making*, vi–xxvii, and editors' introduction in R. W. Home and Sally Gregory Kohlstedt (editors), *International Science and National Scientific Identity: Australia between Britain and America* (Dordrecht, Netherlands: Kluwer, 1991), 1–17.

18. Kathleen G. Dugan, "The Zoological Exploration of the Australian Region and Its Impact on Biological Theory," in Nathan Reingold and Marc Rothenberg (editors), *Scientific Colonialism: A Cross-Culture Comparison* (Washington, D.C.: Smithsonian Institution Press, 1987), 93–97. Elizabeth Dalton Newland, "Dr. George Bennett and Sir Richard Owen: A Case Study in the Colonization of Early Australian Science," in Home and Kohlstedt, *International Science*, 55–74.

19. Crosby, *Ecological Imperialism*, describes the general process. Importations and establishments may be judged from lists and description in Christopher Lever, *Naturalized Mammals of the World* (London: Longman, 1985) and *Naturalized Birds of the World* (London: Longman, 1987). On the acclimatization societies and the movement for importation, Michael A. Osborne, "A Collaborative Dimension of the European Empires: Australian and French Acclimatization Societies and Intercolonial Scientific Co-operation," in Home and Kohlstedt, *International Science*, 97–119. This also relies on Christopher Lever, unpublished acclimatization manuscript, copy from Lever to Dunlap, August 1990.

20. Quoted in H. J. Frith, *Wildlife Conservation* (Sydney: Angus and Robertson, revised edition 1979), 138. The complaint about birds without song and flowers that had no smell was made about America in 1818, twenty years before its appearance in Australia. Richard White, *Inventing Australia* (Sydney: George Allen and Unwin, 1981), 51.

21. To appreciate the cultural dimensions of a suitable quarry, imagine Sir Edwin Landseer's "The Monarch of the Glen" with the stag and the Highlands replaced by a kangaroo and the bush, or a painting entitled "The Emu at Bay" (rather than the stag), or a pack of foxhounds and the local hunt in pursuit of a dingo. This last actually happened, for Australians, lacking foxes, coursed dingoes and emus in the 1850s, but no one thought it quite cricket. The Victorian Acclimatization Society, replying to a questionnaire sent by the British Acclimatization Society in 1864, discussed the deficiencies of the kangaroo. The Melbourne society, on the other hand, recommended kangaroos as valuable and interesting addition to the English parks "and from their speed they might furnish a valuable addition to objects of sport." Lever, unpublished acclimatization manuscript, 181–86, quote from 178. On early hunting see Gordon Inglis, *Sport and Pastime in Australia* (London: Meuthen, 1912), 73.

22. Lever, unpublished acclimatization manuscript.

23. New South Wales's first law was the Game Protection Act, 1866, No. 22; South Australia's "An Act to Prevent the Wanton Destruction of Certain Wild and Acclimatized Animals," 1864, No. 23. Tasmania's first statute was passed in 1865 (29 Vic. 22); Victoria's in 1862 (25 Vic. 161); Queensland's in 1863 (No. 6); and Western Australia's in 1874 (38 Vic. 4). Not surprisingly, legislative discussions of game laws harked back to those at "home"; a review of South Australian wildlife laws noted as a milestone that the debate over the 1912 act was the first in which English game laws did not feature in the legislative discussion. B. C. Newland, "From Game Laws to Fauna Protection Acts in South Australia: The Evolution of an Attitude," *The South Australian Ornithologist* 23 (March 1961), 52–63.

24. Official pest destruction started with New South Wales's "An Act to Facilitate and Encourage

the Destruction of Native Dogs," 1852, No. 44. Private action came first; the Van Dieman Land Company was paying bounties on thylacines as early as 1830, Eric R. Guiler, *Thylacine: The Tragedy of the Tasmanian Tiger* (Melbourne: Oxford University Press, 1985), 16. The irruption of the European rabbit caused all the colonies to take action; Eric Rolls, *They All Ran Wild* (Sydney: Angus and Robertson, 1979) and David Stead, *The Rabbit Menace in New South Wales* (Sydney: Government Printer, 1928). Poisoning was a favorite method; Victoria's "Act for Regulating the Sale and Use of Poisons," 1876, No. 559, justified its restrictions on the grounds that "large quantities of arsenic, strychnine and other poisons are in use in the colony for pastoral, agricultural, and other purposes . . ." Except for the division between native and imported species, the game laws and pest and predator control practices of the other settler colonies were similar, the legal devices and attitudes having come from Great Britain.

25. Only in 1918 did New South Wales, in the Birds and Animals Protection Act, 1918, No. 21, reverse the order and make all species protected unless they were declared game. The Birds and Animals Protection Act, 1922, No. 37, made all protected birds and animals, until lawfully taken, "the property of the Crown." Western Australia declared game property of the Crown in Game Act Amendment, 1913, No. 27. Protection in the other states generally proceeded without such declarations, by expanding the list of species protected, sometimes by administrative action rather than statute, most commonly in the early twentieth century. As with game laws, these are similar to those of the other settler colonies.

26. Rolls, *They All Ran Wild*, 20.

27. T. S. Palmer, "The Danger of Introducing Noxious Animals and Birds," in U.S. Department of Agriculture, *Yearbook of the United States Department of Agriculture for 1898* (Washington, D.C.: Government Printing Office, 1899), 93. Rolls, *They All Ran Wild* and Stead, *The Rabbit Menace*, provide the best overview.

28. Russel Ward, *The Australian Legend* (Melbourne: Oxford University Press, 1958; second edition, 1978). On literary nationalism and the Australian myth see John Barnes (editor), *The Writer in Australia* (Melbourne: Oxford University Press, 1969) and Brenda Niall, *Australia through the Looking-Glass* (Melbourne: Melbourne University Press, 1984). The ordeal of the bush was one of the constants of Australian children's literature, appearing even in books by British authors about Australia. The "new chum" is either made a man or broken by his experience in the outback.

29. Ian Burn, "Beating about the Bush: The Landscapes of the Heidelberg School," in Anthony Bradley and Terry Smith (editors), *Australian Art and Architecture* (Melbourne: Oxford University Press, 1983).

30. Smith, "The Artist's Vision of Australia," 163. For examples see Curry O'Neil, *Classic Australian Paintings* (South Yarra, Victoria: Curry O'Neil Ross, 1983). The gum tree was a symbol for more than artists; the professional journal of the Australian Forest League, which began publication in 1917, was called *The Gum Tree*. Libby Robin, *A Forest Conscience* (Springvale South: Natural Resources Conservation League of Victoria, 1991), 4.

31. Donald MacDonald, *Gum Boughs and Wattle Bloom* (Melbourne: Cassell, 1888). The best analysis of this school is Tom Griffiths, "The Natural History of Melbourne: The Culture of Nature Writing in Victoria, 1880–1945," *Australian Historical Studies*, 23 (October 1989), 339–65.

32. Alec H. Chisholm, *Mateship with Birds* (Melbourne: Whitcombe and Tombs, 1922).

33. Griffiths, "The Natural History of Melbourne," provides an excellent analysis of the patriotic and military aspects of the outdoor movement.

34. One Canadian critic said that animal stories in Great Britain were about humans in fur, in the United States about killing animals, in Canada about being killed as an animal. Margaret Atwood, *Survival* (Toronto: Anansi, 1972), 73. On North American nature literature see Thomas R. Dunlap, *Saving America's Wildlife* (Princeton: Princeton University Press, 1988), Chapter Two. On one of the early Australian authors see Vivienne Rea Ellis, *Louisa Anna Meredith* (Sandy Bay: Blubber Head Press, 1979).

35. Ethel M. Pedley, *Dot and the Kangaroo* (London: Thomas Burleigh, 1899).

36. May Gibbs, *Snugglepot and Cuddlepie* (Sydney: Angus and Robertson, 1918). On Gibbs's botanical exactness see Peter Bernhardt, "Of Blossoms and Bugs: Natural History in May Gibbs'

Art," in Royal Botanic Gardens, Sydney, Gumnut Town, *Botanic Fact and Bushland Fantasy* (Sydney: Royal Botanic Gardens, Sydney, 1992), 5–21.

37. Dorothy Wall, *Blinky Bill* (Sydney: Angus and Robertson, 1933).

38. Elizabeth Ward, "A Child's Reading in Australia," *Washington Post Book World*, 4 November 1990, 17, 22. For a full analysis of this literature see Niall, *Australia through the Looking-Glass*.

39. David Allen, *The Naturalist in Britain* (London: Allen Lane, 1976), 203–04 on the general movement.

40. William Gillies and Robert Hall, *Nature Studies in Australia* (Meburne: Whitecombe and Tombs, 1903). Chisholm, *Mateship with Birds*, 70. Nevillie Caley, *What Bird Is That* (Sydney: Angus and Robertson, 1931), xv–xvi. Alec Chisholm, preface to the 1958 edition (same publisher), v–vii. The state Gould Leagues, which enrolled (by Chisholm's estimate, page 73) 200,000 children by the early 1920s, became a major force for nature education and nature preservation. The papers of Vincent Serventy, a nature writer and filmmaker, active in the Western Australian state group (Serventy collection, National Library, Canberra), contain material from the late 1940s on, and are a source for the postwar development of the Gould Leagues as part of the evolving environmental coalition.

41. Guiler, *Thylacine*, 31. Also D. F. McMichael, "New South Wales," in "Future Case Studies in Selecting and Allocating Land for Nature Conservation," in A. B. Costin and R. H. Groves (editors), *Nature Conservation in the Pacific* (Morges, Switzerland: International Union for the Conservation of Nature, 1973), 53. This may be compared with the "worthless lands" thesis Alfred Runte advanced in *National Parks: The American Experience* (Lincoln: University of Nebraska Press, second edition, 1987), a perspective implicit in W. F. Lothian's *A History of Canada's National Parks* (Ottawa: Environment Canada, 1987).

42. Public outrage at mass slaughter of koalas in the 1920s did provide the push necessary to end this part of the fur trade.

43. A. S. Le Souf, "The Australian Native Animals," in James Barrett (editor), *Save Australia* (Melbourne: Macmillan, 1925), 175. The parallels of vanishing animals and vanishing aboriginals is worth analyzing across the settler colonies.

44. New South Wales provided some protection under the Native Animals Protection Act, No. 18, 1903; Queensland gave the koala protection by proclamation in 1909–1910, but did not follow it up by legislation, see Correspondence re: native animals, 1902–1910, AGE/N350, Queensland State Archives, South Brisbane. General information in J. M. Thomson, J. L. Long and D. R. Horton, "Human Exploitation of and Introductions to the Australian Fauna," in Bureau of Flora and Fauna, *Fauna of Australia*, Volume 1A, 227–49. On sentiment and koala protection, see George R. Wilson, "Cultural Values, Conservation and Management Legislation," in Bureau of Flora and Fauna, *Fauna of Australia*, Volume 1A, 250. U.S. developments in this period are most easily followed in T. S. Palmer, "Chronology and Index of the More Important Events in American Game Protection, 1776–1911," Bulletin 41, Bureau of Biological Survey, U.S. Department of Agriculture (Washington, D.C.: Government Printing Office, 1912).

45. In 1908 a visiting English naturalist commented that he was careful to gather information on the species, because "it will not be very long before it becomes extinct." Geoffry Smith, *A Naturalist in Tasmania* (Oxford: Clarendon Press, 1909), 95. Guiler, *Thylacine*, 25–30. The thylacine's disappearance took Tasmanians by surprise. The Hobart zoo had a total of a dozen specimens up to 1933 but traded them away for more popular animals, apparently believing more could always be trapped. Rumors of survivors persist, but there is no hard evidence, and no biologist I have talked to holds out hope.

46. Guiler, *Thylacine*, 29–31. Michael Sharland, *Tasmanian Wild Life* (Sydney: Melbourne University Press, 1962), 1. M. S. Sharland, "In Search of the Thylacine," *Proceedings of the Royal Society of New South Wales*, 1938–1939, 20–38.

47. J. M. Powell, *Watering the Garden State* (Sydney: Allen and Unwin, 1989), 150–67. J. M. Powell, "Protracted Reconciliation: Society and the Environment," in Roy MacLeod (editor), *The Commonwealth of Science: ANZAAS and the Scientific Enterprise in Australia, 1888–1898* (Melbourne: Oxford University Press, 1988), offers a description of the parallels between North American and Australian conservation, 249–71.

48. On experts and conservation in the United States see Samuel P. Hays, *Conservation and the*

Gospel of Efficiency (Cambridge: Harvard University Press, 1959). A full analysis of the experts hired, the agencies established, and their place in Australian governments would be useful in providing a comparative perspective on the United States conservation movement.

49. Myles Dunphy, "Campcraft and Trailing," 3-volume unpublished manuscript, dated 1930, Box MLK 3335, Dunphy papers, Mitchell Library, Section, "On Walking." The religious appeal is an afterthought that is hardly in evidence in Dunphy's papers, either personal or official. On this see Dick Johnson, *The Alps at the Crossroad* (Melbourne: Victorian National Park Association, 1974), 18. Women's place in bushwalking was ambivalent. The clubs were almost entirely male, but Dunphy's unsolicited correspondence suggests that—at least on the trail—women could find a place as "mates." On this see Mary Lane to Dunphy, 14 August 1931 in Fill Correspondence 1931-42, Box 10, Add on 1823 to Dunphy papers, Mitchell Library.

50. Dunphy, "Campcraft and Trailing," section on "Humping the Swag." The sport is still important in Australian nature preservation efforts. On recent developments and ideas see (Victoria) Land Conservation Council, *Wilderness: Special Investigation, Descriptive Report* (Melbourne: Land Conservation Council, 1990), 13–16. A closer look at the high country and bushwalking is W. K. Hancock's *Discovering Monaro* (Cambridge: Cambridge University Press, 1972), 164–88. On Tasmanian bushwalking, Helen Gee and Janet Fenton (editors), *The South West Book: A Tasmanian Wilderness* (Sydney and Melbourne: Collins and the Australian Conservation Foundation, third edition, 1983), 141–48.

51. Dunphy to E.B. Hawkes, 20 October 1931, File Correspondence, 1931–1932, 1943, Box 10, Dunphy papers, Add on 1823, Mitchell Library. The same sentiments, less articulated, are in the earliest papers, dating from around 1919.

52. See *Conference of Authorities on Australian Fauna and Flora*, Hobart, 7–9 December 1959 (Hobart: Government Printer, 1949). There was a wave of consolidation of wildlife laws at this time. In 1948 New South Wales passed a new general statute on hunting, the Fauna Protection Act, 1948, No. 47. In 1949 Victoria appointed a Director of Fisheries and Wildlife; in 1950 Western Australia repealed its complex of old game laws in favor of a consolidated statute enforced by a new Chief Warden and Fauna Protection Advisory committee under the Fauna Protection Act, 1950, No. 77. In 1952 Queensland followed suit with the Fauna Conservation Act, 1952. Victorian developments are seen in "Fisheries and Wildlife," *Victorian Yearbook, 1984* (Melbourne: Government Printer, 1984), and F. I. Norman and C. A. D. Young, "Short-sighted and Doubly Short-sighted are They," *Journal of Australian Studies* 7 (1980), 2–24. These were limited acts. The Advisory Board set up in New South Wales, for instance, was composed of representatives of ministries concerned with land use. This reduced interdepartmental friction, but did not alter the intent of program under which wildlife laws were made or enforced. In some cases the new laws simply brought the system up to date. Queensland, for the first time, placed its rangers under civil service protection.

53. Between 1935, when the program started, and 1981, Victorian wildlife officials transplanted 8,000 koalas from islands to sixty-five localities in their former range. "Fish and Wildlife," *Victorian Yearbook, 1984*, 5. The public was interested, but this was, occasionally, itself a problem. In Victoria wildlife authorities reported that people were still chopping down trees to get young koalas for pets, despite years of warnings that the species was protected (Victoria, Game Act, 1928, No. 3689). See *Conference on Fauna*, 11, 13.

54. His memoir of this work, *Flying Fox and Drifting Sand* (American edition, New York: McBride, 1938) is a classic of Australian nature writing.

55. Ratcliffe's papers in the National Library, Canberra, are primarily from the 1960s. The best study of his federal wildlife work is in the forthcoming second volume of Boris Schedvin's history of CSIRO. I thank Schedvin for access, in 1990, to rough drafts of two chapters.

56. Schedvin, draft of "Myxomatosis" chapter. Frank J. Fenner and F. N. Ratcliffe, *Myxomatosis* (Cambridge: Cambridge University Press, 1965).

57. On Australian biology see Le Souf, "Australian Native Animals," and Stanbury, "Discovery of the Australian Fauna." Author's interviews, June 1992, with Alan Newsome and John Calaby, CSIRO, Canberra.

58. Schedvin, draft of "Wildlife Research" chapter, 28 and 28–32.

59. J. S. Turner, "The Decline of the Plants," 134–55, and David Pollard and Trevor D. Scott, "Riv-

er and Reef," 112, in J. A. Marshall (editor), *The Great Extermination* (Melbourne: Heinemann, 1966).

60. (Victoria) Land Conservation Council, *Wilderness*, 14. Sarah Bardwell, "For All the People for All Time: A History of National Parks in Victoria," *Royal Historical Society of Victoria*, 56 (1985), 10–18. Johnson, *The Alps at the Crossroad*, is part of this effort.

61. Dorothy Hill, "The Great Barrier Reef Committee, 1922–1985, Part I, The First Thirty Years," and "Part II, The Last Three Decades," *Historical Records of Australian Science* 6, nos. 1 and 2, copies courtesy Dr. Patricia Mather, Curator, Queensland Museum, South Brisbane. First part was unpaged, second is pages 195–211. D. C. Potts, "The Crown of Thorns Starfish—Man-Induced Pest or Natural Phenomenon?" in R. L. Kitching and R. E. Jones (editors), *The Ecology of Pests: Some Australian Case Histories* (Melbourne: CSIRO, 1981), 55–86.

62. Bolton, *Spoils and Spoilers*, 159–68. Vincent Serventy Collection, National Library, Canberra, has many examples and scattered reports and correspondence from various of these battles.

63. Box 5, Ratcliffe papers, National Library, Canberra.

64. Box 5, Ratcliffe papers, National Library, Canberra.

65. Box 1823, Australian Conservation Foundation papers, State Library of Victoria, Melbourne. Dr. Patrica Mather, who had been active in the ACF to that point, provided (personal interview, 5 August 1990) details on the meeting and the scientists' protests.

66. Australian Conservation Foundation, *Conservation Directory* (Melbourne: ACF, 1973). The comparable directory for the United States, the National Wildlife Federation's *Conservation Directory*, indicates a similar trend. The ACF's 1991–1992 *Green Pages* (Fitzroy, Victoria: Australian Conservation Foundation, 1991) lists almost 250 for New South Wales, and several organizations list in addition many branches. The actual numbers are probably not comparable, as the basis for listing may have changed, but the trend is the same.

67. Papers of Judith Wright McKinney and Vincent Serventy, both in the National Library, Canberra, give important information on Queensland and Western Australia, respectively.

68. Gee and Fenton, *The South West Book*, has a large selection of documents and commentary. See also Dick Johnson, *Lake Pedder: Why a National Park Must Be Saved* (Lake Pedder Action Committee of Victoria and Tasmania and the Australian Union of Students, 1972).

69. The records of the Tasmanian Wilderness Society, now the Wilderness Society, in the State Library of Victoria, Melbourne, are the basic source. For events since 1979, they may be supplemented with the society's newsletter, collected in the library of the Tasmanian Department of Parks, Wildlife, and Heritage. See also Gee and Fenton, *The South West Book*.

70. Protest records are in Boxes 2669–70, Wilderness Society records, State Library of Victoria, Melbourne. On the prisons, see "Tasmania and the 'Greenies': Research Note on Prison Crowding," *Australia and New Zealand Journal of Criminology* 17 (March 1984), 41–48.

71. Gee and Fenton, *The South West Book* has a chronology and description to 1983. Current status of parks and work in annual reports of the Australian National Parks and Wildlife Service and of the Tasmanian Department of Parks, Wildlife, and Heritage. On legal issues see D. E. Fisher, *Natural Resources Law in Australia* (Sydney: Law Book Company, 1987), 222–34.

72. Australian Environmental Council, *Guide to Environmental Legislation and Administrative Arrangements in Australia*, Second edition, Report No. 18 (Canberra: Australian Government Publishing Service, 1986).

73. On the kangaroo situation, see H. J. Frith and J. H. Calaby, *Kangaroos* (Melbourne: F. W. Cheshire, 1969), 161–69 and Box 5, Ratcliffe Papers, National Library, Canberra. On the protests see Parliament, House of Representatives, Select Committee on Wildlife Conservation, *Wildlife Conservation* (Report from the House of Representatives Select Committee) (Canberra: Government Publishing Service, 1972), 7 (cited hereinafter as Fox Report) and Box 1822, ACF papers, State Library of Victoria, Melbourne.

74. Fox Report, 7. Box 6, Ratcliffe papers, National Library, Canberra.

75. Fox Report, 1. Numbers in text refer to page numbers in this report.

76. The Council's standing committee is composed of the heads of the national, state, and territorial wildlife and conservation authorities. It has working groups on everything from kangaroos to

cane toads to the management of concessions in national parks. ANPWS's development is best followed through its annual reports and the 1989 review, Department of the Arts, Sports, the Environment, Tourism, and Territories, *Report on the Review of the Australian National Parks and Wildlife Service* (Canberra: Australian Government Printing Service, 1989).

77. Australian Environmental Council, *Guide to Environmental Legislation*, 183–84, 236–37. See also ANPWS annual reports.

78. The Wilderness Society files and newsletters have much on ideological questions. It would be useful to do an analysis of the flyers and newsletters from the United States and Australia, particularly the extent to which more activist organizations like Greenpeace have a different social and political constituency than older organizations.

79. The Hobart (Tasmania) *Mercury*, 20 July 1990. The papers of environmental activists show this quite strikingly in the arguments offered, issues framed, and opposition cited. The Vincent Serventy collection has much illustrative material.

80. Johnson, *Alps at the Crossroad*, 19.

81. Ross Galbreath, "Colonisation, Science, and Conservation," Ph.D. thesis, University of Waikato, 1989. See Royal Forest and Bird Protection Society files, National Archives and Library, Wellington, New Zealand.

82. Tasmanian Wilderness Society publications and records show a striking concern with the "bush." Any commentary on Australian painting will show, as will any trip to a museum, the importance of the gum tree as a landscape symbol. Bernard Smith refers to it as an obsession, "The Artist's Vision of Australia," 163. A recent summary of information on exotics in Australia is R. H. Groves and J. J. Burdon, *Ecology of Biological Invasions* (Melbourne: Cambridge University Press, 1986). On New Zealand see Galbreath, "Colonisation, Science, and Conservation."

83. On Canberra, information from Dan Walton, Australian National Parks and Wildlife Service. On Victoria, Robin, *Forest Conscience*, 49–50.

84. Arthur Bentley, *An Introduction to the Deer of Australia* (Melbourne: The Koetong Trust, revised edition, 1978), 21. By contrast, there is in the United States considerable sentiment for the preservation of wild horses and burros in the West, despite their being introduced and ecologically destructive.

85. Bernard William Smith, *Australian Painting, 1788–1970*, 71–92. On the influence of this school in the interwar period, 196–97. Land use patterns and developments are in Australian Bureau of Statistics, *Year Book, 1988*, 501.

86. J. W. Gregory, *The Dead Heart of Australia* (London: John Murray, 1906). William Ramson, "Wasteland to Wilderness," in D. J. Mulvaney (editor), *The Humanities and the Australian Environment* (Canberra: Australian Academy of the Humanities, 1991), 5–19, provides a history of terms for the center. Edwin J. Brady, *Australia Unlimited* (Melbourne: George Robertson, 1918) was the bible of optimism. The most vocal pessimist was the geographer and climatologist T. Griffiths Taylor. His autobiography, *Journeyman Taylor* (London: Robert Hale, 1958), 139, 202, contains comments on his battle with the irrigation and reclamation faction, as do the documents in Series 4, Taylor Collection, National Library, Canberra, particularly Box labeled Items 359–410.

87. H. H. Finlayson, *The Red Centre* (Sydney: Angus and Robertson, 1935).

88. Quote from Bernard William Smith, "The Artist's Vision of Australia," in Smith, *The Antipodean Manifesto*, 164.

89. Alec M. Blombery, *The Living Centre of Australia* (Kenthurst: Kangaroo Press, 1985). The new popularity of the center might usefully be compared with Canadian attitudes and ideas about the "unconquerable North." An introduction is Louis-Edmond Hamelin, *About Canada: The Canadian North and Its Conceptual Referents* (Ottawa: Minister of Supply and Services, 1988). Comparisions might be made with the first Canadian school of painting, the Group of Seven, which was concerned to paint the "authentic" landscape of Canada, which its members found in the Precambrian Shield in northern Ontario. J. Russell Harmper, *Painting in Canada: A History* (Toronto: University of Toronto Press, second edition, 1977), 265–83. On the Group's relation to the Northern cult, see Roald Nasgaard, *The Mystic North: Symbolist Landscape Painting in Northern Europe and North America, 1890–1940* (Toronto: University of Toronto Press, 1984).

90. The postwar dynamics of American society in Samuel P. Hays, *Beauty, Health, and Permanence* (New York: Oxford University Press, 1988) can usefully be applied to Australia.

Nataraja, *by Stephen J. Pyne*

1. For an introduction to the ecological background of the subcontinent, see M. S. Mani, ed., *Ecology and Biogeography of India* (Dr. W. Junk Publisher, 1974), 2 vols. A quick synopsis of the principal biomes is available in F. R. Bharucha, *A Textbook of the Plant Geography of India* (Oxford University Press, 1983). For forests, consult G. S. Puri et al., *Forest Ecology,* 2nd ed. (Oxford & IBH Publ. Co., 1990), 2 vols. Regrettably no digest of fire practices or summary of Indian fire ecology exists.

2. See M. P. Nayar, "Changing Patterns of the Indian Flora," *Bulletin of the Botanical Survey of India* 19 (1–4) (1977), 145–55.

3. For the influence of caste on shaping the environment, see Madhav Gadgil and Ramchandra Guha, *This Fissured Land: An Ecological History of India* (University of California Press, 1992); Madhav Gadgil, "Deforestation: Problems and Prospects," 13–85, in Ajay S. Rawat, ed., *History of Forestry in India* (Indus Publishing Company, 1991); and for an interesting case study, Madhav Gadgil, "Ecology of a Pastoral Caste: The Gavli Dhangar of Peninsular India," *Human Ecology* 10 (1982).

4. To my knowledge there is no systematic survey of Indian tribal fire practices. The evidence is scattered among travel and ethnographic accounts and especially among reports by British foresters as recorded in working plans and in submissions to the *Indian Forester*. Quotation from Alfred Radcliff-Brown, *The Andaman Islanders* (Free Press, 1948), 258. (Radcliffe-Brown was wrong about the ability of the Andamanders to make fire; they could, but like most aboriginal peoples chose to preserve what they had.) Pearson from "Progress Report of Forest Administration in the Central Provinces, 1863–64" (Calcutta, 1865), 27. For Australian fire practices, see Stephen Pyne, *Burning Bush* (Holt, 1991).

5. The literature on *jhum* cultivation is enormous. For historical references, see H. H. Bartlett, *Fire in Relation to Primitive Agriculture and Grazing in the Tropics: Annotated Bibliography,* 3 vols. (University of Michigan Botanical Gardens, 1955–1961). For a contemporary survey, see P. S. Ramakishnam, *Shifting Agriculture and Sustainable Development,* Man and Biosphere Series Vol. 10 (Parthenon, 1992).

6. See Gadgil and Guha, *This Fissured Land,* 79–82.

7. Agni is everywhere in Vedic theology. A massive inquiry into the fire ceremony is Fritts Staal, *Agni,* 2 vols. (Asian Humanities Press, 1983).

8. Citations from Clarence H. Hamilton, ed., *Buddhism* (Bobbs-Merrill, 1968), 40–41.

9. See Ajay S. Rawat, "Indian Wild Life Through the Ages," 134, in Rawat, ed., *History of Forestry in India.*

10. E. O. Shebbeare, "Fire Protection and Fire Control in India," *Third British Empire Forestry Conference* (Canberra, 1928), 1; Joseph Hooker, *Himalayan Journals,* Vol. 2 (London, 1855), 3; F. B. Bradley-Birt, *Chota Nagpore* (London, 1910), 184; Benjamin Heyne, *Tracts, Historical and Statistical, in India* (London, 1814), 302; Ribbentrop, *Forestry in British India,* 150.

11. Cox quoted in A. A. F. Minchin, "Working Plan for the Ghumsur Forests, Ganjam District" (Madras, 1921), 38–40.

12. For overviews of British forestry, see Edward Stebbing, *The Forests of India,* 3 vols. (Bodley Head, 1928), a fourth volume coming later under the editorship of Harry Champion and F. C. Osmaston (Oxford University Press, 1962); Rawat, ed., *History of Forestry in India;* Gadgil and Guha, *This Fissured Land;* and Richard Haeubaer, "Indian Forestry Policy in Two Eras: Continuity or Change?" *Environmental History Review* 17 (1) (spring 1993), 49–76.

13. D. E. Hutchins, *A Discussion of Australian Forestry* (Perth, 1916), 45; Rudyard Kipling, "In the Rukh," *The Jungle Books* (Oxford, 1987), 326–27, 343.

14. Shebbeare, "Fire Protection and Fire Control in India," 1.

15. For a revealing illustration of how critical fire was in European thinking, see G. Grotenfelt, *Det primitive jordbrukets metoder i Finland under den historiska tiden* (Methods of primitive agriculture in Finland during historical times) (Helsinki, 1899). Even the legendary Linneaus was forced to delete favorable comments on slash-and-burn agriculture in Småland from his Skåne travels, such was the

urgency with which intellectuals argued against fire and in favor of manure; see Lars J. Larsson, "Svedjebruk i Varend och Sunnerbo," 84–86, in Olof Nordstrom et al, *Skogen och Smålanningen*. Historiska foreningens i Kronobergs lan skriftserie 6 (Smp TRYCK AB, 1980). Quotation from C. K. Hewetson, "Fires and Their Ecological Effects in Madhya Pradesh," *Indian Forester* 80 (4) (1964), 238.

16. Pearson quoted in Shebbeare, "Fire Protection and Fire Control in India," 1; Lt. Col. G. F. Pearson, "Report on the Administration of the Forest Department in the Several Provinces Under the Government of India, 1871–72" (Calcutta, 1872), 9; Dietrich Brandis, Memorandum No. 263, *Forest Conference of 1871–72*, 5.

17. Brandis, *Forest Conference of 1875*, 4; Capt. J. C. Doveton, *Forest Conference of 1875*, 5.

18. Pearson, *Forest Conference of 1871–72*, 9; B. H. Baden-Powell, "Report on the Administration of the Forest Department in the Several Provinces Under the Government of India, 1872–73," Vol. 1 (1874), 67.

19. W. Burns et al., "A Study of Some Indian Grasses and Grasslands," *Memoirs of the Dept. of Agriculture in India, Botanical Series 14* (1928), 1–57.

20. An Aged Junior, "Some Remarks on Titles and Tigers," *Indian Forester* 16 (1–3) (1890), 182–84.

21. Shebbeare, "Fire Protection and Fire Control in India." The details of the story unfold over many years in the *Indian Forester* and are best understood in sequence.

22. E. A. Greswell, "The Constructive Properties of Fire in Chil (Pinus longifolia) Forests," *Indian Forester* 52 (1926), 502–05.

23. Political incendiarism, particularly in Kumaon, is well described in Ramachandra Guha, *The Unquiet Woods. Ecological Change and Peasant Resistance in the Himalaya* (Oxford India Paperbacks, 1991). M. D. Chaturvedi, "The Progress of Forestry in the United Provinces," *Indian Forester* XI (1925), 365. For Troup's ecological analysis, see R. S. Troup, "Pinus Longifolia Roxb.: A Silvicultural Study," *The Indian Forest Memoirs*, Silvicultural Series, Vol. 1, Pt. 1 (Calcutta, 1916).

24. For an environmentalist critique of Indian forest policy, see Centre for Science and Environment, *The State of India's Environment 1984–85: The Second Citizens' Report* (New Delhi, 1986), and for a critical review of the policy of state-based resource management, Gadgil and Guha, *This Fissured Land*. The official assessments are available in the Forest Survey of India's "State of Forest" reports, issued annually.

25. Statistics based on Veena Joshi, "Biomass Burning in India," in Joel Levine, ed., *Global Biomass Burning* (MIT Press, 1991), 185–93.

26. R. Saigal, "Modern Forest Fire Control: The Indian Experience," *Unasylva* 162 (41) (1990), 21–27.

Index

adobe, 87–88
African Americans, 196–207
Agni, 293–95
agriculture: in colonial New Mexico, 81; and concept of wilderness, 43; in Concord River meadows, 227–31, 236–40, 242; and English landscape, 54; in India, 193, 303, 306; within *landschaft*, 69; in prehistoric England, 57, 62; role of rice in, 216; threats to, after World War I, 102
Air Quality Act (1967), 114–15, 129–31, 133
air quality standards, 129–43
Albuquerque, N.M., 82–84, 93–96
Alston, Dana, 144–45, 159–60, 199
American Mining Congress, 132–33, 137
American Society for Environmental History, xiv, 196
Anaho Island, 243
Andaman Islands, 293
Arkansas River, 251
Army Corps of Engineers, 106, 125
arroyos, 94, 96
Australia: conservation in, 280–82, 286–88; ecological nationalism of, 277–79, 281–82, 288; economic development in, 287; environmental movement in, 282–87; European settlement in, 273–89; flora and fauna of, 275; game laws in, 276–77, 280; lore of the outback in, 277–78, 281, 288; national parks in, 279; natural history societies in, 275–76; nature consciousness in, 278–80; outdoor recreation in, 280–81; physical description of, 274; rabbit problem in, 277, 282; use of fire in, 293; wildlife policy in, 285–87
Australian Conservation Foundation, 282–84

Baden-Powell, B. H., 301
Bagge, Carl, 128, 138–39
Banks, Joseph, 265
Bathgate, Alexander, 267
Bauer, Catherine, 149

Bean v. Southwestern Waste Management Corp., 205–06
beehive coke oven, 167–71, 178
Been, Vicki, 204
Bellamy, Edward, 74, 76
Bennet Act (1907), 191
Berry, Wendell, 50
Bible, the, 26, 30, 32, 34, 74
Big Outside, the, 44, 47
Billerica Dam, 228, 231–36, 238, 241
biological diversity, 41–42
biota, of civilization, 265; evolution of, 264; of India, 290–91, 302; of Pangaea, 272; portmanteau, 265, 270–72
Bloch, Marc, xii
Bolton, Geoffrey, 273
Bori Forest, 300
Bradley-Birt, F. B., 295
Brandis, Dietrich, 298–300
Braudel, Fernand, xii
Bridenthal, Renaté, 24
Brower, David, 155
Brown, Alice, 206
Brown, Simon, 238–39
Bryant, Bunyan, 198–99
Buddhism, 294–95
Bullard, Robert D., 197, 202
Bureau of Indian Affairs, 245–50, 252, 257–59
bushwalking, 281, 284
"Butterfly Effect," 15
by-product coke oven, 170–77, 179

capitalism, 23, 81
Carnegie Steel Corporation, 173–74, 176–77
Carson, Rachel, 114, 153–56, 159–60, 195
Carter, Jimmy, 143
Cerrell Report, 205
Chacon, Governor Fernando de, 85
Chambless, Edgar, 74–76
chaos theory, 9, 14–16
charcoal deposits, 55–59, 61

cial movements, 156–59; and concept of ecosystems, 8; and consumption, 112–13, 151; decentralization and, 104, 111; demography of, 144, 198–99; diverse interests of, 145–47, 206–07; Earth Day and, 114–15, 117, 133, 146, 156, 158; emergence of, 194; energy issues in, 113–14, 118–19; and environmental history, 195; and environmental justice, 202–06; and equity, 207–09; growth of, after World War II, 104, 110; historiography of, 101–02; and human health issues, 118; and human rights, 210; lifestyle issues in, 119; and natural areas, 116; organizations within, 114, 120, 129, 194–95, 198–99, 206; and political ideology, 123–25; political response to, 115–20; popular support for, 104–05; and protest, 108; racial issues in, 196–200; resource management in, 102–03; response to technology of, 114; role of corporations in, 122–24; and search for meaning, 125; in Tasmania, 284–85; and U.S. standard of living, 109; and wilderness, 28, 31

Environmental Protection Agency (EPA), 127, 133–42, 202
environmental racism, 200–02, 209–10
Environmental Review, xiv–xv
environmental values, 11–12, 107–10, 120, 124
Europe: animals of, in New Zealand, 267–68; descendants of, outside Eurasia, 263–64; diseases of, in New Zealand, 268–69; ecological colonization by, 18, 26; forestry of, in India, 297–99, 302, 307; and Indian fire practices, 295–96, 306; natural history societies of, in Australia, 276; portmanteau biota of, 265, 270–72; settlements in Australia from, 273–77, 289; weed species of, 266–67, 291–92
Everglades National Park, 32

Febvre, Lucian, xii
Federal Water Pollution Control Administration, 131
Fenton, F. D., 269
Finch, Robert, 133–34
Findlayson, H. H., 288
fire, 56, 58–62, 290–307
firesticks, 292–93
Fish and Wildlife Service (U.S.), 117
Flader, Susan, xiv
Flathead Reservation, 254–55
Fleming, Donald, 275
Ford, Gerald, 141–42
Ford, Richard, 191
Foreman, Dave, 43–44, 47
Forest Conference (India), 300
forestry, 297–99, 302, 307
forests: in colonial New Mexico, 87; concept of, after World War II, 106; in Concord River valley, 239–41; Indian, 297–306; mythology of, 70; population biology and, 12; in prehistoric England, 55–61. *See also* timber
Forest Service (India), 298, 301, 304

Forest Service (U.S.), 104, 106, 147–48
French, Judge Henry, 238–39
frontier, 31, 35
frontier myth, 36–38, 40, 44, 47,
Fundamentals of Ecology, The, 6

Gelobter, Michel, 203–04
George, Henry, 74
Georgia, 215, 218–19, 223–24
Gianessi, Leonard P., 203–04
Gibbs, May, 278–79
Gleason, Henry A., 10
Gondwana, 291
Gottlieb, Robert, 198, 208–09
grazing, 83–84, 92, 95, 102
Great Barrier Reef, 283
Great Lakes, 177
Green political parties, 99, 285
Gregg, Josiah, 87
Greswell, E. A., 303–04

Hamilton, Alice, 146, 150–53, 159–60
Hays, Samuel P., xiii
Health, Education, and Welfare, department of, 129, 130–34
Heidelberg School, 278, 288
Hell's Canyon, 105
Hetch Hetchy, 31, 46
Heyne, Benjamin, 295
Hinduism, 290, 293–95
Hitchcock, Ethan Allen, 252
homeostasis, 7
Hooker, Joseph, 295
Horwitz, Morton, 228, 232
Hoxne (England), 55–56
Hughes, J. Donald, xiii–xiv
Hull House, 150–51
Hundley, Norris, 252, 259
Hurley, Andrew, 209–10

India: biota and culture of, 290–91, 302; British rule in, 296–97, 305; European forestry in, 297–99, 302, 307; European weed species in, 291–92; fire in, 290, 292–96, 299–307; monsoons in, 292; population of, 291, 303–05; slash-and-burn agriculture in, 293, 303, 306; social structure of, 292, 295; tiger problems in, 302. *See also* fire
Indian Forester, 292, 302
industrial disease, 150–53
industrialization, 24, 118, 184–86, 190; and English landscape, 53–54; hazards of, 154–55; of New England, 24; Progressive Era and, 156; and utopian fiction, 74
Interior, Department of, 243–44
iron, 163–66. *See also* coke
irrigation, 89–90, 244–46, 248, 250, 253–54, 256–58

James, Henry, 66–68, 73

Club v. Ruckelshaus, 138; and industrial interests, 130–31; and "limits to growth," 142; as national policy, 143; origins of, 127–31; and pollution control technology, 140–41; "protect and enhance" concept and, 134; Senate hearings on, 139; strip mining and, 141

Odum, Eugene P., 6–9, 12, 15
Ohio River Valley, 172
Olerich, Henry, 74–75
Olmsted, Fredrick Law, 68
Olympic National Park, 104
Oñate, don Juan de, 91
Opie, John, xiv–xv

Paiute Indians, 243–44, 247
Pangaea, 272
Panos Institute, 145, 199
patch, 11, 12
Patric, James, 241
Pearson, Colonel, 299–301
peat, 59–63
Peck, Bradford, 74, 76
Pedley, Ethel M., 278
People's Forests, The, 148
Peskin, Henry M., 203–04
pesticides, 117, 154–55
Peterson, Abby, 23
phenols, 171–73, 175–77, 179
Philosophical Society of Australia, 276
Pickett, S. T. A., 11
Pinchot, Gifford, 103
Pittman-Robertson Act (1936), 102
Pittsburgh, 166–67, 169, 178–79
place, sense of, xii–xiii, 51
police power, 173–74, 188
pollution, attempts to control, 123, 177: from coke production, 168; as environmental issue, 104, 108, 116–17, 146; health effects of, 118; in industrial waste disposal, 163; noise as, 182–83, 187–88, 192–93; in Ohio River Basin, 179–80; racial issues in, 201–05
pollution, air: and Air Quality Act (1967), 130; ambient vs. emission standards for, 129–30; from beehive coke ovens, 168–69; from charcoal production, 165; and Clean Air Act, 127; from coke quenching, 178–79; control technology for, 140–41; and human health issues, 128; industrial response to guidelines on, 129, 137; National Conference on, 128; and no-significant deterioration policy, 129–43; racial issues and, 203. *See also* air quality standards, no-significant deterioration
pollution, water, 171–77
population biology, 12
Pratt, Federal Court Judge, 127, 130, 138
Prelude, The, 33
Preston, Porter, 257–58
Prigogine, Ilya, 16
primitivism, 35, 44

production: in capitalist ecological revolution, 25; ecosystems and, 8; and environmental movement, 121–24; focus on, before World War II, 102, 112; human reproduction and, 24; Meillassoux's views on, 23; and nature, 21–22; noise and, 184–86, 190, 193
Progressive Era, 156–58, 209
Public Health Service (U.S.), 175–76, 179–80
Pueblo Indians, use of adobe by, 87–88; interaction with Spanish by, 80, 85, 91, 93, 97; revolt of, 82, 88; settlements of, 83; Tewa group of, 91–92; water use by, 89–90
Pine, Stephen J., xiv
Pyramid Lake, 243–44, 248

railroad: built in India by British, 297, 299; in Henry James, 67; in New Mexico, 95; noise pollution from, 187; and sense of space, 72; in utopian fiction, 76
rain forest, 42
Ratcliffe, Francis, 282–84
Reagan, Ronald, 119–20
Reclamation Act, 244, 248, 252
Reclamation Service: and Bureau of Indian Affairs, 246–48, 257; decline of, 257; growth in jurisdiction of, 247–48, 253; influence of, 259; irrigation projects of, 254–55; and Native Americans, 246–49, 251; and Snake River, 105; water rights policies of, 249–50
recreation: in Australia, 280–81, 287; and environmental movement, 110, 112, 115; and scenic resources, 157; and wilderness, 38, 45
Reich, Charles A., xiii
Reilly, William, 202
reproduction, 22–25
Ribbentrop, Inspector-General, 295, 302
rice production, 215, 126–18, 216–22, 223–25
Richards v. Washington Terminal, 187
Ringlemann Chart, 169
Rio Grande River, 82, 96
Rio Grande Valley, 82–96
rivers, 105, 171–77, 218, 228, 249–52
roads, 71–73, 76, 79
Robbins, William, xv
Roe, Derek A., 56
Rogers, Paul, 136
romanticism, 31–32, 34–35, 39, 48
Roosevelt, Franklin D., 104
Roosevelt, Theodore, 36, 248, 250–51
Rothman, Hal, xv
Ruckelshaus, William, 127, 136, 138

salinization, 90
Santa Cruz de la Canada, 82–84, 96
Santa Fe, 82–84, 93, 96
Schaefer, William, 15
Schnaiberg, Alan, 208
Scientific Revolution, The, 14, 21
scientific revolutions, 19
Sea Around Us, The, 153–54